Solutions Manual

TO ACCOMPANY

UNIVERSITY PHYSICS

REVISED EDITION

HARRIS BENSON

Vanier College

John Wiley & Sons, Inc.

New York Chichester Brisbane Toronto Singapore

ISBN 0-471-14608-0

Printed in the United States of America

10 9 8 7 6 5 4 3 2 1

P R E F A C E

This manual contains brief solutions for UNIVERSITY PHYISCS, Revised Edition, Harris Benson, John Wiley and Sons, N.Y., 1996. The solutions outline the steps that were used to obtain the answers. They are not suitable for student use.

It is a pleasure to acknowledge the help of Phil Eastman, University of Waterloo, who independently solved all the original exercises and problems and Chris Vuille, Embry-Riddle Aeronautical University who independently solved all the additional exercises and problems. As part of their work on the French version of the text, the solutions to all the original exercises and problems were thoroughly checked by Pierre Boucher, Nicole Lefebvre, Marc Séguin, and Benoît Villeneuve. Of course, the responsibility for any errors rests with me. I would appreciate being notified of any that remain.

HARRIS BENSON
Vanier College,
821 Ste. Croix Blvd.,
Montreal, P.Q.
Canada, H4L 3X9.

T A B L E O F C O N T E N T S

CHAPTER 1

Exercises

1. (a) 55 mi/h x 5280 ft/mi x 1 h/3600 s = 80.7 ft/s; (b) 24.6 m/s

2. 13440 furlongs/fortnight

3. 10^{-3} kg/(10^{-2} m)^3 = 10^3 kg/m^3.

4. 3.156 x 10^7 s

5. (a) 9.47 x 10^{12} km; (b) 7.2 AU/h

6. (a) 1.0073 u; (b) 1.67493 x 10^{-27} kg

7. 0.514 m/s

8. (a) 0.984 ft/ns; (b) 1.86 x 10^5 mi/s

9. 134 in^2

10. 30 mi/gal x 1 gal/3.79 L x 1.6 km/mi = 12.6 km/L or 7.85 L/100 km

11. (a) 5, (b) 3, (c) 4, (d) 2 to 4

12. (a) 6.5 x 10^{-9} s; (b) 1.28 x 10^{-5} m; (c) 2 x 10^{10} W; (d) 3 x 10^{-4} A; (e) 1.5 x 10^{-12} A

13. (a) $A = \pi r^2$ = 55.4 m^2, (b) $A = 4\pi r^2$ = 2.7 m^2, (c) $V = 4\pi r^3/3$ = 52.17 m^3.

14. (a) 2.5 x 10^{-1}; (b) 5.00 x 10^{-3}; (c) 7.6300 x 10^{-4}

15. 3.33 x 10^3.

16. (a) 48.0; (b) 403.2

17. (a) 1.495 x 10^{11} m, (b) 5.893 x 10^{-7} m, (c) 2 x 10^{-10} m, (d) 4 x 10^{-15} m.

18. (a) 15.69; (b) 25.9

19. (a) 91.440 m, (b) 0.40469 hectares

20. (a) 6.2×10^3; (b) 2.73×10^1; (c) 6.00000×10^2

21. (a) 2%, (b) 4%, (c) 6%.

22. 243 ± 4.5 cm^2

23. (a) $4\pi R^2 = 5 \times 10^{14}$ m^2, (b) $4\pi R^3/3 = 3 \times 10^{20}$ m^3
(c) $R_S/R_E = 100$, so $(R_S/R_E)^3 = 1 \times 10^6$

24. 2×10^5

25. 14.5 min. error in one day

26. 1670 km/h

27. For a 2-h movie at 30 frames/second, find 2×10^5 frames.

28. 3×10^{-5} m

29. (a) (2 km/d)(400 d/y)(70 y) $\approx 5 \times 10^4$ km.
(b) At 2 kg/d we find 5×10^4 kg.

30. 10^6

31. 10^9 m^3

32. 10^4 grains

33. 0.1 m^3

34. $M^{-1}L^3T^{-2}$

35. (a) Correct, (b) wrong, (c) correct

36. $[A] = LT^{-2}$, $[B] = LT^{-4}$

37. (a) (2.68 m, 2.25 m), (b) (−1.15 m, −1.38 m),
(c) (−1.80 m, 1.26 m), (d) (1.99 m, −1.67 m)

38. (a) 5.00 m, 53.1°; (b) 3.61 m, 124°; (c) 2.92 m, 329°;
(d) 2.24 m, 206°

39. $[\omega] = T^{-1}$, $[k] = M\,T^{-2}$

40. $V = 4\pi r^3/3$, so $r = 9.14$ cm.

41. (a) $V = \pi r^2 h$, h = 14.5 cm. (b) Area = $2\pi rh + 2\pi r^2$ = 330 cm^2.

42. 1 y = 3.15576 x 10^7 s and π = 3.14159. Thus error is 0.449%.

43. Using 3.786 L per gal. and 1 mi = 1.609 km, find 25.9 m.p.g.

44. For each person Δx = 1.6 m. Circumference = 4 x 10^7 m, so we need 2.5 x 10^7 people.

45. (a) 4.96 x 10^2; (b) 2.6 x 10^4

46. Area = 480 ft^2 = 44.595 m^2, so cost is $32.97.

47. (4 x 10^{-3} m^3)/20 m^2 = 0.2 mm.

48. (a) kg.m^2/s^2; (b) E = (10^{-3} kg)c^2 = 9 x 10^{13} kg.m^2/s^2.

49. θ = s/R in radians, or sin(0.5") = 0.5 AU/1 parsec, give 1 parsec = 2.063 x 10^5 AU.

50. 64.206 cm^2.

51. 1 " = 2.54 cm exactly, so g = 32.1740 ft/s^2

52. 1 m^2 = 10.764 ft^2. Cost is 21.05 per m^2.

Problems

1. With a 0.5 m diameter, the circumference is C ≈ 1.5 m. If the tread lasts for 7 x 10^7 m, the number of revolutions is 7 x 10^7 m/1.5 m ≈ 4.5 x 10^7 rev. The tread depth is about 1 cm, so the loss per revolution is 0.01 m/(4.5 x 10^7) ≈ 2 x 10^{-10} m. This about roughly the size of an atom.

2. $a \alpha v^2/r$

3. $T = k^x m^y = M^x T^{-2x} M^y$. The unit of k is N/m, so [k] = MT^{-2} Equating the exponents: 1 = -2x; 0 = x + y. Thus x = -1/2 and y = + 1/2 and so $T = C(m/k)^{1/2}$

4. $x \alpha at^2$

CHAPTER 2

Exercises

1. (a) 4.1 m along the +x axis; (b) 3.2 m at 70° to the +x axis

2. (a) 3.5 m at 8.2^o below the -x axis;
(b) 5.8 m at 52.5^o above the -x axis

3. (a) 3 m at 20° below the +x axis; (b) 2 m along the +y axis

4. $\underline{D} = \underline{C} - \underline{A} - \underline{B}$ = 3.85 m at 10.5^o below the -x axis

5. The angle between adjacent vectors is (a) 180°; (b) 90°;
(c) Many possibilities, e.g. 60° if equal angles; (d) 0.

6. (a) 83°; (b) 151°

7. 60 m at 25° E of N

8. $R_x = -4 \sin 40^o - 3 \cos 20^o = -5.39$ m,
$R_y = 4 \cos 40^o - 3 \sin 20^o = 2.04$ m,
$\underline{R}$ = 5.76 m at 20.7^o N of W.

9. $R_x = A \cos 45^o + B \sin 60^o - C \sin 30^o$
$= 3.54 + 6.06 - 2 = 7.60$ m,
$R_y = A \sin 45^o - B \cos 60^o - C \cos 30^o$
$= 3.54 - 3.50 - 3.46 = -3.42$ m,
$\underline{R}$ = 8.33 at 24.2^o S of E.

10. $\underline{R} = -12\underline{i} - 5\underline{j}$ m, or $\underline{R}$ = 13 m at 22.6^o S of W

11. $(50^2 - 25^2)^{1/2} = 43.3$ cm

12. $y = 100/\tan 30^o = 173$ km

13. (a) $\underline{A} = -\underline{i} - 1.73\underline{j}$ m, $\underline{B} = 1.53\underline{i} + 1.29\underline{j}$ m, $\underline{C} = -1.73\underline{i} + \underline{j}$ m,
$\underline{D} = 1.64\underline{i} - 1.15\underline{j}$ m;
(b) $\underline{R} = \underline{A} + \underline{B} + \underline{C} + \underline{D} = 0.44\underline{i} - 0.59\underline{j}$ m;
(c) $\underline{R}$ = 0.74 m at 53.3^o below the +x axis.

14. (a) $\underline{A} = -3.63\underline{i} - 1.69\underline{j}$ m, $\underline{B} = 2.57\underline{i} - 3.06\underline{j}$ m,

$\underline{C} = -1.37\underline{i} + 3.76\underline{j}$ m, $\underline{D} = -4\underline{j}$ m;

(b) $\underline{R} = -2.43\underline{i} - 4.99\underline{j}$ m;

(c) $\underline{R} = 5.55$ m, at 64.0^o below the -x axis

15. (a) $\underline{R} = \underline{i} - \underline{j}$ m; (b) $R = (2)^{1/2} = 1.41$ m;

(c) $\underline{R}/R = 1/(2)^{1/2}\ (\underline{i} - \underline{j})$ m $= 0.707\underline{i} - 0.707\underline{j}$ m

16. (a) $\underline{S} = 2\underline{i} + 4\underline{j} - 8\underline{k}$ m;

(b) $S = (84)^{1/2}$ m $= 9.17$ m;

(c) $\underline{S}/S = 0.218\underline{i} + 0.436\underline{j} - 0.873\underline{k}$

17. (a) 0, (b) 180°,

(c) $(6 + 4\cos\theta)^2 + (4\sin\theta)^2 = 9$, so $\cos\theta = -0.896$, and $\theta = 153^o$;

(d) $(6 + 4\cos\theta)^2 + (4\sin\theta)^2 = 64$, so $\cos\theta = 0.25$, and $\theta = 75.5^o$.

18. Know $\underline{R} = \underline{A} + \underline{B} = -6.02\underline{i} + 7.99\underline{j}$ m, and $\underline{B} = 3.61\underline{i} + 4.79\underline{j}$, so $\underline{A} = \underline{R} - \underline{B} = -9.63\underline{i} + 3.2\underline{j}$, or $\underline{A} = 10.1$ m at 18.4^o N of W.

19. (a) $\underline{A} = A\cos45^o\underline{i} + A\sin45^o\underline{j} = 1.41\underline{i} + 1.41\underline{j}$ km,

$\underline{B} = B\sin75^o\underline{i} + B\cos75^o\underline{j} = 1.45\underline{i} + 0.39\underline{j}$ km, and

$\underline{C} = -C\sin15^o\underline{i} - C\cos15^o\underline{j} = -0.39\underline{i} - 1.45\underline{j}$ km.

Since $\underline{A} + \underline{B} + \underline{C} + \underline{D} = 0$, we find $\underline{D} = -2.47\underline{i} - 0.35\underline{j}$ km;

(b) $\underline{D} = 2.49$ km at 8.07^o S of W

20. $\underline{S} = \underline{A} - \underline{B} = S\cos30^o\underline{i} + S\sin30^o\underline{j}$, where $S = A/2 = 3$ m. So, $\underline{B} = \underline{A} - \underline{S} = 3.40\underline{i} - 1.5\underline{j}$ m, or $\underline{B} = 3.72$ m at 23.8^o S of E.

21. The initial and final positions are

$\underline{A} = 3.06\underline{i} + 2.57\underline{j}$ km;

$\underline{B} = -3\underline{i} + 5.2\underline{j}$ km.

Thus, $\underline{B} - \underline{A} = -6.06\underline{i} + 2.63\underline{j}$ km, or 6.61 km at 23.5^o N of W.

The least distance between them is $d = A\sin(63.5^o) = 3.58$ km.

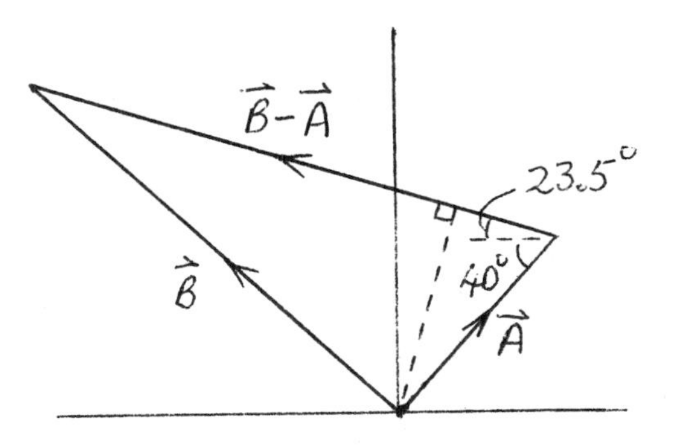

22. $\underline{R} = \underline{A} + \underline{B} + \underline{C}$, where $\underline{A} = 40\underline{j}$ km, $\underline{B} = 30\underline{i}$ km, and $C_x\underline{i} + C_y\underline{j}$.

$\underline{R} = -20\cos30^o\underline{i} - 20\sin30^o\underline{j} = -30\underline{i} + 40\underline{j} + C_x\underline{i} + C_y\underline{j}$, thus

$\underline{C} = 12.7\underline{i} - 50\underline{j}$ km, or $C = 51.6$ km at 75.7^o S of E

23. The vector joining the tip of $\underline{B}$ to the tip of $\underline{A}$ is $\underline{A} - \underline{B}$. Since $\underline{C}$ is to the midpoint of this vector, $\underline{C} = \underline{A} + (\underline{B} - \underline{A})/2 = (\underline{A} + \underline{B})/2$.

24. Since $\underline{A} + \underline{B} = 2\underline{i} + 3\underline{j}$ is at 56.3^o to the x axis, $\underline{C}$ must be at 56.3^o to the y axis. We merely interchange the values of the components. Thus, $\underline{C} = -3\underline{i} + 2\underline{j}$ or $3\underline{i} - 2\underline{j}$.

25. $\underline{S} = -12\underline{i} + \underline{j} + 5\underline{k}$, and $S = (170)^{1/2}$, so
$\underline{S}/S = -0.920\underline{i} + 0.077\underline{j} + 0.383\underline{k}$

26. (a) $A + B = (29)^{1/2} + (13)^{1/2} = 8.99$;
(b) $\underline{A} + \underline{B} = 3\underline{i} - \underline{j}$, so $|\underline{A} + \underline{B}| = (10)^{1/2} = 3.16$;
(c) $\underline{A} - \underline{B} = 7\underline{i} + 5\underline{j}$, so $|\underline{A} - \underline{B}| = (74)^{1/2} = 8.60$;
(d) $A - B = (29)^{1/2} - (13)^{1/2} = 1.78$

27. (a) $\underline{A}' = 2\underline{A} = 12\underline{i} - 4\underline{j} + 6\underline{k}$;
(b) $\underline{A}/A = \underline{A}/7 = -0.857\underline{i} + 0.286\underline{j} - 0.429\underline{k}$
(c) $\underline{B}' = -4\underline{A}/7 = -3.43\underline{i} + 1.14\underline{j} - 1.71\underline{k}$

28. $\underline{C} = 6\underline{B} - 3\underline{A} = -30\underline{i} + 15\underline{j} - 33\underline{k}$

29. Define the plane to be xy, so $\underline{A} = A_x\underline{i} + A_y\underline{j}$, $\underline{B} = B_x\underline{i} + B_y\underline{j}$
Then $\underline{C} = -\underline{A} - \underline{B} = -(A_x + B_x)\underline{i} - (A_y + B_y)\underline{j} + 0\underline{k}$, that is, $\underline{C}$ is also in the xy plane. This doesn't apply to four vectors.

30. $\underline{C} = 3\underline{A} - 2\underline{B} = 9\underline{i} - 5.5\underline{j}$ m, or $C = 10.5$ m at -31.4^o to x axis

31. (a) $\underline{P} = -P\cos 30^o\underline{i} + P\sin 30^o\underline{j} = -4.33\underline{i} + 2.5\underline{j}$ m;
(b) $\underline{Q} = Q\cos 30^o\underline{i} - Q\sin 30^o\underline{j} = 3.12\underline{i} - 1.8\underline{j}$ m

32. (a) $\underline{A} = 3\underline{i} + 2\underline{j}$ m and $\underline{B} = -4\underline{i} + 4\underline{j}$ m, so $\underline{B} - \underline{A} = -7\underline{i} + 2\underline{j}$ m;
(b) 7.28 m at 164° counterclockwise from +x axis

33. $\underline{A} = 4\underline{i} + 3\underline{j}$ and $\underline{A} + \underline{B} = -3\underline{i}$, so $\underline{B} = -7\underline{i} - 3\underline{j}$

34. (a) $\underline{A} = 3\underline{i} + 5.2\underline{j}$ m, $\underline{B} = 5.2\underline{i} - 3\underline{j}$ m, $\underline{B} - \underline{A} = 2.2\underline{i} - 8.2\underline{j}$ m
(b) $\underline{A} = 5.2\underline{i} + 3\underline{j}$ m, $\underline{B} = -5.8\underline{i} + 1.55\underline{j}$ m,
$\underline{B} - \underline{A} = -11.0\underline{i} - 1.45\underline{j}$ m

35. (a) $\underline{W} = W \sin 30^o \underline{i} - W \cos 30^o \underline{j} = 10\underline{i} - 17.3\underline{j}$;
$\underline{T} = -T \cos 37^o \underline{i} + T \sin 37^o \underline{j} = -24\underline{i} + 18\underline{j}$;
$\underline{F} = F \cos 30^o \underline{i} + F \sin 30^o \underline{j} = 8.66\underline{i} + 5\underline{j}$
(b) $\underline{W} + \underline{T} + \underline{F} = -5.34\underline{i} + 5.7\underline{j}$

36. $\underline{A} = A \sin 30^o \underline{i} + A \cos 30^o \underline{j} = -2.50\underline{i} + 4.33\underline{j}$ m,
$\underline{B} = B \sin 15^o \underline{i} + B \cos 15^o \underline{j} = 1.04\underline{i} + 3.86\underline{j}$ m,
$\underline{C} = -2.00\underline{k}$ m. $\underline{R} = \underline{A} + \underline{B} + \underline{C} = -1.46\underline{i} + 8.19\underline{j} - 2.00\underline{k}$ m.

37. (a) $\underline{B} = \pm 5\underline{k}$;
(b) Since $\underline{A}$ is at 56.3^o to the x axis, $\underline{B}$ is 56.3^o to y axis. Thus $B_y/B_x = \pm 2/3$, that is, $\underline{B} = C(3\underline{i} - 2\underline{j})$ or $C(-3\underline{i} + 2\underline{j})$, which means $B = (13)^{1/2}C$. But B = 5 m, so $C = 5/(13)^{1/2}$ and $\underline{B} = -4.16\underline{i} + 2.77\underline{j}$ m, or $4.16\underline{i} - 2.77\underline{j}$ m.

38. $-181\underline{i} - 84.5\underline{j} + 100\underline{k}$ m

39. $\cos\theta = \underline{A}.\underline{B}/AB = (-4)/(5x13)^{1/2}$, so $\theta = 120^o$

40. (a) $\underline{A}.\underline{B} = -10 + 2 + 3 = -5$; (b) $A^2 - B^2 = -16$

41. $\cos\theta = \underline{A}.\underline{B}/AB = -4/15$, thus $\theta = 105^o$

42. $\cos\theta = \underline{A}.\underline{B}/AB = -21.2/(4.82)(4.79) = -0.918$, so $\theta = 157^o$.

43. $\underline{A}.\underline{B} = (3.2)(2.4)\cos 115^o = -3.25\ m^2$

44. (a) $\underline{A} + \underline{B}$, and $\underline{A} - \underline{B}$ or $\underline{B} - \underline{A}$; (b) $(\underline{A} + \underline{B}).(\underline{A} - \underline{B}) = 0$

45. $\underline{A}.\underline{i} = A_x = A \cos\alpha$, thus $\cos\alpha = \underline{A}.\underline{i}/A$, etc.
$\cos\alpha = \underline{A}.\underline{i}/A = 3/(14)^{1/2}$, thus $\alpha = 36.7^o$;
$\cos\beta = \underline{A}.\underline{j}/A = 2/(14)^{1/2}$, thus $\beta = 57.7^o$;
$\cos\gamma = \underline{A}.\underline{k}/A = 1/(14)^{1/2}$, thus $\gamma = 74.5^o$.

46. (a) $\underline{C}.(\underline{A} + \underline{B}) = (-2\underline{j}).(4\underline{i} - 4\underline{j}) = +8$; (b) Not allowed;
(c) $C + (\underline{A}.\underline{B}) = 2 + 3 = 5$; (d) 6; (e) $-6\underline{j}$

47. $\cos\theta = \underline{A}.\underline{B}/AB = 1/(6x25)^{1/2}$, thus $\theta = 85.3^o$. The component of $\underline{A}$ along $\underline{B}$ is $A \cos\theta = 0.2$ m.

48. $\underline{A} \times \underline{B} = (\underline{i} + 2\underline{j} - 4\underline{k}) \times (3\underline{i} - \underline{j} + 5\underline{k}) =$
$(-\underline{k} - 5\underline{j}) + (-6\underline{k} + 10\underline{i}) + (-12\underline{j} - 4\underline{i}) = 6\underline{i} - 17\underline{j} - 7\underline{k}$

49. (a) $\underline{A}.(\underline{A} \times \underline{B}) = 0$; (b) $\underline{A}$ is perpendicular to $\underline{A} \times \underline{B}$.

50. $\underline{A} \times \underline{B} = (3.6)(4.4) \sin 135^{o}\ \underline{k} = 11.2\underline{k}\ m^2$

51. Area = Base x Height = $(B)(A \sin\theta) = |\underline{A} \times \underline{B}|$

52. (a) $3(8\underline{k}) = 24\underline{k}$; (b) 0; (c) Not allowed;
(d) $3\underline{i} \times 8\underline{k} = -24\underline{j}$; (e) $3\underline{i} + 8\underline{k}$

53. $\underline{A} = 2.83\underline{i} + 2.83\underline{j}$, $\underline{B} = 1.5\underline{j} + 2.6\underline{k}$.
$\underline{A} \times \underline{B} = 7.36\underline{i} - 7.36\underline{j} + 4.25\underline{k}$.

54. The unit vector $\underline{n}$ normal to $\underline{A}$ and $\underline{B}$ is given by

$$\underline{n} = (\underline{A} \times \underline{B})/(AB \sin\theta)$$
$$= (14\underline{i} + 19\underline{j} - \underline{k})/(23.6)$$

Thus, $\underline{C} = 5\underline{n} = 2.96\underline{i} + 4.02\underline{j} - 0.21\underline{k}$

55. $\underline{A} = 4.10\underline{i} + 2.87\underline{j}$, $\underline{B} = 1.84\underline{i} - 2.62\underline{j}$, so
$\underline{A} - \underline{B} = 2.26\underline{i} + 5.49\underline{j}$ m.

56. $\underline{A} = -2.5\underline{i} + 4.33\underline{j}$, $\underline{B} = -2.82\underline{i} - 1.03\underline{j}$, so
$\underline{A} + \underline{B} = -5.32\underline{i} + 3.30\underline{j}$ m

57. $\underline{A} = 3.28\underline{i} + 2.29\underline{j}$, $\underline{B} = -2.35\underline{i} + 0.855\underline{j}$, so
$\underline{B} - \underline{A} = -5.63\underline{i} - 1.44\underline{j}$ m.

58. $\underline{A} = 1.766\underline{i} + 0.939\underline{j}$, $\underline{B} = -0.634\underline{i} + 1.359\underline{j}$, $\underline{C} = 2.5\underline{i}$, so
$\underline{D} = -2.00\underline{i} + 3.66\underline{j}$ m.

59. $\underline{A} = 3.857\underline{i} + 4.596\underline{j}$, $\underline{C} = -2.867\underline{i} + 2.008\underline{j}$, so
$\underline{B} = -6.72\underline{i} - 2.59\underline{j}$ m, or 7.20 m at 21.1^{o} S of W.

60. $A^2 = A_x{}^2 + A_y{}^2 + A_z{}^2$ leads to $A_z = \pm 3.12$ m.

61. $\underline{A} + \underline{B} = -2.00\underline{i} + 3.464\underline{j}$, so $\underline{B} = -4\underline{i} - 0.464\underline{j}$ m.

62. $\underline{B} = -4\underline{i} + 6\underline{j}$ m.

63. $-\underline{i} \pm 1.73\underline{j}$ m.

64. $(1.732 - B\cos 50^{o})^2 + (1 + B\sin 50^{o})^2 = (2.6)^2$. Find B = 2.04 m and $\theta = 80.7^{o}$ N of E. Graphically, about 2 m and 80°.

65. (a) $A^2 + B^2 = (2.12)^2$, so A = B = 1.50 m;

(b) $\underline{C} = (A\cos 30 - B\sin 30)\underline{i} + (A\sin 30 + B\cos 30)\underline{j} =$
$= 0.549\underline{i} + 2.05\underline{j}$ m

(c) $2.05\underline{i} - 0.549\underline{j}$ m.

66. $B^2 = A^2 + C^2 - 2AC\cos 80^o$, so C = 3.24 m. Law of Sines gives $\theta = 35.7^o$ N of W.

67. (a) $4\underline{i} - 3\underline{j} + 6\underline{k}$ m; (b) 7.81 m; (c) 3.74 + 4.58 = 8.32 m.

68. (a) $-3\underline{i} + 5\underline{j}$ m; (b) 5.83 m.

69. (a) $6\underline{i} - 4\underline{j} + 3\underline{k}$ m; (b) 7.81 m; (c) 4.583 - 3.742 = 0.841 m.

70. $2\underline{R} = 6\underline{i} - 2\underline{j} + 4\underline{k}$ m.

Problems

1. Since $\underline{A}\cdot\underline{B} = 0$, we have $3B_x + 6B_y = 0$, i.e. $B_x = -2B_y$, which means $\underline{B} = C(2\underline{i} - \underline{j})$ or $C(-2\underline{i} + \underline{j})$, where C is a constant. Since $B = C(5)^{1/2} = 5$ m, we find that $C = (5)^{1/2}$, thus $\underline{B} = 4.46\underline{i} - 2.23\underline{j}$ m, or $-4.46\underline{i} + 2.23\underline{j}$ m.

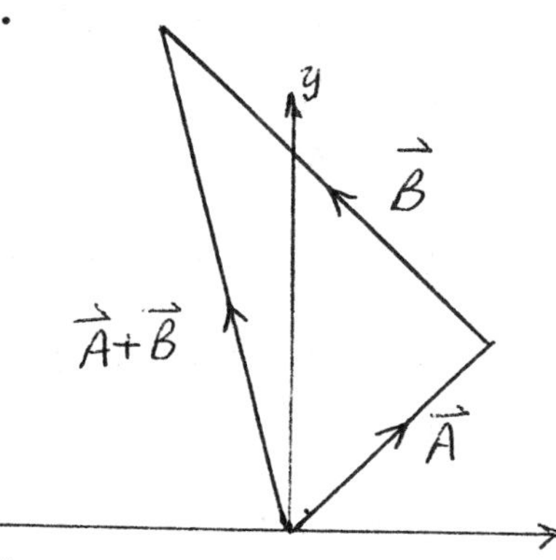

2. Given $\underline{A} = 1.41\underline{i} + 1.41\underline{j}$ m.

(a) $\underline{B} = -3\underline{A} = -4.24\underline{i} - 4.24\underline{j}$ m;

(b) $\underline{B} = \underline{A} = 1.41\underline{i} + 1.41\underline{j}$ m;

(c) From the figure we find $B = (4^2 - 2^2)^{1/2} = 3.46$ m, so
$\underline{B} = -B\cos 45^o\underline{i} + B\sin 45^o\underline{j}$
$= -2.45\underline{i} - 2.45\underline{j}$ m;

(d) $\underline{A} + \underline{B} = -4\underline{j}$, thus $\underline{B} = -1.41\underline{i} - 5.41\underline{j}$ m.

3. (a) $\underline{R} = \underline{A} + \underline{B}$ is one diagonal. One half of R is $A \cos\theta/2$, thus $|\underline{A} + \underline{B}| = 2A \cos(\theta/2)$.
(b) $\underline{S} = \underline{A} - \underline{B}$, is the other diagonal. One half of S is $A \sin\theta/2$, thus $|\underline{A} - \underline{B}| = 2A \sin(\theta/2)$.

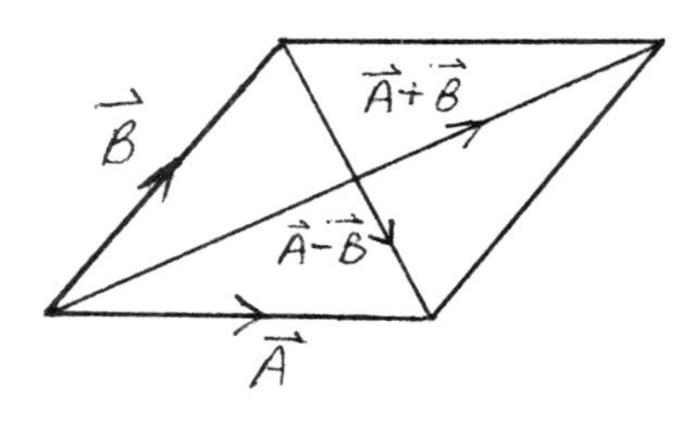

4. (a) $\underline{r} = r \cos\phi\ \underline{i} + r \cos\phi\ \underline{j}$,
$\underline{r}' = r \cos(\phi - \theta)\ \underline{i} + r \sin(\phi - \theta)\ \underline{j}$
(b) $x' = r \cos\phi \cos\theta + r \sin\phi \sin\theta = x \cos\theta + y \sin\theta$
$y' = r \sin\phi \cos\theta - r \cos\phi \sin\theta = y \cos\theta - x \sin\theta$

5. (a) $\underline{D} = (\underline{i} + \underline{j} + \underline{k})L$ and $D = (3)^{1/2}L$. The angle γ between $\underline{D}$ and the z axis is given by $\cos\gamma = \underline{D}.\underline{k}/D = L(\underline{i} + \underline{j} + \underline{k}).\underline{k}/(3)^{1/2}L$, thus $\gamma = 54.7^o$.
(b) $\underline{A} = (\underline{i} + \underline{j})L$, $\underline{B} = (\underline{i} + \underline{k})L$, thus $\underline{A}.\underline{B} = L^2$. The angle is given by $\cos\theta = 1/2$, so $\theta = 60^o$.
(c) $\cos\delta = \underline{A}.\underline{D}/AD = 2/(6)^{1/2}$, so $\delta = 35.3^o$.

6. $\underline{A} = 1.73i + \underline{j} + 1.2\underline{k}$ km, $\underline{B} = 0.707\underline{i} - 0.707\underline{j} + 0.8\underline{k}$ km.
$\underline{B} - \underline{A} = -1.02\underline{i} - 1.71\underline{j} - 0.40\underline{k}$ km

7. From E2.51, we know that Area of base = $|\underline{B} \times \underline{C}|$, where the vector $\underline{B} \times \underline{C}$ is along $\underline{n}$, the unit vector normal to the plane of $\underline{A}$ and $\underline{B}$. The height of the figure is $H = A \cos\theta$, where θ is the angle between $\underline{A}$ and $\underline{n}$, thus $H = \underline{A}.\underline{n}$. The volume is V = Height x Area of base = $(\underline{A}.\underline{n})(|\underline{A} \times \underline{B}|) = \underline{A}.(\underline{B} \times \underline{C})$

8. Let $\underline{B} = B_x\underline{i}$, then show $\underline{A} \times (B_x\underline{i} \times \underline{C}) = B_x\underline{i}(\underline{A}.\underline{C}) - \underline{C}(A_xB_x)$

9. Place vectors on a circle of unit radius; take components:
$\underline{r} = \cos\theta\ \underline{i} + \sin\theta\ \underline{j}$, $\underline{\theta} = -\sin\theta\ \underline{i} + \cos\theta\ \underline{j}$

10. $\cos\alpha = \underline{A}.\underline{i}/A = x/r$, $\cos\beta = y/r$, $\cos\gamma = z/r$.
$\cos^2\alpha + \cos^2\beta + \cos^2\gamma = (x^2 + y^2 + z^2)/r^2 = 1$.

11. $A_x = 10 \cos65^o = 4.23$ m, $A_z = 10 \cos40^o = 7.66$ m,
$A_y = (A^2 - A_x{}^2 - A_z{}^2)^{1/2} = 4.84$ m.

CHAPTER 3

Exercises

1. 10.1 m/s

2. (a) 24 km/h, 18.7 km/h; (b) 1.75 h

3. (a) Δx = 320 km; 3.2 h, thus 0.8 h = 48 min.

4. (a) Δx = 720 m + 1920 m = 2640 km, so v = 2640 m/180 s = 14.7 m/s;
   (b) Δx = 720 m - 1920 m = -1200 m, so v = -6.67 m/s

5. 38.5 km/2.817 h = 13.7 km/h or 3.8 m/s

6. (a) 5 m/s; (b) 2 m/s

7. (a) +5 m/s; (b) +2.5 m/s; (c) -5 m/s; (d) 0

8. Time for Mansell to cover the 315 km distance:
   T = 315 km/(208 km/h) = 1.514 h = 5.452×10^4 s
   In this time De Cesaris travels 57.53 m/s x T = 313.66 km
   Thus he loses by 315 km - 313.66 km = 1.34 km

9. B covers 485 km when A finishes, thus its average speed is
   v_B = 485 km/4 h = 121.25 km/h. Time to complete 500 km is
   T_B = 500 km/v_B = 4.124 h = 4 h 7 min 25 s.

10. Distance = 82.67 h x 65.5 km/h = 5415 km. For the Queen Mary
    v = 5415 km/92.7 h = 58.4 km/h

11. (a) 6 m/s, (b) 3.5 m/s; (c) -7 m/s

12. (a) 5 m/s; (b) 0; (c) -10 m/s; (d) -5 m/s; (e) 0

13. (a) 550 m/30 s = 18.3 m/s;
    (b) 50 m/30 s = 1.67 m/s;
    (c) -45 m/s/(30 s) = -1.5 m/s^2

14. (a) 2 m/s^2; (b) -12.5 m/s^2; (c) 200 m/s^2

15. a_{av} = -70 m/s/(0.04 s) = -1750 m/s^2

16. (a) -3 m/s; (b) -1.5 m/s^2

17. (a) 0.4 km/5.4 s = 74.1 m/s; (b) (410 km/360 s)/(5.4 s) = = 21.1 m/s^2.

18. Time for total trip should be 4 h. The first leg took 1.8 h = 108 min. Thus second leg must take 240 - 108 - 12 = 120 min. Therefore, v = 120 km/2 h = 60 km/h

19. (a) 2.83 m/s^2; (b) 1.88 m/s^2; (c) 1.06 m/s^2; (d) -6.94 m/s^2

20. (b) +1 m/s; (c) -5 m/s; (d) -5 m/s

21. (a) x = 1.04 m to 4.98 m; (c) ≈5 m/s; (d) 5 m/s
    (b) 0.425 m/0.1 s = 4.25 m/s
    (c) 2.25 m/0.55 s = 5 m/s
    (d) x(0.5) = 4.3301 m, x(0.501) = 4.3354 m.
    Thus, $\Delta x/\Delta t$ = 0.0052 m/0.01 s = 5.25 m/s

22. (a) -1.21 m/s; (b) -1.35 m/s; (c) -1.34 m/s

23. (a) v = -5 + 6t = 13 m/s, a = 6 m/s^2;
    (b) v = 0 at t = 5/6 = 0.83 s

24. (a) (20 m/s)/5 s = +4 m/s^2; (b) 5 m/s^2; (c) 2 s

25. (a) 0; 3 s; (b) 3 s; (c) 2 m/s^2; (d) 4 m/s^2

26. (a) 3-4 s; (b) 0, 7 s; (c) 1-2 s, 5-6 s; (d) 4-5 s;
    (e) 6-7 s; (f) 0-1 s; 7-8 s; (g) 2-3 s

27. (a) 15 m; (b) 11.7 m/s

28. 27 m, 9 m/s

29. (a) 2.5 m/s; (b) 5.83 m/s

30. (a) 0; (b) 2.4 m/s

31. (c) 1.67 m/s^2; (d) 5 m/s^2

32. (c) 2 m/s^2; (d) 4 m/s^2

35. (b) 14.3 mph/s; (c) 8.4 s; (d) 13 s at 105 mph

36. (a) 5 s, 40 m; (b) v_A = 8 m/s, v_B = 16 m/s

37. (a) 9.2 mph/s, 4.8 mph/s, 3 mph/s;
(b) 9.2 mph/s = 13.5 ft/s^2. $\Delta x = 1/2(13.5\ ft/s^2)(17\ s)^2 \approx 2000$ ft;
(c) (72 + 240 + 504) mph.s ≈ 1200 ft

38. (a) $6.75 \times 10^5\ m/s^2$; (b) 1.33×10^{-3} s

39. $\Delta x = v_0t + 0.5at^2$: 6 = -2x5 + 12.5a, so a = 1.28 m/s^2, E

40. Use Eq. 3.12 with v = 0 and v_0 = 31.1 m/s.
(a) 4.11 s, -7.56 m/s^2;
(b) 6.4×10^{-3} s, -484 m/s^2

41. 96 km/h = 26.7 m/s so a = 5.34 m/s^2.
(a) $\Delta t = (10\ m/s)/(5.34\ m/s^2) = 1.87$ s,
$\Delta x = (v^2 - v_0^2)/2a = (300\ m^2/s^2)/2(5.34\ m/s^2) = 28.1$ m;
(b) Δt = 1.87 s, $\Delta x = (500\ m^2/s^2)/2a = 46.8$ m.

42. (a) 33 s, 5440 m; (b) 3.11 h, 6.27×10^6 m;
(c) 34.7 d, 4.5×10^{13} m

43. (a) Using $x = 1/2\ at^2$: t_F = 20 s, t_R = 24 s, thus Δt = 4 s.
(b) From v = at: v_F = 10 m/s; v_R = 12 m/s.

44. (a) x_C = 60t, x_T = 1 + 40t. Equate x's: t = 0.05 h, and x = 3 km;
(b) x_C = 60t, x_T = 1 - 40t give t = 0.01 h and x = 0.6 km

45. (a) In 0.5 s, he moves 15 m, so Δx = 70 - 15 = 55 m. Time needed to stop is t = (30 m/s)/(8 m/s^2) = 3.75 s. In this time he travels $\Delta x = 30t - 4t^2$ = 112.5 - 56.3 = 56.3 m. Needs 1.3 m more.

46. (a) From Eq. 3.12, a = -3.2 m/s^2. With v = 0 and v_0 = 16 m/s, we find Δx = 40 m;
(b) From Eq. 3.9: t = 7.5 s

47. (a) From Eq. 3.10: 32 = 48 + 8a, so a = -2 m/s^2;
(b) v = 12 m/s - (2 m/s^2)(4 s) = -4 m/s.

48. From Eq. 3.10: $30 = 20t - 2.5t^2$, thus t = 2 s, 6 s. Reject 6 s since car stops at t = 4 s. So $v_{av} = 30$ m/2 s = 15 m/s

49. From $\Delta v = a\,\Delta t$, we have $-16 = 4a$, so $a = -4$ m/s^2.
(a) From $v = v_o + at$ we find $v_o = 16$ m/s.
Eq. 3.11 then leads to $x_o = -28$ m.
(b) From $0 = v_o + at$ we find that the particle stops and turns around at t = 4 s, which means x(4) = 4 m. At t = 7 s, we find x(7) = -14 m. The total distance traveled between 3 s and 7 s is 20 m, so the average speed is 5 m/s
(c) $v_{av} = \Delta x/\Delta t = (-16\text{ m})/4\text{ s} = -4$ m/s.

50. (a) $v_0^2 = 2gy$ leads to $v_0 = 7.92$ m/s; (b) $t = 2v_0/g = 1.62$ s

51. From $v_o^2 = 2g\,\Delta y$, $v_0 = 22.1$ m/s. Then $\Delta y = (v_o^2/2g) = 150$ m.

52. (a) $\Delta y = 1/2\ gt^2 = 11$ m; (b) 14.7 m/s

53. $v^2 = 2g\,\Delta y$ leads to v = 6.93 m/s

54. (a) $v_0^2 = g$, so $v_0 = 3.13$ m/s; (b) $v_0^2 = 12g$, $v_0 = 10.8$ m/s

55. (a) From $0 = 7^2 - 2a(0.6)$, we find a = 40.8 m/s^2;
(b) From $0 = 7^2 - 2a(0.1)$, we find a = 245 m/s^2.

56. (a) Fall: $\Delta y = 4.9(4^2) = 78.4$ m; (b) $t = 2v_0/g$, so $v_0 = 39.2$ m/s.

57. (a) $v^2 = 20^2 - 2g(-24)$ so v = -29.5 m/s. From $v = v_o - gt$, we find t = 5.05 s.
(b) $0 = 20^2 - 2g\Delta y$, so $\Delta y = 20.4$ m and H = 44.4 m.
(c) 29.5 m/s.

58. (a) $v^2 = 900 - 2(9.8)(25)$, so $v = \pm 20.2$ m/s;
(b) Since $v = \pm 15$ m/s, we find t = 1.53 s, 4.59 s;
(c) $40 = 30t - 4.9t^2$ leads to t = 1.96 s, 4.16 s

59. (a) $10.2 = 20t - 4.9t^2$, gives t = 0.6 s, 3.48 s;
(b) $v = \pm 10$ m/s leads to t = 1.02 s, 3.06 s

60. From Eq. 3.12: $30 = 2v_o - 4.9\text{x}4$, so $v_o = 24.8$ m/s.
From Eq. 3.10: v = 24.8 - 9.8x2 = 5.2 m/s.

Thus, $t = (24.8 + 5.2)/g = 3.06$ s

61. From $v_0^2 = 2g\Delta y$, we find $v_0 = 5.94$ m/s. The time to reach the maximum height is found from $0 = 5.94 - 9.8t$, thus $t = 0.606$s. The time for one round trip is $T = 2(0.606) + 0.3 = 1.512$ s, thus the time between balls is $1.512/3 = 0.504$ s. When one ball is at the top, the other two are at $y = 1.8\text{ m} - 4.9(0.504)^2 = 0.555$ m

62. Find $v_1 = -9.9$ m/s and $v_2 = +7.92$ m/s. $a = \Delta v/\Delta t = +495\text{ m/s}^2$

63. From $\Delta y = 20$ m we find $v_o = 19.8$ m/s.
(a) From Eq. 3.11: $-50 = 20t - 4.9t^2$, so $t = 5.84$ s
(b) From Eq. 3.12 with $\Delta y = -50$ m, we find $v = -37.4$ m/s.
(c) From Eq. 3.11: $-20 = 19.8t - 4.9t^2$, so $t = 4.88$ s.

64. (a) $-40 = 4v_0 - 4.9(4^2)$, so $v_0 = 9.6$ m/s.
From $0 = 9.6^2 - 19.6\Delta y$ we find $\Delta y = 4.7$ m, so $H = 44.7$ m;
(b) $v^2 = 9.6^2 - 19.6(-15)$ leads to $v = -19.7$ m/s

65. From Eq. 3.12: $v_o^2 = 2g'\Delta y = (3.0)(2.8\text{x}10^5)$, so $v_o = 917$ m/s. Then $0 = 917\text{ m/s} - 1.5t$, so $t = 10$ min.

66. (a) $-13 = v_0 - (9.8)(0.8)$, so $v_0 = -5.16$ m/s;
(b) $13^2 = 5.16^2 - 19.6H$, so $H = 7.26$ m;
(c) $-13 = -5.16 - 9.8t$, so $t = 1.85$ s

67. $(0.4v_0)^2 = v_0^2 - 19.6(4.2)$, thus $v_0 = 9.9$ m/s.
Then $0 = v_0^2 - 19.6H$, so $H = 5$ m.
The time of flight is $t = 2v_0/g = 2.02$ s

68. If H is the maximum height:

$$H/2 = v_0(2) - 4.9(2^2) \qquad \text{(i)}$$

but, we know $0 = v_0^2 - 19.6$ H. Substituting for H into (i):

$$v_0^2 - 78.4v_0 + 768 = 0$$

So $v_0 = 11.5$ m/s, 66.9 m/s. The 2nd solution would apply to the object moving downward at 2 s. The maximum height is $H = v_0^2/2g = 6.75$ m, 228 m

69. (a) $0.7v_0 = v_0 - 9.8(1.8)$, so $v_0 = 58.8$ m/s.

The maximum height is $H = v_0^2/19.6 = 176$ m, and the time of flight is $t = 2v_0/g = 12$ s

70. If H is the maximum height:

$$30^2 = v_o^2 - 2g(3H/4)$$
$$0 = v_o^2 - 2gH$$

Thus, $900 = gH/2$ and $H = 184$ m. Time of flight $t = 2v_0/g = 12.3$ s

Problems

1. (a) $\Delta v = 96$ km/h $= 26.7$ m/s, $\Delta t = 4.6$ s, so $a = 5.80$ m/s^2. $x_M = 12.5\,t$ whereas $x_C = (5.8)t^2/2$. Set $x_M = x_C$ to find $t = 4.31$ s and then $x = 53.9$ m.
   (b) $v_M = 12.5$ and $v_C = 5.8 \times 4.31 = 25$ m/s.

2. (a) Equate $x_T = 0.5t^2$ and $x_C = (t - 10)^2$, to find $t = 5.9$ s, 34.1 s. Must reject $t < 10$ s, so $x = 0.5(34.1)^2 = 581$ m.
   (b) $v_T = 34.1$ m/s, $v_c = a_C(t - 10) = 48.2$ m/s

3. $x_T = 38 + 20t$ and $x_C = 20t + t^2 = x_T + 11$ m. Thus $t = 7$ s and $\Delta x_T = 178 - 38 = 140$ m.

4. $x_A = 6t$, $x_B = 10 - 4t$, so they meet at $t = 1$ h. In this time the bird flies 20 km

5. The positions of the fronts of A and B are

$$x_A = 1500 + 50t; \qquad x_B = 3t^2/2$$

   (i) B reaches 60 m/s at $t_1 = 20$ s, at which time $x_A = 2500$ m and $x_B = 600$ m.
   (ii) If they travel at constant speed for time t_2:

$$x_A = 2500 + 50t_2; \qquad x_B = 600 + 60t_2$$

   so $t_2 = 190$ s. Finally, $t_1 + t_2 = 210$ s, and $x_A = (50 \text{ m/s})(210 \text{ s}) = 10.5$ km.

6. (a) Equate $x_T = 60 + 10t$ and $x_C = 30t - 2.5t^2$ to find t is not a real number, thus there is no collision.
   (b) $x_T = 65 + 10t$, $x_C = 15 + 30t - 1/2\,at^2$. Set $x_T = x_C$ to find t is not a real number if $a > 4.0$ m/s^2.

7. (a) $y_A = 100 + 5t - 4.9t^2$, $y_B = 100 - 20(t-2) - 4.9(t-2)^2$.
Set $y_A = y_B$ to find t = 3.78 s, and then y = 48.9 m.
(b) $v_A = 5 - 9.8(3.78) = -32$ m/s,
$v_B = -20 - 9.8(1.78) = -37.4$ m/s

8. (a) From Eq. 3.11:

$$300 = v_0(10) + 1/2\ a(100) \qquad \text{(i)}$$

$$600 = v_0(25) + 1/2\ a(625) \qquad \text{(ii)}$$

From (i) and (ii) we find $a = -0.8$ m/s^2 and $v_0 = 34$ m/s.
At t = 25 s, $v = 34 - 0.8(25) = 14$ m/s.
From Eq. 3.12: $0 = 14^2 - 2(0.8)\Delta x$, thus $\Delta x = 123$ m;
(b) $0 = 14 - 0.8t$, so t = 17.5 s

9. (a) $25^2 = 15^2 + 2a(100)$, so a = 2 m/s^2.
$v^2 = 25^2 + 2a(100) = 1025$, so v = 32 m/s
(b) $32 = 25 + 2t$, so t = 3.5 s

10. (a) $x_P = -9 + 4.5t$, $x_B = 0.5t^2$, so t = 3 s, 6 s. At t = 3 s, x = 4.5 m. (b) In the x vs t graph, the line for the person is tangent to the parabola for the bus, thus $v_P = v_B = 4.5$ m/s. The bus reaches this speed in 4.5 s at x = 10.1 m. If D is the initial distance of theperson from the busstop, then $x_p = -D + 4.5(4.5) = 10.13$, so D = 10.1 m. The extra time is (10.13 m - 9 m)/(4.5 m/s) = 0.25 s

11. $x_A = 20t + 1/2\ a_A t^2$; $x_B = 12t + 1/2\ a_B t^2$
Since $v_B = 36 = 12 + a_B t$, $a_B = 2$ m/s^2, so $x_B = 12t + t^2$.
At t = 12 s: $20t + 1/2\ a_A t^2 = 12t + t^2$ so $a_A = 2/3$ m/s^2.
$v_A = 20 + (2/3)(12) = 28$ m/s.

12. (a) The sum of the areas of the v vs t graph is $1/2\ (at_1^2) + (at_1)t_2 + 1/2\ (a/2)(2t_1^2) = 270$ (i)
We know $v_1 = at_1 = 20$ m/s, and that $3t_1 + t_2 = 21$ s. Using this in (i) we find $t_1 = 5$ s so that a = 20/5 = 4 m/s^2. 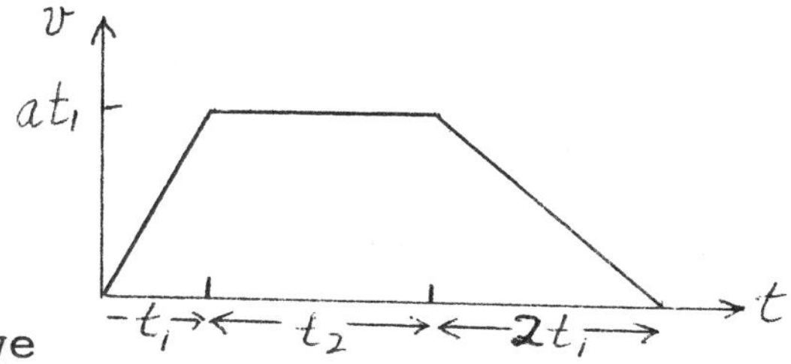
(b) $t_2 = 6$ s, so $\Delta x_2 = (20\text{ m/s})(6\text{ s}) = 120$ m

13. From the v vs t graph: $(vt_1)/2 = 200$, so $v = 400/t_1$. The deceleration phase must take $t_1/4$, thus the time during which the speed is constant is $t_2 = (33 - t_1 - t_1/4)$, $v(33 - 5t_1/4) = 160$. Using $v = 400/t_1$ we find $t_1 = 20$ s, and finally $t_2 = 8$ s.

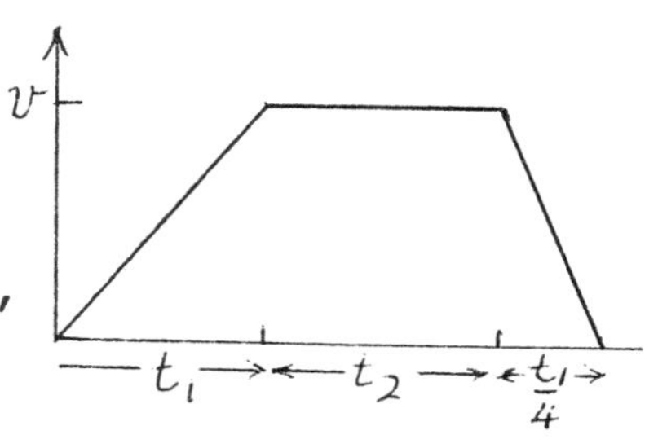

14. $x_A = 16t - 1.2t^2$, $x_B = 48 - 8t + 2t^2$. Setting $x_A = x_B$ leads to $t < 0$, which implies there is no colliison. However, we must consider the fact that B stops in 2 s at x = 40 m and we must find when A reaches this point. The equation $40 = 16t - 1.2t^2$ leads to t = 3.33 s.

15. (a) Note that 105 km/h = 29.2 m/s and 90 km/h = 25 m/s.
$a_C = 14.6$ m/s^2, $a_A = 12.5$ m/s^2.
(i) C accelerates until t = 2 s, while A starts at 0.5 s and accelerates until t = 2.5 s. At t = 2.5 s:
$x_C = 1/2\ (14.6)(2^2) + (29.2)(0.5) = 43.8$ m
$x_A = 100 + 1/2\ (12.5)(2^2) = 125$ m
Thus A is still ahead
(ii) At constant speed: $x_C = 43.8 + 29.2t$; $x_A = 125 + 25t$.
$x_A = x_C$ leads to t = 19.3 s which means C is already resting, so A is not caught.
(b) The closest approach occurs at t = 15 s, just before C stops. At this time
$x_C = 43.8 + 29.2(13) = 423.4$ m,
$x_A = 125 + 25(13) = 450$ m.
Thus the minimum distance is 26.6 m.

16. (a) $y_A = H + 15t - 4.9t^2$; $y_B = H - 4.9(t - 2)^2$
Set $y_A = y_B$ to find t = 4.26 s and y = H - 25 m, i.e. 25 m below the roof.
(b) $v_A = 15 - 9.8(4.26) = 26.7$ m/s,
$v_B = -9.8(2.26) = -22.1$ m/s

17. (a) $y_A = 25t - 4.9t^2$; $y_B = 95 - 15(t - 1) - 4.9(t - 1)^2$.
Set $y_A = y_B$ to find t = 3.48 s and y = 27.7 m

(b) $v_A = 25 - 9.8t = -9.1$ m/s,

$v_B = -15 - 9.8(t - 1) = -39.3$ m/s

18. (a) $\Delta y_1 = 1/2\ (4\ m/s^2)(64\ s^2) = 128$ m. At this time $v = 32$ m/s. With this initial velocity, the extra height is $\Delta y_2 = 52.2$ m. The total height is 180 m.

(b) The time in free-fall is given by

$$0 = 128 + 32t - 4.9t^2$$

which yields $t = 9.33$ s. The total time of flight is 17.3 s.

19. $y_B = 3 + 2t - 4.9t^2$ and $y_E = 2t = 1.56$ m. Thus $t = 0.78$ s. $0 = 2^2 - 2g\ \Delta y$ leads to $\Delta y = 0.2$ m. The distance traveled is $d = 0.4 + (3 - 1.56) = 1.84$ m.

20. (a) $0 = 20^2 - 19.6\Delta y$ gives $\Delta y = 20.4$ m, so the maximum height is 45.4 m

(b) $y_B = 25 + 20t - 4.9t^2$, $y_E = 25 + 7t$. The ball meets the floor when $y_B = y_E$, which occurs at $t = 2.65$ s

21. $1.25 = H - 4.9t^2$ and $0 = H - 4.9(t + 0.1)^2$.

Thus $t = 1.225$ s and $H = 8.6$ m

22. It lands at t, so $0 = H - 4.9t^2$ and falls 0.36H in $(t - 1)$. $0.36H = 1/2\ g(t - 1)^2$. From these $t = 2.5$ s and $H = 30.6$ m

23. (a) $H = 1/2\ gt^2 = 30.6$ m

(b) If t_1 is the time to fall and t_2 is the time for the sound to travel then $H = 4.9t_1^2$ and $H/330 = t_2$, where $t_1 + t_2 = 2.5$ s.

$$330(2.5 - t_1) = 4.9t_1^2$$

Thus $t_1 = 2.41$ s and $H = 28.5$ m.

CHAPTER 4

Exercises

1. (a) $\underline{v} = (6t - 2)\underline{i} - 3t^2\underline{j} = 10\underline{i} - 12\,\underline{j}$ m/s;
   (b) $\underline{a} = 6\underline{i} - 6t\underline{j} = 6\underline{i} - 24\underline{j}$ m/s^2,
   (c) $\underline{a}_{av} = 6\underline{i} - 12\underline{j}$ m/s^2.

2. (a) $0.5\underline{i} - 1.5\underline{j} + 2\underline{k}$ m/s;
   (b) $\underline{v}_1 = \underline{v}_2 - \underline{a}\Delta t = 40\underline{i} - 10\underline{j} - 2\underline{k}$ m/s

3. (a) $\underline{v}_{av} = 4\underline{i} + 7\underline{j}$ m/s;
   (b) $\underline{v}_i = 12\underline{i} + 9\underline{j}$ and $\underline{v}_f = 18\underline{i} + 24\underline{j}$; so $\underline{a}_{av} = 1.2\underline{i} + 3\underline{j}$ m/s^2.

4. (a) 350 m; (b) 250 m, 37° N of E; (c) 10 m/s, 37° N of E;
   (d) 14 m/s; (e) $\underline{a}_{av} = (10\underline{j} - 20\underline{i})/25 = -0.8\underline{i} + 0.4\underline{j}$ m/s^2

5. (a) $\Delta\underline{r} = (0.707R\underline{i} + 0.707R\underline{j}) - R\underline{j} = -0.9\underline{i} - 0.37\underline{j}$ m;
   (b) $\Delta\underline{r} = (0.707R\underline{i} - 0.707R\underline{j}) - R\underline{j} = 0.3\underline{i} - 0.71\underline{j}$ m/s;
   (c) $\Delta\underline{v} = -(2)^{1/2}\underline{i}$, $\underline{a}_{av} = \Delta\underline{v}/\Delta t = -0.71\underline{i}$ m/s^2

6. The 5810 km distance is along the curved surface of the earth. The angular displacement is $\Delta\theta = 5810$ km/6370 km $= 0.912$ rad. The magnitude of the displacement is the length of the straight line joining the initial and final points: $2\tan(\Delta\theta/2)$ = 5611 km. Thus, $v_{av} = 5611$ km/33.5 h = 167 km/h

7. (a) From Eq. 4.11, $0 = 100 + 20t - 4.9t^2$, thus $t = 7$ s.
   (b) From Eq. 4.12, $0 = 400 - 2gH$, so $H = 120$ m
   (c) $R = v_x\, t = 105$ m.
   (d) $v_y = 20 - gt = -48.6$ m/s, so $\underline{v} = 15\underline{i} - 48.6\underline{j}$ m/s

8. (a) From Eq. 4.11: $0 = 40 - (v_o \sin37^o)2 - 4.9(2^2)$, so $v_o = 16.9$ m/s. Then, $R = (v_o \cos37^o)\, t = 27.0$ m;
   (b) $v_y = -v_o \sin37^o - 9.8t = -29.8$ m/s, $v_x = 13.5$ m/s.
   $\tan\theta = v_y/v_x = 2.21$, thus $\theta = 65.6^o$ below the horizontal

9. From Example 4.2, $R = v_o^2 \sin2\theta/g = 20.3$ m
   From Eq. 4.11: $0 = 0 + (v_o \cos45^o)t - 4.9t^2$, so $t = 2.03$ s.
   Person must run at $v = (30\text{ m} - 20.3\text{ m})/2.03\text{ s} = 4.77$ m/s toward the thrower.

10. From Example 4.2: $\sin 2\theta = Rg/v_o^2$ thus $\theta = 31.4^o$ or 58.6^o

11. (a) $t = (18.3\text{ m})/(30\text{ m/s}) = 0.61\text{ s}$, so $\Delta y = -4.9t^2 = -1.82\text{ m}$
(b) $x = 18.3 = 30\cos\theta\, t$; $y = 0 = 30\sin\theta\, t - 4.9t^2$.
Substitute $t = 18.3/(30\cos\theta)$ into the equation for y to find:
$$\sin 2\theta = 9.8(0.61)^2/18.3$$
Thus $\theta = 5.75^o$ or 84.2^o (reject). $D = 18.3\tan 5.75^o = 1.84\text{ m}$
OR: From Example 4.2, $\sin 2\theta = Rg/v_o^2 = 0.199$, so $\theta = 5.75^o$.

12. (a) From $x = 3 = v_o t$ and $\Delta y = -0.05 = -4.9t^2$, find
$v_o = 29.7\text{ m/s}$;
(b) From Example 4.2, $\sin 2\theta = Rg/v_o^2 = 0.0333$, so $\theta = 0.95^o$

13. From Example 4.2, $R = v_o^2 \sin 2\theta/g$, we find $v_o = 31.3\text{ m/s}$
From Eq. 4.12, $H = v_o^2/2g = 50\text{ m}$

14. (a) $0 = 200 - v_o \sin 37^o(4) - 4.9(4^2)$, so $v_o = 50.5\text{ m/s}$;
(b) $x = v_o \cos 37^o\, t = 161\text{ m}$

15. (a) From $v_o^2 = Rg/\sin 2\theta$, we find $v_o = 9.81\text{ m/s}$.
(b) 10.9 m/s.

16. $x = v_o t$, $\Delta y = -10^{-4}\text{ m} = -4.9t^2$, so $x = v_o(10^{-4}/4.9)^{1/2}$
$= 13.6\text{ km}$

17. $x = 4 = v_o \cos\theta\, t$; $y = 0.8 = v_o \sin\theta\, t - 4.9t^2$. Substitute
$t = 4/(v_o\cos\theta)$ into the second equation to find $v_o = 7\text{ m/s}$.

18. Use $R = v_o^2 \sin 2\theta/g$ with $\theta = 15^o$ to find $v_o = 34.3\text{ m/s}$

19. (a) $50 = 27\cos(32^o)t$, so $t = 2.18\text{ s}$, then
$y = 1 + 27\sin(32^o)t - 4.9t^2 = 8.9\text{ m}$.
(b) $R = v_o^2 \sin 2\theta/g = 66.9\text{ m}$, and the time of flight is
$t = R/(v_o\cos\theta) = 2.92\text{ s}$. Player has $\Delta t = 2.92 - 0.5 = 2.42\text{ s}$
to run 16.9 m, so $v = 16.9\text{ m}/2.42\text{ s} = 7\text{ m/s}$.

20. $1.6 = v_o t$, and $-1 = -4.9t^2$, so $t = 0.452\text{ s}$ and $v_o = 3.54\text{ m/s}$

21. (a) Speed of rocket at 6.5 s is 52 m/s and it has risen to a height of $(26\text{ m/s})(6.5\text{ s})\sin 70^o = 169\sin 70^o = 159\text{ m}$. The rest of the trip is in free-fall. To find the highest point,

$0 = (52\ \sin 70^o)^2 - 2g\ \Delta y$, yields $\Delta y = 122$ m.

Thus H = 122 + 159 = 281 m.

(b) In the initial phase the horizontal displacement is $\Delta x_1 = 169\ \cos 70^o = 57.8$ m. During free-fall:

$x = 52\cos\theta\ t$; $y = 0 = 159 + 52\sin\theta\ t - 4.9t^2$,

which yields t = 12.55 s and then $\Delta x_2 = 223.2$ m.

The total horizontal range is R = 223.2 + 57.8 = 281 m.

22. (a) Given $v_x = 24$ m/s, $v_y = -8$ m/s for $\Delta y = 9.8$ m. From Eq. 4.12 we find $v_{oy} = 16$ m/s, thus $\underline{v}_o = 24\underline{i} + 16\underline{j}$ m/s.

(b) The maximum height is $H = v_{oy}^2/2g = 13.1$ m

23. (a) Given R = 8.3 m and $v_o = 9.7$ m/s. Using $\sin 2\theta = Rg/v_o^2$, we find $\theta = 29.9^o$.

(b) $0 = (v_o\ \sin\theta)^2 - 2g\Delta y$, leads to $\Delta y = 1.19$ m;

(c) $t = 2\ v_o\ \sin\theta/g = 0.99$ s

24. (a) $v_o\ \cos\theta = 15$ and $v_o\ \sin\theta = 29.4$, thus $v_o = 33.0$ m/s and $R = v_o^2\ \sin 2\theta/g = 90$ m

(b) $\tan\theta = v_y/v_x = -29.4/15$, so $\theta = 63^o$ below horizontal

25. Initial angle is given by $\tan\theta = 10/40$, so $\theta = 14^o$.

$x = (20\cos 14^o)t$; $y = 0 = 10 - (20\sin 14^o)t - 4.9t^2$, thus t = = 1.02 s and $R = v_x\ t = 19.8$ m. The ball misses by 20.2 m.

26. (a) $v_x = 20$ m/s, $v_{oy} = 25.4$ m/s so $v_o = 32.3$ m/s and $\theta = 51.8^o$

Then, $R = v_o^2\ \sin 2\theta/g = 104$ m;

(b) $H = v_{oy}^2/2g = 32.9$ m

27. $x = v_o\ t$, so $t = 8.33 \times 10^{-9}$ s.

$y = 1/2\ at^2 = 1/2\ (4\text{x}10^{14})(69.4\text{x}10^{-18}) = 1.39$ cm.

$\tan\theta = at/v_o = 1.39$, so $\theta = 54.2^o$.

28. $\sin 2\theta = Rg/v_o^2 = 0.907$, thus $\theta = 32.6^o$, 57.4^o

29. (a) $x = 42 = v_o\ \cos 30^o\ t$, $y = 0 = 2 + v_o\ \sin 30^o\ t - 4.9t^2$.

Using $t = 42/(v_o\ \cos 30^o)$ in the second equation, $v_o = 21.0$ m/s

(b) t = 2.31 s.

(c) $\Delta y = (v_o\ \sin\theta)^2/2g = 5.63$ m, so maximum height is 7.63 m.

30. $70 = (20\cos\theta)(4)$, so $\theta = 29^o$,
$0 = H - (20\sin 29^o)(4) - 4.9(4^2)$, so $H = 117$ m

31. (a) and (b) $x = 200 = (50\cos 30^o)t$, so $t = 4.62$ s. Using this t in $y = 50\sin 30^o\, t - 4.9t^2 = 10.9$ m. It would hit the wall.
(c) $v_y = 25 - g(4.62) = -20.3$ m/s.
$\tan\theta = v_y/v_x$ so $\theta = 25.1^o$ below the horizontal.

32. $H = (v_o \sin\theta)^2/2g = 2$ m, thus $v_o = 12.5$ m/s

33. $x = 17 = v_o \cos 42^o\, t$; $y = 0 = 2.1 = v_o \sin 42^o\, t - 4.9\, t^2$.
Substituting $t = 17/(v_o \cos 42^o)$ into the second equation and solving: $v_o = 12.1$ m/s. Next, $\theta = 40^o$, so $v_o \sin\theta = 7.78$ m/s:
$y = 0 = 2.1 + 7.78t - 4.9t^2$, so $t = 1.82$ s and $R = 16.9$ m
So the second range is 0.1 m less

34. $t = 12/30 = 0.4$ s; $y = 2.4 - 4.9(0.4)^2 = 1.62$ m, which is 0.72 m above the net

35. (a) and (b) $t = 64/(v_o \cos 35^o)$, thus $t = 1.85$ s. Use this in $y = 28 = (v_o \sin 35^o)t - 4.9t^2$, to find $v_o = 42.2$ m/s;
(c) $v_x = v_o \cos 35^o = 34.6$ m/s,
$v_y = (v_o \sin 35^o) - 9.8(1.85) = 6.07$ m/s
Thus, $\underline{v} = 34.6\underline{i} + 6.1\underline{j}$ m/s

36. $v_x = 16.1$ m/s, $v_{oy} = 19.2$ m/s
At the given point, $v_y = \pm v_x \tan 30^o = \pm 9.28$ m/s
$\pm 9.28 = 19.2 - 9.8t$, thus $t = 1.01$ s, 2.91 s

37. Use $a = 4\pi^2 r/T^2$ for all.
(a) 3.37×10^{-2} m/s^2; (b) 5.9×10^{-3} m/s^2; (c) 3×10^{-10} m/s^2

38. (a) 9.13×10^{22} m/s^2; (b) 7.90×10^5 m/s^2

39. (a) $a = (27.8\text{ m/s})^2/50\text{ m} = 1.58g$.
(b) $a = (417\text{ m/s})^2/5000\text{ m} = 3.55g$
(c) $a = 16.1g$; (d) $T = 1.8$ s, $a = 0.187g$
(e) $T = 2$ ms, $a = 1.51\times 10^5$ g

40. With $r = 35{,}800 + 6400 = 42{,}200$ km, and $T = 24 \times 3600$ s, $a = 0.223$ m/s^2

41. $T^2 = 4\pi^2 r/a$, so $T = 142$ s

42. Use $r = 6370$ km and $a = 9.8$ m/s^2 to find $T = 84.4$ min

43. $r = 1.2\sin 20^o = 0.41$ m, so $a = (1.21)^2/(0.41) = 3.57$ m/s^2.

44. $v^2 = ar$ leads to $v = 19.4$ m/s

45. $v = 40$ m/s, $r = 1.27$ m, so $a = v^2/r = 1260$ m/s^2

46. $v = 7.35$ m/s, $T = 2\pi r/v = 0.513$ s. For 8 revs, $t = 4.1$ s

47. $r = v^2/a = (83.3 \text{ m/s})^2/(0.05g) = 14.2$ km

48. (a) $\underline{v}_{BA} = \underline{v}_B - \underline{v}_A = -3\underline{i} + 4\underline{j}$ m/s
(b) $\underline{v}_{AB} = \underline{v}_A - \underline{v}_B = 3\underline{i} - 4\underline{j}$ m/s
Both $\underline{v}_A$ and $\underline{v}_B$ are relative to the ground.

49. $\underline{v}_{BS} = \underline{v}_{BL} + \underline{v}_{LS}$, or in simpler notation $\underline{R} = \underline{W} - \underline{V}$
(x): $8 = W_x - (-5)$, so $W_x = 3$ m/s;
(y): $-6 = W_y - 0$, $W_y = -6$ m/s.
$\underline{W} = 3\underline{i} - 6\underline{j}$ or 6.72 m/s at 63.5^o S of E.

50. (a) $t = 100 \text{ m}/(4 \text{ m/s}) = 25$ s;
(b) See Fig. 4.20 in text: $\sin\theta = 3/4$, so $\theta = 48.6^o$ W of N, then $t = 100 \text{ m}/(2.65 \text{ m/s}) = 37.8$ s

51. The component of A's velocity perpendicular to the river is $(10^2 - 5^2)^{1/2} = 8.66$ m/s, so $t_A = 200/8.66 = 23.1$ s.
$t_B = 100/15 + 100/5 = 26.7$ s.

52. $\underline{v}_{RC} = \underline{v}_{RG} + \underline{v}_{GC}$, $\tan\theta = v_{RG}/v_{GC} = 10/20$, so 26.6^o above horizontal

53. (a) From the law of sines,
$200/\sin 45^o = 50/\sin\alpha$, thus $\alpha = 10.2^o$.
The velocity of the plane relative to the ground, $\underline{R}$, is found from the components of the velocities along $\underline{R}$:
$R = 200\cos\alpha + 50\cos 45^o = 232.2$ km/h.
The heading is $45^o + \alpha = 55.2^o$ N of E.

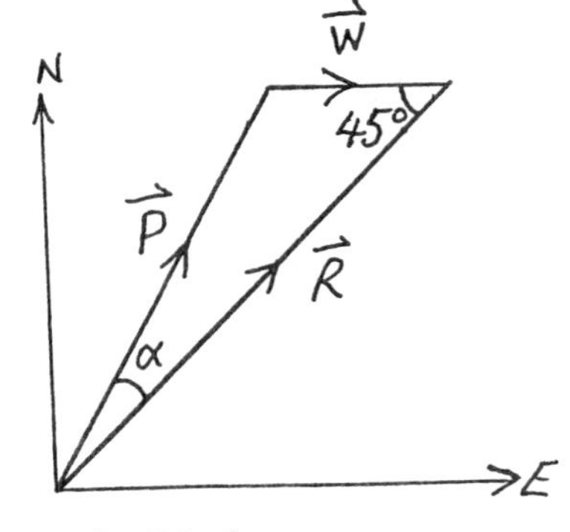

(b) The time taken is 600 km/232.2 km/h = 2.58 h

54.

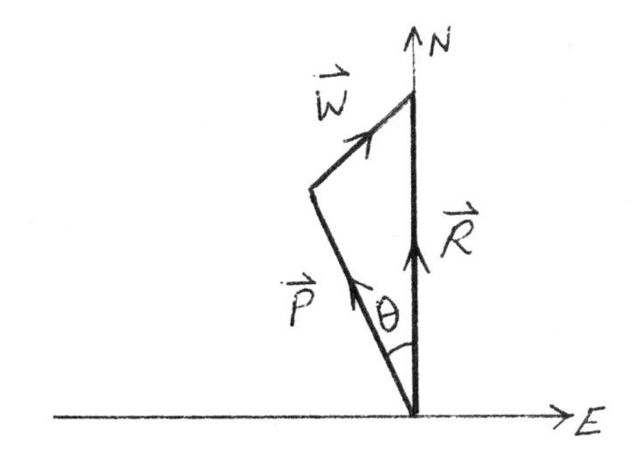

$\underline{v}_{PL} = \underline{v}_{PW} + \underline{v}_{WL}$, or $\underline{R} = \underline{P} + \underline{W}$. Taking components:

$0 = P_x + 80\cos 37^o$, so $P_x = -63.9$

$400 = 48.1 + P_y$, so $P_y = 351.8$

The airspeed is $(63.9^2 + 351.8^2)^{1/2}$ = 358 km/h. For the direction $\tan\theta$ = 63.9/351.8, so $\theta = 10.3^o$ W of N

55. From the law of sines $5/\sin(63.4^o) = 4/\sin\beta$, so $\beta = 45.7^o$. The angle $\theta = 180^o - 63.4^o - 45.7^o = 70.9^o$ N of W.

(b) The time taken is $t = 100/(5\sin 71.9^o) = 21$ s.

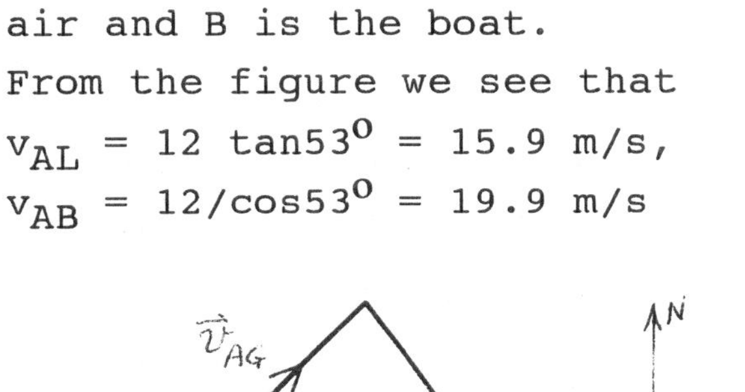

56.

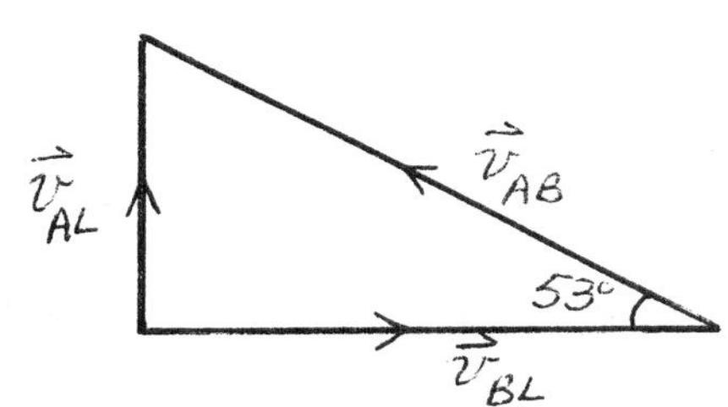

$\underline{v}_{AL} = \underline{v}_{AB} + \underline{v}_{BL}$, where A is the air and B is the boat.

From the figure we see that

$v_{AL} = 12\tan 53^o = 15.9$ m/s,

$v_{AB} = 12/\cos 53^o = 19.9$ m/s

57. $\underline{v}_{PG} = \underline{v}_{PA} + \underline{v}_{AG}$. The angle between $\underline{v}_{PA}$ and $\underline{v}_{AG}$ is 75°. From the law of cosines, $v^2 = 40^2 + 180^2 - 2(40)(180)\cos 75^o$, so $v_{PG} = 174$ km/h. Using the law of sines: $174/\sin 75^o = 40/\sin\alpha$, thus $\alpha = 12.8^o$ and $\theta = 30 + \alpha = 42.8$ or 47.2^o W of N.

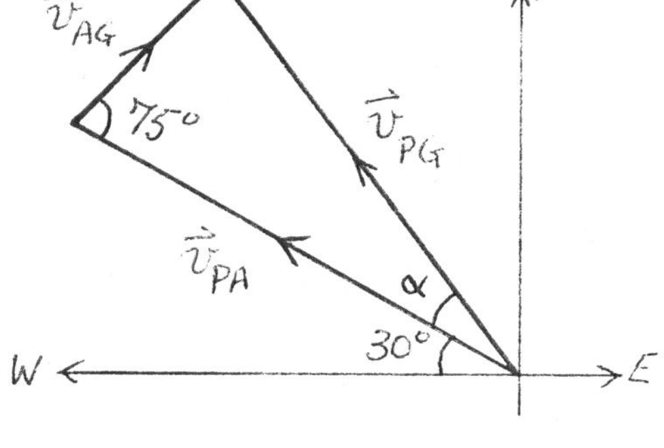

58. (a) After 8 s, $\underline{v}_A$ and $\underline{v}_B$ are 72° above and below the x axis (east) respectively. Thus

$\underline{v}_B - \underline{v}_A = 5\cos 72^o\underline{i} - 25\sin 72^o\underline{j}$ = $1.54\underline{i} - 23.8\underline{j}$ m/s

(b) $10t + 50 = 15t$, so $t = 10$ s

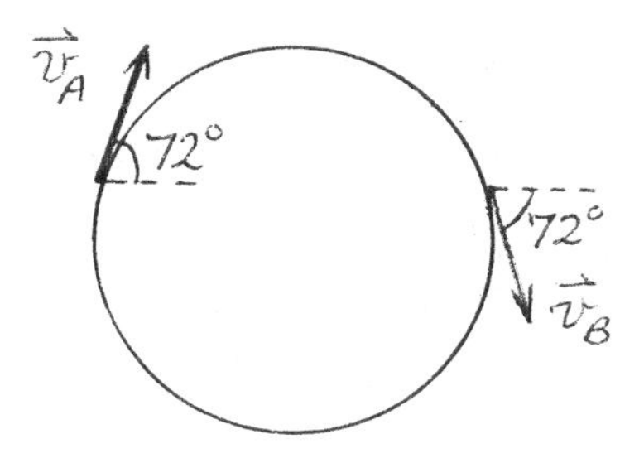

59. If the motion of A relative to B is not along $\underline{v}_{AB}$, the apparent position (direction) changes with time.

60. The shortest time occurs when the velocity of the penguin relative to the fish lies along the line joining their initial positions, i.e. due north. The northerly component of the penguin's velocity is 29.85 km/h. Thus, t = 50 m/(8.29 m/s) = 6.03 s

61. (a) Speeding up; (b) $a_c = v^2/r = 2 \tan60^o = 1.155$, so $v = 11.8$ m/s.

62. (a) $a_t = 10 \cos37^o = 7.99$ m/s^2, $a_c = 10 \sin37^o = 6.02$ m/s^2;
(b) At the given point, the speed is $v_o = (ar)^{1/2} = 17.3$ m/s Since the tangential acceleration is constant, the speed after one revolution is given by $v^2 = v_o^2 + 2as = (17.3)^2 + 16(100\pi)$, which leads to v = 73 m/s. For the time, $v = v_o + at$ yields t = 6.96 s.

63. (a) $a = (4 + 86)^{1/2} = 6.32$ m/s^2.
(b) $v^2 = ar = 24$, so v = 4.9 m/s.

64. From $C = 2\pi r$, find r = 429.7 m. Then $a = v^2/r = 3.02$ m/s^2.

65. $C = 2\pi r$, find r = 382 m. Also v = 30 m/s, so a = 2.36 m/s^2.

66. (a) $y = y_o + v_o \sin\theta t - 4.9t^2$, gives $\theta = 53.1^o$.
(b) $0 = v_{oy}^2 + 2a(y - 0)$, leads to y = 19.6 m.

67. (a) $y = y_o - 4.9t^2$, leads to t = 3.03 s
(b) Using $\tan\theta = v_y/v_{ox}$, $\theta = -56.1^o$
(c) $v^2 = v_x^2 + v_y^2$ leads to v = 35.8 m/s.

68. (a) $y = 0.5t^2 = -0.992$ m; (b) $\tan\theta = v_y/v_x$ gives $\theta = -6.29^o$.

69. From $0 = v_{oy}^2 + 2a(20)$, $v_{oy} = 19.8$ m/s. With $v_x = 10$ m/s, find $v_o = 22.2$ m/s. From $\tan\theta = v_{oy}/v_{ox} = 1.98$, find $\theta = 63.2^o$

70. (a) $0 = 60 - v_o \sin40^o(2) - 4.9(2)^2$, so $v_o = 31.4$ m/s;
(b) $x = v_o \cos\theta t = 48.1$ m; (c) $\mathrm{Tan}\theta = v_y/v_x$, so $\theta = -58.8^o$.

71. $r = R\cos\theta = 4.83 \times 10^6$ m, then $a = 4\pi^2 r/T^2 = 2.55 \times 10^{-2}$ m/s^2.

72. From $C = 2\pi r$ find $r = 5.73$ m, then $a = v^2/r = 157$ m/s^2.

73. $v_x = 40\cos 60^o = 20$ m/s, so $v_y^2 = v_o^2 - v_x^2$ and $v_y = \pm 34.6$ m/s. From $v_y = v_{oy} + at$ find $t = 1.25$ s, 5.82 s.

74. $0 = 9.2 - 4.9t^2$, so $t = 1.37$ s. From $\tan 30^o = h/d$, find $d = 15.9$ m. Finally, $x = v_o t$ gives $v_o = 11.6$ m/s.

75. $x = v_o\cos\theta t$ gives $v_o\cos\theta = 16$ m/s and $0 = 44 - v_o\sin\theta t - 4.9t^2$ gives $v_o\sin\theta = 12.2$. Square and add to find $v_o = 20.1$ m/s. Then, $\theta = 37.3^o$.

76. Since $0 = v_{oy}^2 - 2gH$ and $-v_{oy} = v_{oy} - gT$, we see $T^2 = 8H/g$.

77. (a) Since $v_x = 19.0$ m/s, $v_y = -40.8 = -8.875 - 9.8t$, thus $t = 3.26$ s. (b) $0 = y_o - 8.875t - 4.9t^2$ gives $y_o = 81.0$ m.

78. $t = 7.2$ m$/12\cos 35^o = 0.732$ s. Then $\Delta y = 6.883t - 4.9t^2 = 2.41$ m, and $y = 4.01$ m. (b) $v_y = 6.883 - 9.8t = -0.29$ m/s. So, $\tan\theta = v_y/v_x = -0.0295$ leads to $\theta = -1.69^o$.

79. Substitute $t = 46/v_o\cos 15^o$ into $0 = 14 + v_o\sin 15^o t - 4.9t^2$ to find (a) $v_o = 20.5$ m/s; and then (b) $t = 2.32$ s.

80. $r = 6630$ km, so $a = 4\pi^2 r/T^2 = 9.05$ m/s^2.

81. (a) $v^2 = ar$, so $v = 1.65$ km/s. (b) $v = 2\pi r/T$, so $T = 108$ min.

Problems

1. $R_1 = (v_o \cos\theta)\ t_1;\quad 0 = H + (v_o \sin\theta)\ t_1 - 1/2\ gt_1^2$
   $R_2 = (v_o \cos\theta)\ t_2;\quad 0 = H - (v_o \sin\theta)\ t_2 - 1/2\ gt_2^2$
   Substitute the t's from the x equations into the y equations. Add to find $(R_1 + R_2)\tan\theta = g(R_2^2 - R_1^2)/(2vo^2 \cos^2\theta)$, thus

$$R_2 - R_1 = v_o^2 \sin 2\theta/g$$

2. $R = v_o \cos\theta\ t$, and $0 = 2R + v_o \sin\theta\ t - 1/2\ gt^2$
   Substitute $t = R/(v_o \cos\theta)$ into the second equation to find

$$0 = 2R + R\tan\theta - (g/2)(R/v_o \cos\theta)^2$$

   Simplify to find $R = 2v_o^2 \cos^2\theta(2 + \tan\theta)/g$

3. For the horizontal motion, $R = vt$ and for the vertical motion: $0 = H - 1/2\ gt^2$. Thus, $H = gR^2/2v^2$

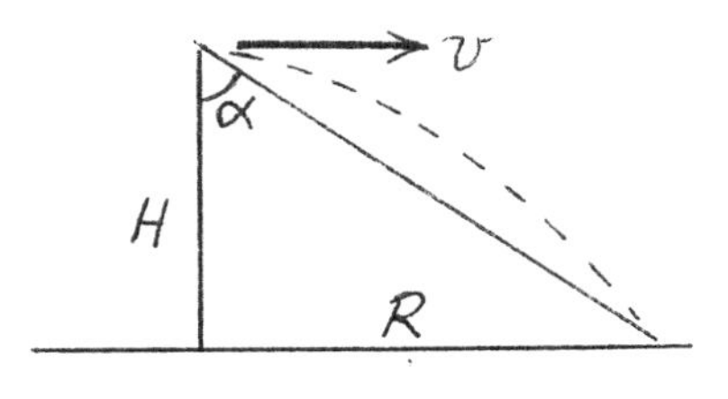

(a) $\tan\alpha = R/H = (2v^2/gH)^{1/2}$;

(b) $R = (2v^2H/g)^{1/2}$

4. $10 = (15\cos\theta)t$, and $3 = (15\sin\theta)t - 4.9t^2$. Square and add expressions for the trigonometric functions to find

$$24t^4 - 195.6t^2 + 109 = 0$$

which yields $t^2 = 7.548\ s^2$, $0.602\ s^2$. So, $t = 2.75$ s, 0.776 s. The corresponding angles are 76° and 30.8^o. [One could also solve a quadratic in one of the trigonometric functions.]

5. (a) $H = (v_o \sin\theta)^2/2g$ and $R = v_o^2 \sin 2\theta/g$. If $R = H$, then $\tan\theta = 4$, and $\theta = 76^o$.

(b) If $R = H/2$, then $\tan\theta = 8$ and $\theta = 82.9^o$.

6. (a) See Fig. 4.42. $x = R\cos\alpha = (v_o\cos\theta)t$;

$$y = R\sin\alpha = (v_o\sin\theta)t - 1/2\ gt^2$$

Substitute $t = (R\cos\alpha)/(v_o\cos\theta)$ into the equation for y, then solve to find R.

(b) Recast, R has the form

$R = v_o^2[\sin(2\theta - \alpha) - \sin\alpha]/(g\cos^2\alpha)$

Maximum R occurs when $(2\theta - \alpha) = \pi/2$. To obtain the given expression use $\cos^2\alpha = (1 - \sin\alpha)(1 + \sin\alpha)$.

7. In the x direction: $R\cos\alpha = v_o \cos(\theta)\ t$ (i)

In the y direction: $-R\sin\alpha = -v_o \sin(\theta)\ t - 1/2\ gt^2$ (ii)

Substitute t from (i) into (ii) and rearrnge to find:

$$R = 2v_o^2 \sin(\theta + \alpha)\cos\theta/(g\cos^2\alpha)$$

Using $2\sin A\cos B = \sin[(A + B)/2] + \sin[(A - B)/2]$, we find

$$R = v_o^2[\sin(2\theta + \alpha) + \sin\alpha]/(g\cos^2\alpha)$$

The maximum value occurs when $2\theta + \alpha = 90^o$. Use $\cos^\alpha = (1 + \sin\alpha)(1 - \sin\alpha)$ to find $R_{max} = v_o^2/[g(1 - \sin\alpha)]$

8. $T = 2v_o\sin\theta/g$ and $R = v_o\cos\theta\ T$.

$$x = T_1/T_2 = \sin\theta_1/\sin\theta_2 = \cos\theta_2/\cos\theta_1$$

Thus,

$$x^2 = s_1^2/(1 - c_2^2) = s_1^2/(1 - x^2c_1^2)$$
$$c_1^2x^4 - x^2 + s_1^2 = 0$$

which yields,

$$x^2 = [1 + (1 - 4c_1^2s_1^2)^{1/2}]/2c_1^2 = [1 + \cos 2\theta_1]/(2\cos^2\theta_1)$$
$$= \tan^2\theta_1,\ 1 \text{ (reject)}$$

Thus, $T_1/T_2 = \tan\theta_1$

9. From Eq. (iii) of Example 4.2, the time of flight is $T = 2v_o \sin\theta/g$, so,
$T_1 = 2\ v_o \sin(45 - \alpha)/g$; and $T_2 = 2\ v_o \sin(45 + \alpha)/g$
Expand the sine functions to obtain
$$T_2 - T_1 = 2(2)^{1/2}\ v_o \sin\alpha/g$$

10. $v_x^2 = (v_o\cos\theta)^2$ and $v_y^2 = (v_o \sin\theta)^2 + 2gH$ (Note $\Delta y = -H$)
Thus, $v^2 = v_x^2 + v_y^2 = v_o^2 + 2gH$, which is independent of θ.

11. Substitute $t = x/30$ into $y = 60 = 40t - 4.9t^2$ to obtain $4.9x^2 - 1200x + 54 \times 10^3 = 0$ which yields $x = 59.4$ m, 185.5 m.

12. (a) $r^2 = x^2 + y^2 = A^2$, so the path is a circle
(b) $\underline{v} = d\underline{r}/dt = -\omega A \sin(\omega t)\underline{i} + \omega A\cos(\omega t)\underline{j}$
$\underline{a} = d\underline{v}/dt = -\omega^2 A\cos(\omega t)\underline{i} - \omega^2 A\sin(\omega t)\underline{j}$
(c) $v^2 = v_x^2 + v_y^2 = \omega^2A^2$, so $v = \omega A$
(d) Compare (b) to $\underline{r}$ to see that $\underline{a} = -\omega^2\underline{r}$

13. (a) $\underline{v}_{BA} = \underline{v}_B - \underline{v}_A = -3\underline{i} - 4\underline{j}$ m/s, i.e. 5 m/s at 36.9° W of S.
(b) Choose a frame in which A is at rest. Thus v_{BA} is the velocity of B in this frame. The minimum distance between A and the path of B is 100 sin8.1° = 14.1 km.
(c) Draw a circle centered at A with a radius of 20 km. The length of the chord that intersects the path of B is $2(20^2 - 14.1^2)^{1/2} = 28.4$ km. The time B takes to travel this distance is 28.4 km/(5 m/s) = 94.7 min.

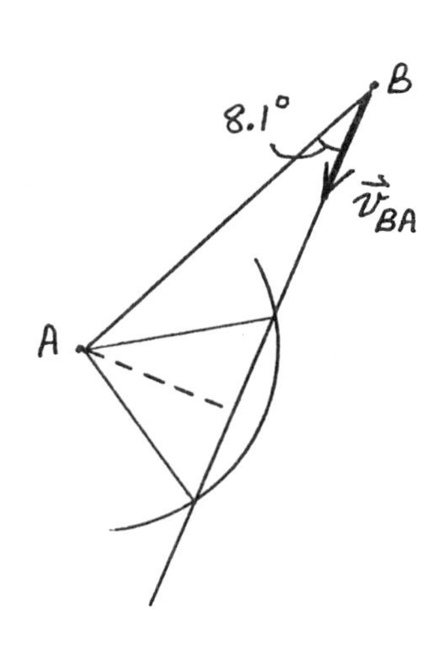

14. $x = R\cos\alpha = (v_o\cos\theta)t$, $\Delta y = -R\sin\alpha = (v_o\sin\theta)t - 1/2\ gt^2$

Substitute for t from the first equation into the second and use a trigonometric identity to find:

$$R = (2v_o^2\cos\theta)\sin(\alpha + \theta)/(g\cos^2\alpha) = 102 \text{ m}$$

15. $R = 15^2 \sin 2\theta/g = 15 \cos(\theta)\ t = 25 - vt$, so $\theta = 20^o$.

$$x_B = 15 \cos(\theta)\ t; \quad x_P = 25 - vt$$

Since $T = 2v_o \sin\theta/g$, we find

$$(15 \cos\theta + v)(30 \sin\theta/g) = 25$$

which yields v = 9.78 m/s

16. (a) $L = v_o \cos\theta\ t$, $h = v_o \sin\theta\ t - 1/2\ g\ t^2$. Substitute for t into the second to find

$$v_o^2 = gL[2 \cos^2\theta(\tan\theta - h/L)]^{-1}$$

(b) Substitute for t into $\tan\alpha = (v_o \sin\theta - gt)/(v_o \cos\theta)$ to find $\tan\alpha = 2h/L - \tan\theta$.

17. The x and y coordinates of the landing spot are:

$$R = v_o \cos(\theta)\ t; \quad -h = -v_o \sin(\theta)\ t - 1/2\ gt^2$$

Substituting for t from the x equation into the y equation:

$$v_o = (R/\cos\theta)\ (g/2(h - R\tan\theta)^{1/2})$$

18. $R = v_o \cos\theta\ t$, $\Delta y = -h = v_o\sin\theta\ t - 1/2\ gt^2$, i.e.

$$gt^2 - 2v_o \sin\theta\ t - 2h = 0$$

Thus

$$t = [2v_o\sin\theta + (4v_o^2\sin^2\theta + 8gh)^{1/2}]/2g$$

Substitute this t into $R = v_o \cos\theta\ t$ to find

$$R = v_o^2 \sin 2\theta[1 + (1 + 2gh/v_o^2\sin^2\theta)^{1/2}]/2g$$

19. (a) $\underline{v}_{BA} = \underline{v}_B - \underline{v}_A = 30\underline{i} - 50\underline{j}$ km/h, or 58.3 km/h at 31° W of N

(b) Take A to be at rest. The angle between $\underline{v}_{BA}$ and the line (of length $10(2)^{1/2}$ km) joining the initial positions of A and B is 14°. The minimum distance between A and B is $D = 10(2)^{1/2}\sin 14^o = 3.42$ km

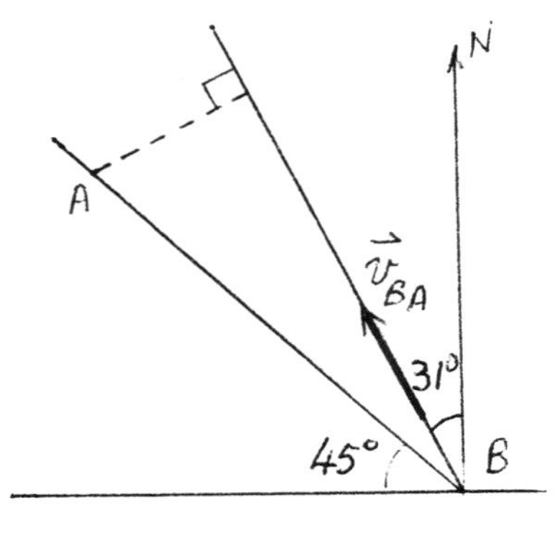

(c) The time taken to reach the point

of closest approach is $t = 10(2)^{1/2}\cos 14^{o}/(58.3) = 0.235$ h. In this time B moves 11.75 km, and A moves 7.05 km. Thus, their actual positions relative to the intersection are
$r_B = 1.75$ km N; $r_A = -2.95$km. W.

20. 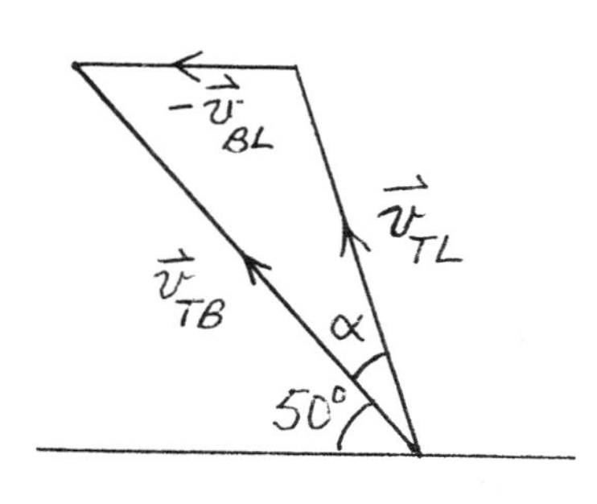

Take the ship to be at rest. The shortest time to reach it requires the velocity of the tug relative to the ship to lie along the line joining their initial positions.

$\underline{v}_{TB} = \underline{v}_{TL} - \underline{v}_{BL}$.

From the law of sines: $6/\sin 50^o = 3/\sin\alpha$, thus $\alpha = 22.5^o$. Considering just the northerly components, the time taken is given by $t = (2000 \cos 40^o)/(6 \sin 72.5^o) = 6$ min 20 s

21. In $R = v_o^2 \sin 2\theta/g$ we use $\theta_1 = 45^o - \alpha$ and $\theta_2 = 45^o + \alpha$:
$R_1 = v_o^2 \sin(90^o - 2\alpha)/g$ and $R_2 = v_o^2 \sin(90 + 2\alpha)/g$.
Thus $R_1 = R_2$.

CHAPTER 5

Exercises

1. $\Sigma F_x = T_1 \cos 40^o - T_2 \cos 60^o = 0$
   $\Sigma F_y = T_1 \sin 40^o + T_2 \sin 60^o - 7g = 0$
   Find $T_1 = 34.8$ N; $T_2 = 53.4$ N.

2. $\Sigma F_x = T_1 \cos 30^o - T_2 = 0$; $\Sigma F_y = T_1 \sin 30^o - 3g = 0$
   $T_1 = 58.8$ N, $T_2 = 50.9$ N

3. $\Sigma F_x = F - T \sin 37^o = 0$; $\Sigma F_y = T \cos 37^o - mg = 0$.
   (a) $F = 14.7$ N; (b) $T = 24.5$ N

4. $\Sigma F_y = 2T\cos\theta - mg = 0$, where $\cos\theta = 1.5/50$, thus $T = 11.4$ kN

5. $\underline{F}_1 + \underline{F}_2 = m\underline{a}$, thus $-\underline{i} + 2\underline{j} + 3\underline{k} + \underline{F}_2 = 8\underline{i} - 6\underline{j}$, and
   $\underline{F}_2 = 9\underline{i} - 8\underline{j} - 3\underline{k}$ N.

6. $\underline{F}_1 = \underline{i} + 2\underline{j}$ N and $\underline{F}_2 = 3.19\underline{i} + 2.41\underline{j}$ N,
   $\underline{a} = (\underline{F}_1 + \underline{F}_2)/(0.2 \text{ kg}) = 21.0\underline{i} + 22.1\underline{j}$ m/s^2

7. $\underline{a} = (4\underline{i} - \underline{j})/(1.5 \text{ kg}) = 2.67\underline{i} - 0.67\underline{j}$ m/s^2.
   $\underline{v} = \underline{v}_o + \underline{a}t = (2\underline{i} + 3\underline{j}) + (5.34\underline{i} - 1.34\underline{j}) = 7.34\underline{i} + 1.66\underline{j}$ m/s.

8. $a = (400)^2/(0.06 \text{ m}) = 2.67 \times 10^6$ m/s^2.
   $F = ma = 2.67 \times 10^4$ N, about 45 times the person's weight

9. (a) $v^2 = v_o^2 + 2a\Delta x$, so $a = (6\text{x}10^{13})/(0.04) = 1.5 \times 10^{15}$ m/s^2.
   $F = ma = 1.37 \times 10^{-15}$ N
   (b) $t = (6 \times 10^6)/(1.5 \times 10^{15}) = 4 \times 10^{-9}$ s.

10. $a = v^2/2\Delta x = 1/6 = 0.167$ m/s^2.
    $\Sigma F_x = mg \sin\theta - f = ma$, so $f = 109$ N

11. (a) $v^2 = 2a\Delta x$, thus $a = 22.5$ m/s^2; (b) $F = (50 \text{ kg})a = 1125$ N.

12. (a) $a = 2.67$ m/s^2, $F = ma = 3.27$ kN;
    (b) $a = -7.56$ m/s^2, $F = ma = -9.26$ kN
    In both cases, F is the frictional force exerted by the road.

13. (a) $a = 31.6$ m/s^2, so $F = 3.95 \times 10^5$ N
    (b) $a = -31.25$ m/s^2, so $F = -3.90 \times 10^5$ N.

14. $F - mg = ma$, so $F = m(g + a) = (0.215\ kg)(13.8\ m/s^2) = 2.97\ N$

15. To rise by 0.4 m, the initial speed v required is given by $0 = v^2 - 19.6x0.4$, so $v^2 = 7.84\ m^2/s^2$. This v is the final speed while the feet are in contact with the ground, thus $v^2 = 2a(0.15)$ and $a = 26.1\ m/s^2$ is the acceleration of the torso. The force exerted is $F - mg = ma$, so $F = 1795$ N.

16. (a) $\Sigma F_x = F\cos\theta - mg\sin\theta = ma$, so $a = -1.90\ m/s^2$ (down);
(b) $\Delta x = (4)(2) - 1/2\ (1.9)(2^2) = +4.2$ m (up incline)

17. With $v_o = 4.17$ m/s, $0 = (4.17)^2 - 2(g\sin 10^o)\Delta x$, so $\Delta x = 5.1$ m.

18. (a) $F\cos\theta - f = 0$, $f = 69.3$ N;
(b) $F'\cos\theta - f = ma$, $F' = 103$ N

19. From $x = 1/2\ at^2$, $a = 14\ m/s^2$ and $F = ma = 840$ N.

20. $v_y = a_y t = (F/m)t = -40$ m/s (S). Thus $\underline{v} = 10\underline{i} - 40\underline{j}$ m/s or 41.2 m/s at 76° S of E. Path is parabolic.

21. (a) $a = v^2/2\Delta x = 416.7\ m/s^2$, so $F = 3.13 \times 10^4$ N
(b) $a = 1250\ m/s^2$, so $F = 9.38 \times 10^4$ N.

22. $a = 1.4 \times 10^4\ m/s^2$, $F = (0.15\ kg)a = 2.1$ kN

23. (a) $a = F/m = 12.8\ m/s^2$; $(61.1)^2 = 2(12.8)\Delta x$, so $\Delta x = 146$ m.
(b) $F = (1.6 - 1.225)x10^5$ N, so $a = 3\ m/s^2$.
From $\Delta x = 1/2\ at^2$, we find $t = 25.8$ s.

24. $F - mg = ma$, thus $a = 2.42\ m/s^2$

25. $F - mg = ma$ leads to $a = 4.49\ m/s^2$
For t = 1 min., $\Delta y_1 = 1/2\ at^2 = 1/2(4.49)(60^2) = 8.08$ km.
When the engine stops $v = at = 269.4$ m/s.
The added height in free-fall is $\Delta y_2 = v^2/2g = 3.70$ km.
The total height is 11.8 km

26. The landing speed is $v = (2gh)^{1/2} = 4.43$ m/s.
(a) $a = v^2/2\Delta y$, where $\Delta y = 0.3$ m then $F = ma = 1.31$ kN;
(b) 9.8 kN

27. (a) $a = v^2/2\Delta x = 156\ m/s^2$, $F = ma = 23.4\ N$;
(b) $a = (35 + 25)/(5x10^{-3}) = 1200\ m/s^2$, $F = 1800\ N$
(c) $a = v^2/2\Delta x = 1330\ m/s^2$, $F = 200\ N$

28. From $R = v_o^2 \sin2\theta/g$, find $v_o = 12.52$ m/s. Then $a = v_o^2/2\Delta x = 52.3\ m/s^2$ and $F = ma = 379\ N$

29. Moving down: $Mg - F_B = Ma$,
Moving up: $F_B - (M - m)g = (M - m)a$, where m is the ballast.
Adding these equations: $mg = (2M - m)a$, so $m = 2Ma/(a + g)$

30. $mg - T = ma$, leads to $a = 1.8\ m/s^2$

31. (a) $a = 20/5 = 4\ m/s^2$; (b) $20 - F_{BA} = 12$, so $F_{BA} = 8$ N.
(c) 12 N; (d) $F_{BA} = 12$ N.

32. (a) $F_{BG} = (0.9\ kg)a = 0.45$ N (forward); (b) $F_{CG} = 0$ (given);
(c) $F_{BC} = (0.2\ kg)a = 0.1$ N (backward)

33. (a) $60g = 588$ N; (b) $67g = 657$ N.

34. M = mass of engine, m = mass of a railcar.
$a = F/(10m) = 2\ m/s^2$.
(a) $F - T_1 = ma$, $T = 7.2 \times 10^5$ N; (b) $T_{10} = ma = 8 \times 10^4$ N;
(c) $f = (M + 10m)a = 1.24 \times 10^6$ N

35. $a = 4.44\ m/s^2$, $T = 20a = 88.8$ N

36. $6g\sin60^2 - T = 6a$, $T - 5g\sin30^o = 5a$. From these we find $a = 2.4\ m/s^2$ and $T = 36.5$ N

37. (a) $T_1 = (m_A + m_B)g = 4.9$ N, $T_2 = m_Bg = 2.94$ N;
(b) Same as (a);
(c) $T_1 - (m_A + m_B)g = (m_A + m_B)a$, so $T_1 = 5.9$ N
$T_2 - m_Bg = m_Ba$, so $T_2 = 3.54$ N;
(d) $(m_A + m_B)g - T_1 = (m_A + m_B)a$, so $T_1 = 3.9$ N, $T_2 = 2.34$ N.
(e) $10 - 0.5g = 0.5a$, so $a = 10.2\ m/s^2$.

38. $8Mg - T_1 = 8Ma$, $T_1 - 2Mg\sin\theta - T_2 = 2Ma$, $T_2 - 4Mg\sin\theta = 4Ma$.
Adding these: $T_1 - T_2 = 2M(a + g\sin\theta)$
(a) $a = 2.63\ m/s^2$; (b) $T_1 - T_2 = 19.1$ N

39. (a) $2T - 9g = 0$, so $T = 9g/2 = 44.1$ N;
(b) $a = 0$, so $T = 44.1$ N;
(c) $2T - mg = ma$, $T = 46.4$ N.

40. The monkey climbs up: $T - 10g = 10a$. To lift the bananas, we need $T = 12g$, so $a = 2g/10 = 1.96$ m/s^2.

41. (a) $F_o - 10g = 10a$; thus $F_o = 118$ N;
(b) $F = ma = 4$ N;
(c) $T - 4g = 4a$, so $T = 47.2$ N.

42. (a) $m_1g - T = m_1a$, $T = m_2a$, so $m_2 = m_1(g - a)/a = 2.9$ kg;
(b) $T = m_1m_2g/(m_1 + m_2)$, $m_2 = m_1T/(m_1g - T) = 1.38$ kg

43. $F \cos20^o - T = m_1a$; $T - m_2g = m_2a$. Adding these equations:

$$10 \cos20^o - 3g \sin30^o = 8a$$

Thus $a = -0.66$ m/s^2 and $T = 12.7$ N.

44. (a) $N - mg = ma$, $N = 826$ N;
(b) $W = mg = 686$ N;
(c) From the 3rd law, reaction is 686 N, and acts on the earth

45. (a) $N = mg$, so v = constant;
(b) $mg - N = ma$, so $a = 1.8$ m/s^2, downward
(c) $N - mg = ma$, $a = 2.2$ m/s^2, upward

46. (a) Acceleration is downward. $\Sigma F = mg - T = ma$, so $T = 17.4$ N;
(b) Acceleration is downward. $\Sigma F = mg - T = ma$, so $T = 23.4$ N.

47. $N_y - mg = ma \sin\theta$, thus $N_y = 950$ N; $N_x = ma_x = 96$ N
$N = (950^2 + 96^2)^{1/2} = 955$ N at 5.77^o to the vertical

48. (a) $a = 38.1$ m/s^2. $\Sigma F = F + 160$ kN $= ma$, thus $F = 316$ kN;
(b) $\underline{N} = ma\underline{i} + mg\underline{j} = 2670\underline{i} + 686\underline{j}$, or $N = 2.75$ kN at 14.4^o to the horizontal.

49. Rocket: $T - Mg/6 = Ma$ so $a = 1.6$ m/s^2.
Person: $N - mg/6 = ma$, so $N = 226$ N

50. $\underline{F}_1 = 8.66\underline{i} + 5\underline{j}$ N, $\underline{F}_2 = 4.38\underline{i} - 12\underline{j}$ N so $\underline{F}_3 = -13\underline{i} + 7.0\underline{j}$ N

51. (a) $-5.67\underline{i} + 8.83\underline{j}$ N; (b) $-12.6\underline{i} + 19.6\underline{j}$ m/s^2

52. (a) $m_Ca = +960$ N; (b) -960 N; (c) $(m_T + m_C)a = +2560$ N; (d) $m_Ta = 1600$ N.

53. $m_1g - T = m_1a$; $T - F\cos\theta = m_2a$. So $a = 1.54$ m/s^2; $T = 19.8$ N.

54. (a) $T\cos\theta = ma$, so $T = 9.06$ N.
(b) $N + T\sin\theta - mg = 0$, so $N = 65.2$ N.

55. From $m_1g - T_A = m_1a$, $T_A - T_B = m_2a$, and $T_B - m_3g = m_3a$, find $a = 1.96$ m/s^2, $T_A = 35.3$ N, $T_B = 32.9$ N.

56. $mg\sin\theta - f = ma$, so $f = 1.2$ N

57. $m_1g - T = m_1a$, $T + m_2g\sin\theta = m_2a$. So $a = 5.5$ m/s^2, $T = 4.3$ N.

58. From $12 + F_2x = ma_x$ and $F_2y = ma_y$; $F_2 = 15.9$ N at 19.1^o N of W.

59. $F\cos\theta - T = m_1a$, $T = m_2a$. Find $a = 4.7$ m/s^2 and $T = 6.11$ N.

60. $F = 48$ N for both. (a) 0.6 m/s^2; (b) 0.8 m/s^2.

61. $m_1g - 22 = (m_1 + m_2)(1)$ and $44 - m_1g = (m_1 + m_2)(1.75)$.
Find $m_1 = 3.06$ kg and $m_2 = 4.94$ kg.

62. $F\cos\alpha - m_2g\sin\theta - T = m_2a$ and $T = m_2a$ lead to $a = 1.42$ m/s^2 and $T = 2.84$ N.

63. $v^2 = 2a\Delta x$, so $a = 10^5$ m/s^2 and $F = ma = 1$ kN.

64. $F_1\cos\theta_1 + F_2\cos\theta_2 = ma$; $-F_1\sin\theta_1 + N - mg + F_2\sin\theta_2 = 0$.
Find $a = 4.75$ m/s^2 and $N = 21.3$ N.

65. $\Sigma F_x = 9 = 2a_x$; $\Sigma F_y = -8 = 2a_y$: $a = 6.02$ m/s^2 at 41.6^o S of E

66. $m_1g - m_2g\sin\theta = (m_1 + m_2)a$ and $m_2g\sin\theta - m_1g = (m_1 + m_2)a$ lead to $m_2 = 3.2$ kg and $\theta = 17.5^o$.

67. $N - mg = ma$, so $m = N/(g + a) = 60.8$ kg and $mg = 596$ N.

68. (a) $0 = 4 - (g\sin\theta)t$, so $t = 0.816$ s.
(b) $0 = 4^2 - 2(g\sin\theta)\Delta x$, so $\Delta x = 1.63$ m.

69. $F\cos\alpha - mg\sin\theta = ma$; $N - F\sin\alpha - mg\cos\theta = 0$.
Find $a = 1.15\ m/s^2$ and $N = 24.7$ N.

Problems

1. (a) $2T = 90g$, so $T = 441$ N
(b) $2T - 90g = 90a$, so $T = 459$ N
(c) $T = 90g$. Rope breaks.

2. $m_1g - T = m_1a_1$, $2T - m_2g = m_2a_2$.
Note that $m_1 = m_2$ and that $a_1 = 2a_2$.
$a_1 = 3.92\ m/s^2$, $a_2 = 1.96\ m/s^2$, $T = 5.88$ N

3. Since the pulley is massless, $100 - 2T = 0$, so $T = 50$ N. Say A is the upward acceleration of the pulley and a is the acceleration relative to the pulley.
$$5g - T = 5(a - A) = 5a_1$$
$$T - 2g = 2(A + a) = 2a_2$$
Thus, $3T = 20a$, and so $a = 7.5\ m/s^2$, then $A = 7.7\ m/s^2$.
(a) $a_2 = 15.2\ m/s^2$ upward, $a_5 = 0.2\ m/s^2$ upward; (b) 50 N

4. $a = (m_2 - m_1)g/(m_1 + m_2) = 2.45\ m/s^2$. Thus the speed of the 3 kg block just as the 5 kg block lands is $v = (2ah)^{1/2} = 4.43$ m/s. The is the initial speed for the free-fall of the 3 kg mass. The additional height is given by $\Delta y = v^2/2g = 1$ m. The maximum height is 5 m.

5. The length of the chord is $L = 2R\cos(45^o - \theta/2)$; and the acceleration along the chord is $a = g\cos(45^o - \theta/2)$.
$L = 1/2\ at^2$, so $t^2 = 2L/a = 4R/g$, thus $t = 2(R/g)^{1/2}$.

6. Let A be the acceleration relative to the ground and a′ be the downward acceleration relative to the wedge. We know $F = 5(M + m)$.
$\Sigma F_x = mg\sin\alpha = m(a' + A\cos\alpha)$
$\Sigma F_y = N - mg\cos\alpha = mA\sin\alpha$
Thus, $a' = g\sin\alpha - A\cos\alpha = 1.9\ m/s^2$.

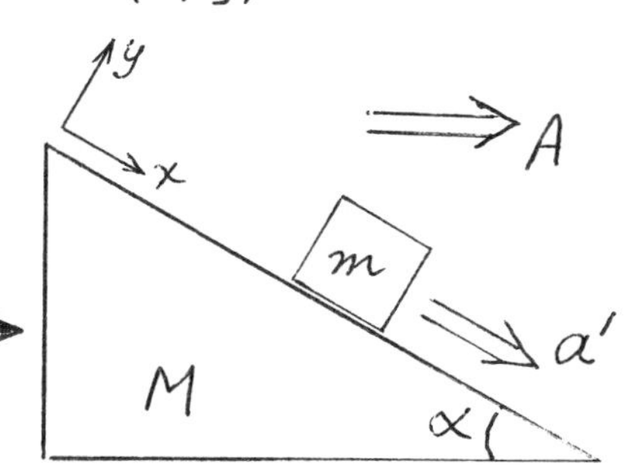

7. First show $a = mg/(2M + m)$. Then $v = D/t = [2mgh/(2M + m)^{1/2}]$
Solve for g:

$$g = (2M + m)D^2/(2mHt^2)$$

8. Acceleration a′ is relative to the wedge.

(W) $\Sigma F_x = -N_1 \sin\theta = -MA$ (i)

(B) $\Sigma F_x = N_1 \sin\theta = m(-A + a' \cos\theta)$ (ii)

From (i) and (ii),

$$A = ma'\cos\theta/(M + m) \quad \text{(iii)}$$

(B) $\Sigma F_y = N_1 \cos\theta - mg = -ma'\sin\theta$ (iv)

Use (i) and (iii) in (iv) to obtain

$a' = (M + m)g\sin\theta/(M + m\sin^2\theta)$

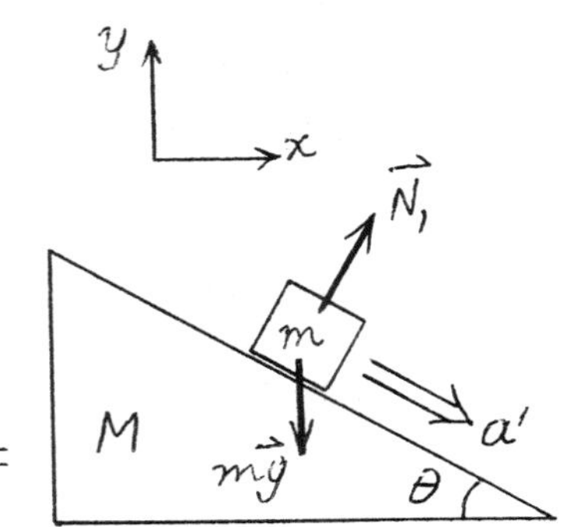

Then from (iii)

$A = mg\sin\theta \cos\theta/(M + m\sin^2\theta)$ (v)

The horizontal and vertical comps. of the acceleration of m:

$$a_H = -A + a'\cos\theta = Mg\sin\theta \cos\theta/(M + m \sin^2\theta)$$
$$a_V = -a' \sin\theta = -(M + m)g \sin^2\theta/(M + m \sin^2\theta)$$

From (i) and (v): $N_1 = MA/\sin\theta = Mmg \cos\theta/(M + m \sin^2\theta)$

9. (a) F = ma takes the form $M\, d^2y/dt^2 = (Mg/L)y$, so

$$d^2y/dt^2 = dv/dt = gy/L$$

(b) $dv/dt = (dv/dy)(dy/dt) = (dv/dy)\, v$, but we know

$$dv/dt = (g/L)\, y, \text{ so } v\, dv = (g/L)\, y\, dy.$$

Integrating this:

$$\int_0^v v\, dv = (g/L) \int_{y_o}^{L} y\, dy$$

$$v^2 = (g/L)\,(L^2 - y_o^2)$$

10. Let A be the acceleration of m_1 and a the acceleration of m_2 and m_3 relative to their pulley. The tensions are related by $T = 2P$.

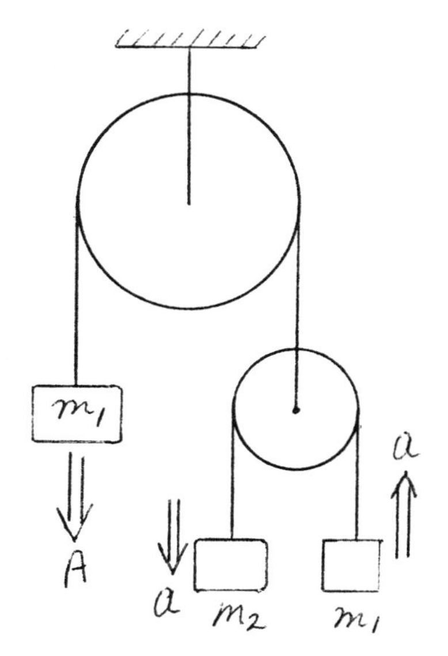

$$m_1g - T = m_1A \qquad \text{(i)}$$

$$m_2g - P = m_2(a - A) \qquad \text{(ii)}$$

$$P - m_3g = m_3(a + A) \qquad \text{(iii)}$$

Eliminate a from (ii) and (iii) to obtain

$$2m_2m_3g - T(m_2 + m_3)/2 = -2m_2m_3A \qquad \text{(iv)}$$

Multiply (i) by m_2m_3 and (iv) by m_1 to eliminate A and rearrange:

$$T = 8m_1m_2m_3g/[m_1(m_2 + m_3) + 4m_2m_3]$$

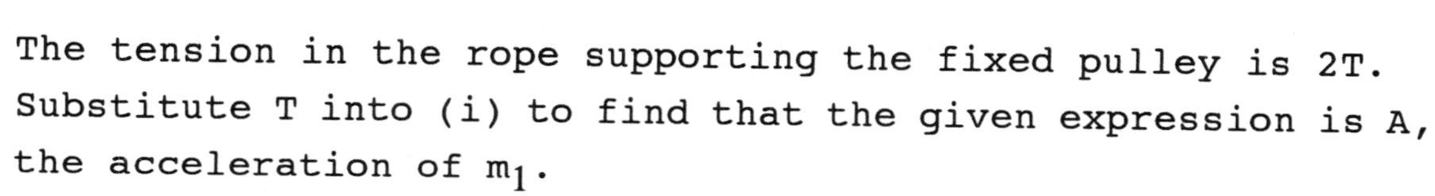

The tension in the rope supporting the fixed pulley is $2T$. Substitute T into (i) to find that the given expression is A, the acceleration of m_1.

CHAPTER 6

Exercises

1. (a) From $v^2 = v_o^2 + 2a\Delta x$, with $a = -f/m$, we find $f = 0.135$ N.
   (b) $f = \mu mg$, so $\mu = 0.153$

2. (a) $f = \mu mg/2 = ma$. Thus, $a = \mu g/2 = 3.92$ m/s^2;
   (b) All four wheels contribute, so $f = \mu mg$, $a = \mu g = 7.84$ m/s^2

3. (a) $5g\sin\theta - T - \mu(5g\cos\theta) = 0$, $T - \mu(2g) = 0$. Adding these: $5g\sin\theta = \mu(5g\cos\theta + 2g)$. Thus, $\mu = 0.395$
   (b) $T = 2\mu g = 7.74$ N.

4. (a) $\Sigma F_x = mg\sin\theta - \mu(mg\cos\theta)$. Since $\mu_s > \tan\theta$, block does not start to move.
   (b) $\Sigma F_y = N - mg\cos\theta - F\sin\theta$; and $f = \mu N$.
   $\Sigma F_x = F\cos\theta - mg\sin\theta - \mu_k(mg\cos\theta + F\sin\theta) = ma$, $a = 6.86$ m/s^2

5. (a) $\mu_s Mg - \mu_k mg = ma$; leads to $a = 27.4$ m/s^2
   (b) Need the force exerted by the person on the crate:
   $F = \mu_s Mg = 627$ N

6. (a) Maximum $f_s = \mu_s mg = 34.3$ N. Thus, $f = F = 30$ N;
   (b) $30 + \mu_k mg = ma$, so $a = 10.9$ m/s^2 (to the right);
   (c) $30 - \mu_k mg = ma$, so $a = 1.1$ m/s^2 (to the right)

7. (a) $F\cos 37^o - \mu_s(mg + F\sin 37^o) = 20 - 0.5(44.4) < 0$, so it does not start to move.
   (b) $20 - 0.2(44.4) = 3a$ so $a = 3.71$ m/s^2.

8. (a) $F\cos 37^o - \mu_s(mg - F\sin\theta) = 20 - 0.5(14.4) > 0$, so it does move.
   (b) $20 - 0.2(14.4) = 3a$, so $a = 5.71$ m/s^2

9. (a) $\Sigma F_x = mg\sin\theta - \mu_s(mg\cos\theta) > 0$, so the body does move.
   $\Sigma F_x = 19.6 - \mu_k(14.7) = 2.5a$, so $a = 6.37$ m/s^2, downward
   (b) $-mg\sin\theta - \mu_k(mg\cos\theta) = ma$; thus $a = 9.31$ m/s^2 (downward)
   (c) $mg\sin\theta - \mu_k\, mg\cos\theta = ma$; thus $a = 6.37$ m/s^2 (downward)

10. $f = ma = \mu mg$, so $\mu = a/g = 0.612$

11. Given $v_o = 27.8$, from $0 = v_o^2 - 2a\Delta x$, we find $a = \mu g = 6.43\ m/s^2$, so $\mu = 0.656$.
(a) $g \sin\theta - \mu(g\cos\theta) = a = -4.63\ m/s^2$, so $\Delta x = 83.3$ m.
(b) $g \sin\theta + \mu(g\cos\theta) = a = 8.03\ m/s^2$, so $\Delta x = 48.1$ m.

12. (a) $mg\sin\theta - \mu(mg\cos\theta) = ma$, so $a = 5.55\ m/s^2$;
(b) $v = v_o + at$, so $t = 4.0$ s;
(c) $\Delta x = v^2/2a = 44.5$ m

13. $a = g \sin\theta + \mu g \cos\theta = 2.67\ m/s^2$.
From Eq. 3.12, $0 = (22.2)^2 - 2a\ \Delta x$, so $\Delta x = 92.3$ m.

14. (a) $t_R/t_F = 1.5$; (b) $\Delta x_R/\Delta x_F = 1$

15. (a) $F - \mu(m_A + m_B + m_C)g = 100a$; $a = 1.02\ m/s^2$
(b) $F - \mu m_A g - T_1 = m_A a$, thus $T_1 = 140$ N.
(c) $T_2 - \mu m_C g = m_C a$, so $T_2 = 40$ N.

16. $x = 1/2\ at^2$ yields $a = 0.533\ m/s^2$.
$mg\sin 20 - \mu(mg\cos\theta) = ma$, so $\mu = 0.306$

17. $\Sigma F_x = F \cos\theta - \mu(mg + F \sin\theta) = F(\cos\theta - \mu\sin\theta) - \mu mg$
(a) Need $F > \mu_s mg/(\cos\theta - \mu \sin\theta)$ for block to move.
(b) If $\mu_s = \cot\theta$, $F = \infty$;
(c) $a = -\mu_k g$

18. (a) $\Sigma F_x = F\cos\theta - \mu(mg\cos\theta + F\sin\theta) - mg\sin\theta = ma$,
$20.0 - 0.1(39.1 + 15.0) - 29.5 = 5a$, so $a = -2.98\ m/s^2$ (downward)
(b) $\Delta x = 6x2 - 1/2(2.98)(2^2) = 6.04$ m

19. (a) 0; (b) For m_A: $f = ma$ gives $\mu(2g) = 2a$, so $a = 2.45\ m/s^2$.
The maximum force on system is $F = 7a = 17.2$ N.

20. $\Sigma F_x = N = m_A a$, $\Sigma F_y = f - m_A g = 0$. Since $f = \mu N$, find $\mu = g/a$.
$F = (m_A + m_B)a$, so $a = 12\ m/s^2$ and then $\mu = 0.82$

21. (a) Add $\Sigma F_x = ma$ for both blocks to find:
$m_1 g - \mu m_2 g \cos\theta - m_2 g \sin\theta = (m_1 + m_2)a$
Thus $a = 0.973\ m/s^2$.

(b) $m_1g + \mu m_2g \cos\theta - mg \sin\theta = (m_1 + m_2)a$, so $a = 2.93 \text{ m/s}^2$.
m_1 moving down: $m_1 = \mu m_2 \cos\theta + m_2 \sin\theta = 4.8$ kg.
m_1 moving up: $m_1 = m_2(\sin\theta - \mu \cos\theta) = 2.4$ kg.

22. $m_1g - 2T = m_1a$ (i), $T - f_2 = m_2(2a)$ (ii).
Use given a's in (i): $T_1 = (3g - 1.8)/2$; $T_2 = (4g - 6.4)/2$
Use these T's and the a's in (ii) and s olve to find
$m_2 = 1.3$ kg, $f_2 = 12.2$ N

23. $F - \mu m_1g - m_2g \sin53^o - \mu m_2g \cos53^o = (m_1 + m_2)a$
(a) $F - 19.6 - 39.13 - 14.74 = 9$, so $F = 82.5$ N;
(b) $F - T - \mu m_1g = m_1a$, so $T = 58.9$ N
(c) $F + \mu m_1g - m_2g \sin\theta + \mu m_2g \cos\theta = (m_1 + m_2)a$, so
$10 + 19.6 - 39.13 + 14.74 = 9a$, $a = 0.58 \text{ m/s}^2$ upward.

24. $f_1 = \mu(8g)$, $f_2 = \mu(2g)$. $\Sigma F_x = F - \mu(10g) = 6a$, so $\mu = 0.061$

25. $\Sigma F_x = T \sin\theta = 0.58a$ (i), $\Sigma F_y = T \cos\theta - 0.58g = 0$ (ii),
From (ii), $\cos\theta = 0.947$, so $\theta = 18.7^o$. From (i) $a = 3.31 \text{ m/s}^2$

26. Use $\Delta x = v^2/2\mu g$. (a) 51 m; (b) 153 m

27. $F\cos25^o - mg\sin15^o - \mu(mg\cos15^o - F\sin25^o) = ma$; so $\mu = 0.197$.
With $F = 0$: $\Sigma F_x = mg\sin15^o - \mu mg\cos15^o = ma$, so $a = 0.67 \text{ m/s}^2$.

28. $\Sigma F_y = mg = mv^2/r$ (Note $N = 0$.) Thus, $v^2 = rg$, so $v = 2.8$ m/s

29. (a) $\Sigma F_y = mg = mv^2/R$, so $v^2 = Rg$, which gives $v = 14$ m/s.
(b) $\Sigma F_y = N - mg = mv^2/R$, so $N = 2mg = 1470$ N.

30. $\Sigma F_x = f = mv^2/r$, but $f = \mu N = \mu mg$, so $\mu = v^2/rg = 0.472$

31. Since a is horizontal, pick x axis horizontal.
$\Sigma F_x = N\sin\theta = mv^2/r$; $\Sigma F_y = N \cos\theta - mg = 0$. Eliminate N to find $\tan\theta = v^2/rg$.

32. (a) Let P be the lift acting normal to the wings.
$\Sigma F_x = P\sin\theta = Mv^2/r$; $\Sigma F_y = P\cos\theta - Mg = 0$, thus $\tan\theta = v^2/rg$.
Thus $\theta = 32.2^o$.
(b) From $\Sigma F_y = 0$, $N = mg/\cos\theta = 811$ N

33. (a) $\Sigma F_x = mg \sin\theta - \mu N = ma_x = 0$; $\Sigma F_y = N - mg\cos\theta = mv^2/r$. Thus, $g \sin\theta = \mu(g \cos\theta + v^2/r)$.
(b) With values: $\sin\theta = 0.75\cos\theta + 0.302$. Use $s = (1 - c^2)^{1/2}$: $1.563c^2 + 0.453c - 0.909 = 0$, $\cos\theta = 0.631$, so $\theta = 50.9^o$

34. $\Sigma F_x = N = mv^2/r$, $\Sigma F_y = f - mg = 0$, thus $\mu = rg/v^2 = 0.8$.

35. $\Sigma F_y = N_y - mg = 0$; $\Sigma F_x = N_x = mv^2/r$. Thus, $\underline{N} = 686\underline{j} + 394\underline{i}$ N, or $N = 791$ N at 29.9^o to the vertical.

36. (a) Given $T = 2.5$ s. $\Sigma F_x = N = mv^2/r$; $\Sigma F_y = f - mg = 0$. Apparent weight $= (N^2 + m^2g^2)^{1/2} = m(v^4/r^2 + g^2)^{1/2} = 640$ N
Note $f_s < \mu_s N$.

37. (a) $T = 2\pi r/v = 1.51 \times 10^{-16}$ s.
(b) $F = mv^2/R = 8.3 \times 10^{-8}$ N.

38. $\Sigma F_x = N\sin\theta = mv^2/r$, $\Sigma F_y = N\cos\theta - mg = 0$. Given $N = 1.4mg$, so $\cos\theta = 1/1.4$ and $\sin\theta = 0.70$. Using $v = 222$ m/s in ΣF_x: $r = v^2/1.4g\sin\theta = 5.13$ km

39. From $g = 4\pi^2R/T^2$, find $T = 84.4$ min.

40. (a) $T + mg = mv^2/r$, $T = m(v^2/r - g) = 0.707$ N;
(b) $T - mg = mv^2/r$, $T = m(v^2/r + g) = 12.4$ N;
(c) $T = mv^2/r = 6.57$ N

41. (a) $\Sigma F_y = mg = mv^2/R$, so $v^2 = Rg$, thus $v = 14$ m/s.
(b) $\Sigma F_y = N - mg = mv^2/R$, thus $N = m(v^2/R + g) = 1180$ N

42. $\Sigma F_x = f = mv^2/r$, $\Sigma F_y = N - mg = 0$, thus $\mu = v^2/rg = 0.34$

43. (a) $v^2 = 2a\Delta x = 2\mu gD$, thus $D = v^2/2\mu g$.
(b) $mv^2/R = \mu mg$; $R = v^2/\mu g$ (= 2D)
(c) Slow down as much as possible before turning.

44. $g/5 = 4\pi^2r/T^2$, so $T = 142$ s

45. (a) $\Sigma F_y = mg = mv^2/R$, so $v^2 = Rg$. Need $v > 7.98$ m/s.
(b) $\Sigma F_y = N + mg = mv^2/R$, thus $N = m(v^2/R - g) = 163$ N

46. (a) Kepler's 3rd law: $M_E = 4\pi^2r^3/GT^2 = 6.02 \times 10^{24}$ kg.

(b) Similarly, M_S = 1.97 x 1030 kg

47. Use Kepler's 3 rd law: $M_J/M_E = (r_J{}^3/T_J{}^2)(T_E{}^2/r_E{}^3) = 323$.

48. (a) $M = 4\pi^2 r^3/GT^2 = 1.32 \times 10^{41}$ kg
(b) $M/(2\times10^{30}$ kg) $\approx 6.6 \times 10^{10}$ stars. Spherically symmetric distribution of mass.

49. $\Sigma F_r = GmM/r^2 = mv^2/r$, so $M = 4\pi^2/(rGT^2) = 4.74 \times 10^{24}$ kg.

50. (a) Orbital speed $v = (GM/R)^{1/2} = (gR)^{1/2}$. (b) 84.4 min

51. (a) $(R_2/R_1)^3 = (T_2/T_1)^2$, thus $R_2 = 6.71 \times 10^5$ km.
(b) $T^2 = 4\pi^2 r^3/GM$, so $M = 4\pi^2 r^3/GT^2 = 1.90\times10^{27}$ kg.

52. $v = (GM/R)^{1/2} = 0.408$ mm/s

53. (a) $T^2 = 4\pi^2 r^3/GM$, thus T = 114 min.
(b) $F = ma = m\,(4\pi^2 r/T^2) = 1.52 \times 10^4$ N

54. Since $M\ \alpha\ \rho R3$ and $r \approx R$, Kepler's 3rd. law yields $T^2\ \alpha\ 1/\rho$. Thus $T_A/T_B = (\rho_B/\rho_A)^{1/2}$, or $T_A = (2)^{1/2}T_B$.

55. $(T_1/T_2)^2 = (r_1/r_2)^3 = (6685/6730)^3$, so $T_1/T_2 = 0.99$.
$nT_2 = (n + 0.5)T_1$, so $n/(n + 0.5) = 0.99$ and $n = 49.5$ orbits.

56. (a) $mg = kv_T{}^2 = 196$ N; (b) $F_D = 196$ N/4 = 49 N

57. $mg = \gamma v_T$, thus $F_D = \gamma v = (mg/v_T)v = mg/2 = 2.45 \times 10^{-2}$ N.

58. $a = dv/dt = (\gamma v_T/m)\exp(-\gamma t/m) = ge^{-\gamma t/m}$, since $mg = \gamma v_T$

59. $T\cos\theta = mg$, $T\sin\theta = ma'$, so $\tan\theta = a'/g$, and $a' = 1.38$ m/s^2.

60. (a) $T\cos\theta = mg$, $T\sin\theta = ma'$, thus $\tan\theta = a'/g$ and $\theta = 14.9^o$.
Then, $d = L\sin\theta = 20.6$ cm;
(b) $T = mg/\cos\theta = 4.06$ N

61. $\tan\theta = a'/g = v^2/rg = 1/3$, thus $v^2 = rg/3$ and $v = 11.4$ m/s

62. $a' = 2\omega v'$, $d = 1/2\ at^2 = 1/2(2\omega v')(R/v)^2 = \omega R^2/v' = 0.54$ m

63. $m_1 g - \mu m_2 g = (m_1 + m_2)a$, so $a = 2.38$ m/s^2

64. $m_1 g - \mu m_2 g = 0$, so $\mu = m_1/m_2 = 0.586$.

65. $N - mg = mv^2/L$, so $N = 445$ N.

66. $T - mg\sin\theta - \mu(mg\cos\theta) = ma$, leads to $a = 1.74$ m/s^2

67. $F_1\cos\theta_1 + F_2\cos\theta_2 - \mu N = ma$; $N - F_1\sin\theta_1 - mg + F_2\sin\theta_2 = 0$.
Find $N = 34.55$ N and then $a = 5.37$ m/s^2.

68. (a) $mg\sin\theta + \mu(mg\cos\theta) = ma$, thus $a = 6.91$ m/s^2 down the slope
(b) $mg\sin\theta - \mu(mg\cos\theta) = ma$, thus $a = 4.34$ m/s^2 down the slope

69. $\Sigma F_x = T\cos\alpha - mg\sin\theta - \mu N = ma$; $\Sigma F_y = N - mg\cos\theta + T\sin\alpha = 0$.
Find $a = 1.74$ m/s^2.

70. $\mu mg = ma$, so $a = 5.39$ m/s^2. Then, $0 = 24^2 - 2(5.39)\Delta x$, so
$\Delta x = 53.4$ m.

71. (a) $F - \mu(m_1 + m_2)g = (m_1 + m_2)a$, so $a = 0.54$ m/s^2.
(b) $F_{21} - \mu m_2 g = m_2 a$, thus $F_{21} = 7.50$ N.

72. From $v = (GM/r)^{1/2}$ find r, then $T = 2\pi r/v = 2\pi GM/v^3 = 88$ min.

73. $\Sigma F_y = f - mg = 0$, where $f = \mu F$, so $F = mg/\mu = 1.47$ N

74. Centripetal $f = mv^2/r$, where $f = \mu mg$. From $C = 2\pi r$ find r,
then $\mu = v^2/rg = 0.123$.

75. $F\cos\alpha - \mu(m_1 g - F\sin\alpha) - \mu(m_2 g\cos\theta) - m_2 g\sin\theta = (m_1 + m_2)a$,
leads to $a = 1.60$ m/s^2

76. (a) 0.13 μs; (b) $a = v^2/r = 1.45 \times 10^{15}$;
(c) $mv^2/r = 2.42 \times 10^{-12}$ N

77. (a) $f = v/2\pi r = 6.59 \times 10^{15}$ s^{-1}; (b) $mv^2/r = 8.26 \times 10^{-8}$ N

78. $F\cos\theta - \mu(m_1 g - F\sin\theta) - \mu m_2 g = (m_1 + m_2)a$, thus $a = 2.71$ m/s^2.
$T - \mu m_2 g = m_2 a$, so $T = 9.29$ N

Problems

1. $F\cos\theta = \mu(mg - F\sin\theta)$, so $F = \mu mg/(\cos\theta + \mu\sin\theta)$.
From $\partial F/\partial\theta = 0$ we find $\tan\theta = \mu$. Substitute this to find
$F_{min} = mg\sin\theta$.

2. (a) $T - \mu mg = ma$; $F - T - \mu(M + 2m)g = Ma$. (M experiences friction at both surfaces.) Set $a = 0$ and eliminate T to find $F = \mu(M + 3m)g = 19.6$ N;
(b) $F - \mu(10g) = 6a$, so $F = 31.6$ N

3. (a) Downhill: $mg\sin\theta - f = 0$; Uphill: $F - mg\sin\theta - f = 0$, so $F = 2mg\sin\theta = (6000g)\sin 5^o = 5.12$ kN.
(b) Uphill: $F - mg\sin\theta - f = 0$; so $F = mg\sin\theta + f$.
Level road: $F - f = ma$, thus $a = g\sin\theta = 1.70$ m/s^2.

4. For minimum speed, friction is outward.
$\Sigma F_x = N\sin\theta - f\cos\theta = mv^2/r$;
$\Sigma F_y = N\cos\theta + f\cos\theta - mg = 0$
Eliminate N to find,
$v_{min}{}^2 = rg(\sin\theta - \mu\cos\theta)/(\cos\theta + \mu\sin\theta)$
Similarly,
$v_{max}{}^2 = rg(\sin\theta + \mu\cos\theta)/(\cos\theta - \mu\sin\theta)$
With the given values, $v_{min} = 9.6$ m/s, $v_{max} = 24.5$ m/s

5. $T + \mu mg = mv^2/R_1$ and $T - \mu mg = mv^2/R_2$, where $T = Mg$. Thus
$$R_2/R_1 = (M + \mu m)/(M - \mu m)$$

6. (a) $a_r = v^2/L = 2.25$ m/s^2, $a_t = g\sin\theta = 3.35$ m/s^2,
(b) $T - mg\cos\theta = mv^2/L$, so $T = 18.4 + 4.5 = 22.9$ M

7. $\Sigma F_y = (T_1 - T_2)\cos\theta = mg$, so $T_1 - T_2 = 4.9$ N
$\Sigma F_x = (T_1 + T_2)\sin\theta = (mv^2/R)$, so $T_1 + T_2 = 11$.
Find $T_1 = 7.94$ N and $T_2 = 3.04$ N.

8. (a) Relation for frictionless banked curve is $\tan\theta = v^2/rg$.
With the given values $v^2 > rg\tan\theta$, so f is down the slope.
(b) $\Sigma F_x = N\sin\theta + \mu N\cos\theta = mv^2/r$;
$\Sigma F_y = N\cos\theta - \mu N\sin\theta - mg = 0$
Eliminate N to find
$$\mu = (v^2 - rg\tan\theta)/(rg + v^2\tan\theta)$$
With $v = 30$ m/s, $r = 80$ m and $\theta = 15^o$, we find $\mu = 0.673$.

9. $8g\sin\theta - (\mu_1 m_1 g\cos\theta + \mu_2 m_2 g\cos\theta) = (m_1 + m_2)a$, so $a = 2.04$ m/s^2
$T + m_1 g\sin\theta - \mu_1 m_1 g\cos\theta = m_1 a$, thus $T = 1.6$ N.

10. $T\sin\theta = mv^2/r$; $T\cos\theta - mg = 0$ thus $\tan\theta = v^2/rg$, where $r = L\sin\theta$. $T = 2\pi r/v = 2\pi(L\cos\theta/g)^{1/2}$

11. (a) From $mg\sin\theta - \mu mg\cos\theta = ma$, we find $\tan\theta_s = \mu_s$.
(b) $d = 1/2\ at^2$, so $t^2 = 2d/a = 2d/[g(\sin\theta_s - \mu_k\cos\theta_s)$, thus, $t = [2d/(g\cos\theta_s(\mu_s - \mu_k)^{1/2})]$
(c) v = constant, $\tan\theta_k = \mu_k$.

12. When F_o is a maximum, the frictional force is downward
$\Sigma F_x = N\sin\theta + \mu N\cos\theta = ma$,
so $a = N(s + \mu c)/m$;
$\Sigma F_y = N\cos\theta - \mu N\sin\theta - mg = 0$,
so $N = mg/(c - \mu s)$
$F_o = (M + m)a =$
$= (M + m)g(s + \mu c)/(c - \mu s) = 71$ N
When F_o is a minimum, f is upward: $a = N(s - \mu c)/m$, and $N = mg/(c + \mu s)$, thus $F_o = (M + m)a = 3.9$ N

13. $\Sigma F_x = N(\sin\theta + \mu\cos\theta) = mv^2/R$;
$\Sigma F_y = N(\cos\theta - \mu\sin\theta) - mg = 0$.
Thus, $v^2 = Rg(\tan\theta + \mu)/(1 - \mu\tan\theta)$

14. $mg - \gamma v = m\ dv/dt$, so $\int dv/g(1 - \gamma v/mg) = \int dt$

$$-(m/\gamma)\ln(1 - \gamma v/mg) = t$$

$$1 - \gamma v/mg = \exp(-\gamma t/m)$$

So, using $v_T = mg/\gamma$, we find

$$v = v_T(1 - e^{-\gamma t/m})$$

CHAPTER 7

Exercises

1. $W = Fs\cos\theta = 24$ J

2. $W = (2\underline{i} - 3\underline{j} + \underline{k}).(2\underline{i} - 2\underline{j} - 4\underline{k}) = 6$ J

3. 40 runs of 10 m = 400 m, thus W = (40N)(400 m) = 16 kJ.

4. (a) $W_P = Fs = 240$ J;
   (b) $W_g = -mg\Delta y = -147$ J;
   (c) $W_f = -fs = -66$ J

5. $W = fs = (\mu N)(12x2\pi r) = (8\ N)(12)(2\pi r) = 24.1$ J

6. $\Sigma F_x = F\cos\theta - f = 0$, $\Sigma F_y = N + F\sin\theta - mg = 0$. Thus, $F\cos\theta = f = \mu(mg - F\sin\theta)$ or $F = \mu mg/(\cos\theta + \mu\sin\theta) = 4.99$ N
   (a) $W_F = Fs\cos\theta = 7.06$ J
   (b) $W_f = -fs = -7.06$ J;   (c) $W_g = 0$

7. $\Sigma F_x = F\cos\theta - f = 0$, $\Sigma F_y = N - F\sin\theta - mg = 0$. Using $f = \mu N$, find $F = \mu mg/(\cos\theta - \mu\sin\theta) = 8.32$ N.
   (a) $W_f = Fs\cos\theta = 11.8$ J;
   (b) $W_f = -fs = -11.8$ J;   (c) $W_g = 0$.

8. $W_g = -mg\Delta y = -29.4$ J is independent of the path taken

9. (a) $W_g = -mg\Delta y = +mgs\sin\theta = 49.7$ kJ; (b) $W_f = -fs = -4$ kJ.

10. F = 157.5 N, then $W_F = mg\Delta y = 1.89$ kJ

11. (a) $W_{motor} = -mgs\sin\theta = -134$ J; (b) +134 J

12. $W = (mg)(v\Delta t) = 784$ J

13. (a) Net area = +30 J (Displacement and force are positive);
    (b) Area = -10 J (Displacement is negative, force is positive)

14. (a) $+F_oA/2$; (b) $-F_oA/2$

15. (a) $+F_oA/2$; (b) $-F_oA/2$

16. (a) $-F_oA/2$; (b) $F_oA/2$

17. (a) $W_{ext} = 1/2\ k(x_f^2 - x_i^2) = 0.2$ J; (b) $W_{ext} = 0.6$ J

18. $k = \Delta F/\Delta x = 73.5$ kN/m (Assume weight shared equally by all springs)

19. $W_{ext} = 1/2\ kx^2$. (a) $W_1/W_2 = k_1/k_2$;
(b) $W = 1/2\ k(F/k)^2 \propto 1/k$. Thus, $W_1/W_2 = k_2/k_1$

20. $\Delta K = 1/2\ m(v_f^2 - v_i^2) = 1(29 - 13) = 16$ J

21. $K = 1/2\ mv^2 = 1/2\ m(2\pi r/T)^2$, where $T = 1$ y and $r = 1.5\text{x}10^{11}$ m. Find $K = 2.66\text{x}10^{33}$ J $= 6.3\text{x}10^{23}$ tons

22. $K = 1/2\ mv^2$, $W = F\Delta x = \Delta K$, so $\Delta x = \Delta K/F$
(a) 120 J, 0.15 m; (b) 26.2 kJ, 3.28 m;
(c) 3.62 MJ, 4.53 km; (d) $3.48\text{x}10^{10}$ J, $4.36\text{x}10^{7}$ m

23. (a) $\Delta K = -\ 1/2\ (0.2)(400) = -\ 40$ J;
(b) $W_g = -mg\Delta y = -\ 35.3$ J; (c) No, because of air resistance

24. $K = 1/2\ mv^2$, $W = F\Delta x = \Delta K$, so $F = \Delta K/\Delta x$.
(a) $9.34\text{x}10^{6}$ N; (b) $1.31\text{x}10^{7}$ N; (c) $2.79\text{x}10^{7}$ N

25. (a) $W = Fs = 5\text{x}10^{4}$ J;
(b) $W = 1/2\ mv^2 + f(100\text{ m}) = 2.25\text{x}10^{5}$ J;
(c) $W = \Delta K + fs = 1/2\ (10^3)(1200) + (250)(300) = 6.75\text{x}10^{5}$ J

26. (a) $K_H = 5.47$ kJ, $K_W = 109$ kJ;
(b) $(mg\sin\theta - f)s = \Delta K$, so $s = 234$ m

27. (a) $W_{eng} = Fs = 3\text{x}10^{7}$ J;
(b) $W_f = -fs = -(1.96\text{x}10^{4}\text{ N})(10^3\text{ m}) = -1.96\text{x}10^{7}$ J;
The net work done on the railcars ($m = 40\text{x}10^{4}$ kg) is
$W = Fs - fs = 1.04\text{x}10^{7} = 1/2\ mv^2$, thus $v = 7.21$ m/s.

28. (a) $T - mg = ma$, $W_C = Ts = m(a + g)s = 12.0$ kJ;
(b) $W_g = -mg\Delta y = -11.76$ kJ;
(c) $\Delta K = W_C + W_g = 240$ J

29. $W_{net} = W_F + W_f = \Delta K$, thus
$W_F = \Delta K + fs = 1/2\ (1100)(2.5^2) + (200)(30) = 9.44$ kJ

30. $\Delta K = 1/2\ mv^2 = 4.34$ kJ, $\Delta K = Fs$, so $F = \Delta K/(0.6\ m) = 7.23$ kN

31. $\Delta K = -1/2\ (10^{-2})(16x10^4) = -800$ J, $\Delta K = -Fs$, thus
$F = 3.2x10^4$ N

32. (a) $W = \Delta K = -1.09x10^{10}$ J;
(b) $\Delta K = -Fs$, so $F = 2.17x10^6$ N;
(c) $K_f = 0.5K_i = 0.5(1.085x10^{10}) = 1/2\ mv_f^2$,
so $v_f = 10.8$ m/s or 38.9 km/h.

33. $\Sigma F_x = F\cos\theta - mg\sin\theta - \mu N = 0$; $\Sigma F_y = N - mg\cos\theta - F\sin\theta = 0$.
Substituting for N we find F = 51.7 N
(a) $W_F = Fs\cos\theta = 179$ J;
(b) $W_g = -mgs\ \sin\theta = -117.6$ J
(c) $W_f = -fs = -\mu Ns = -61.4$ J;
(d) $W_{NET} = 0$.

34. (a) $\Delta K = +3$ J, $W_g = +mgd\ \sin\theta = 4.9$ J;
(b) $W_f = -fd = -1.9$ J;
(c) $W_f = -\mu(mg\cos\theta)d$, so $\mu = 0.224$

35. $\Delta K = -9$ J; $W_f = -\mu mg\cos\theta\ d = -3.79d$; $W_g = -mg\sin\theta d =$
$-5.07d$. Since $\Delta K = W_{NET}$, we find d = 1.02 m.

36. $1/2\ mv^2 = 1/2\ kA^2$, so $v = A(k/m)^{1/2} = 15$ m/s

37. $-1/2\ mv^2 = -\mu mgd$, so $d = v^2/2\mu g = 57.4$ m.
(a) $W_F = Fd\ \cos\theta = 398$ J;
(b) $f = \mu(mg\cos\theta + F\sin\theta)$, $W_f = -fs = -204$ J;
(c) $W_g = -mgd\sin\theta = -68.3$ J;
(d) $W_{net} = 126$ J $= 1/2\ mv^2$, so v = 2.51 m/s

38. $-1/2\ mv^2 = -\mu mgd$, so $d = v^2/2\mu g = 57.4$ m

39. (a) Net area = +10 J (Note displacement is negative.)
(b) $K_i = 1/2(0.25)(20^2) = 50$ J, thus $K_f = 50 + 10 = 60$ J

40. (a) $W_{sp} = 1/2\ kA^2 = 1.60$ J
(b) $W_f = -\mu mgA = -0.392$ J
(c) $1/2\ mv^2 = 1.21$, so v = 2.2 m/s
(d) $-1.21 = -\mu mgs$, thus s = 0.617 m

41. $N - mg + F\sin\theta = 0$, so $N = 8.62$ N and $f = \mu N = 0.86$ N
(a) $W_F = F s\cos\theta = 4.8$ J;
(b) $W_f = -fs = -0.34$ J
(c) $W_{sp} = -1/2\ (20)(0.4)^2 = -1.6$ J
(d) $W_{NET} = 2.86$ J $= 1/2\ mv^2$ thus $v = 1.38$ m/s.

42. $P = (20$ hp$)(746$ W/hp$) = 14920$ W, but $P = fv$, so $f = 671$ N.

43. $\Delta T = 200g$; $P = \Delta T v = 784$ W

44. $P = \Delta E/\Delta t = -mgh/\Delta t = -\ 1.23\text{x}10^{-5}$ W

45. $P = Tv = (mg\sin\theta + f)v = (886$ N$)(0.5$ m/s$) = 443$ W

46. $P = Fv = (5.8\text{x}10^4$ N$)(278$ m/s$) = 1.61\text{x}10^7$ W $= 21{,}600$ hp

47. $P = Fv$, so $P_E = 2.35\text{x}10^7$ W; $P_G = 4.92\text{x}10^7$ W

48. (a) $P = \Delta E/\Delta t = mg\Delta y/\Delta t = 1715$ W $= 2.30$ hp;
(b) $P = 10(mg\Delta y/\Delta t) = 39.2$ W.

49. $P = (mg)(55$ m/s$)/746 = 43.4$ hp

50. $E = (0.25$ hp$)(0.746$ kW/hp$)(2$ h$) = 0.373$ kW.h, so
cost = 2.6 cents

51. $P = 600$ kcal/1 h means 600/9 g/h.
Thus $t = (100$ g$)(9/600$ h/g$) = 1.5$ h.

52. $F = P/v = 3.58$ kN

53. $P = 373$ W $= (18.5$ N$)v$, so $v = 20.2$ m/s; $\Delta x = v\Delta t = 12.1$ km

54. (a) $P = \Delta E/n\Delta t$, where $n = 2.2\text{x}10^8$ and $\Delta t = 3.156\text{x}10^7$ s.
$P = 7.2$ kW;
(b) 7.2 kW$/(0.2$ kW/m$^2) = 36$ m^2

55. $F = ma = (3\text{x}10^{-3}$ kg$)(145$ m/s$^2) = 0.435$ N; $P_{av} = Fv = 0.74$ W

56. (a) $f = P/v = 8.952$ kW$/(22.2$ m/s$) = 403$ N;
(b) $P = (mg\sin\theta + f)v = 4.86\text{x}10^4$ W $= 65$ hp

57. $R = v_o{}^2\sin 2\theta/g$, so takeoff speeds are $v_S = 14$ m/s; $v_J = 29.7$

m/s. $P_S = \Delta K/\Delta t = (\Delta K/\Delta x)(v_{av}) = 3.3$ kW; $P_J = 2.4$ kW.

58. $\Delta E = P\Delta t = (0.33\text{ hp})(169\text{ min}) = 2.496$ MJ
At 22%, 1 g of fat provides 8.272 kJ.
Thus 2.496 MJ/(8.272 kJ/g) = 302 g

59. $(3.4\times10^7\text{ J/L})(100\text{ km}/3600\text{ s})(1\text{ L}/12\text{ km}) = 7.87\times10^4$ W
25 hp = 1.87×10^4 W, thus efficiency = 23.7%

60. $P = (2F\cos\theta)v = 1300\text{ W} = 1.74$ hp

61. Energy available = 4.6×10^{14} J, Use = 5×10^{19} J/y
Thus, $t = 9.2\times10^{-6}$ y = 4.84 min

62. $W = mgd$, so $d = 34.0$ m

63. (a) $\Delta K = W = 2300$ J, so $v = 2.09$ m/s; (b) $Fd = W$, so $F = 129$ N.

64. $\Delta K = 0.5m(38 - 14) = 2.4$ J

65. (a) $W_g = -mgd\sin\theta = -42.9$ J;
(b) $\Delta K = 0.5m(1.5^2 - 0.5^2) = 14$ J;
(c) Net work = ΔK = 14 J; (d) $W_g + W_p = \Delta K$, so $W_p = 56.9$ J.

66. (a) $W_g = -mgd\sin\theta = -56.3$ J; (b) $W_f = -\mu(mg\cos\theta)d = -30.9$ J;
(c) $W_p + W_g + W_f = 0$, so $W_p = 87.2$ J.

67. (a) $W_g = +mgd\sin\theta = 1956$ J; (b) $W_f = -\mu(mg\cos\theta)d = -1104$ J;
(c) $W_{NET} = 852\text{ J} = 0.5mv^2$, so $v = 5.33$ m/s.

68. $P_{av} = mgd/\Delta t = 399\text{ W} = 0.535$ hp.

69. $F = \mu mg = 78.4$ N, $P = Fv = 56.6$ W.

70. (a) $T\cos\alpha - mg\sin\theta - \mu(mg\cos\theta - T\sin\alpha) = 0$, thus $T = 11.7$ N;
(b) $W_g = -mgd\sin\theta = -12.9$ J; (c) $W_f = -\mu(mg\cos\theta - T\sin\alpha)d$
$= -4.11$ J; (d) $Td\cos\alpha = +17$ J.

71. (a) $W_p = Fs = 223$ J; (b) $\Delta K = 145\text{ J} = W_p + W_f$, thus
$W_f = -78\text{ J} = -fd$, so $f = 21.7$ N.

72. $\Delta K = -(0.5)(1250)(15.6)^2 = 1.52\times10^5$ J; $\Delta K/\Delta t = -680$ hp.

73. $P = Fv = (mg\sin\theta)v = 297$ W $= 0.399$ hp

74. (a) $W_g = -mg\Delta y = 0.157$ J; (b) $W_{sp} = -0.5kx_f^2 = -0.04$ J;
(c) $\Delta K = 0.117 = 0.5mv^2$, so $v = 1.71$ m/s.

75. $W_{EXT} = 0.5k(x_f^2 - x_i^2)$ leads to $k = 8$ N/m.

76. 1 N.m = (0.2248 lb)(3.28 ft) = 0.7376 ft.lb.

77. (a) $W_g = -mg\Delta y = +20.6$ J; (b) $\Delta K = W_g = 20.6$ J;
(c) 8.28 m/s.

78. (a) $W_g = -mg\Delta y = +753$ J; (b) $\Delta K = W_g = 753$ J; (c) 7.92 m/s.

79. (a) $W_g = -mg\Delta y = -59.7$ J; (b) 59.7 J;
(c) $W_p = Fd$, $F = 74.6$ N.

80. (a) $\Delta K = 0.5(1200)(14.5)^2 = 1.26 \times 10^5$ J,
$P_{av} = \Delta K/\Delta t = 33.8$ hp
(b) $P = Fv = (ma)v = 67.7$ hp.

Problems

1. $\Delta K = 1/2\ m(36 - 64) = -\mu mg(2.4\pi)$; thus $\mu_k = 0.19$.
$\Delta K = -1/2\ m(36) = -\mu_k(19.6)(2.4\pi N)$, thus $N = 1.28$ rev.

2. (a) $W_f = -fA = -\mu mgA$, $W_{sp} = +1/2\ kA^2$, thus $W_{net} = 0.134$ J
$W_{net} = \Delta K = 1/2\ mv^2$, thus $v = 0.946$ m/s
(b) $-1/2\ mv^2 = -fx - 1/2\ kx^2 = -0.134$ J. With the given values $6x^2 + 0.53x - 0.134 = 0$, so $x = 0.112$ m (compression)

3. $P = d(0.5mv^2)/dt = (m/2)dv^2/dt$;

$m/2 \int dv^2 = \int P\,dt$, thus $mv^2/2 = Pt$, so

$$v = dx/dt = (2Pt/m)^{1/2}$$

$\int dx = (2P/m)^{1/2} \int t^{1/2}\,dt$, thus $D = (8Pt^3/9m)^{1/2}$

4. $W = \int F\,dx = [8x^2 + x^4/8] = 25.9$ J

5. (a) $W_{sp} = -1/2\ kx^2 = -1.6$ J; (b) $W_f = -\mu mg \sin\theta\ s = +0.79$ J
(c) $W_g = +mg \sin\theta\ s = +6.27$ J
(d) $\Delta K = W_{net}$ leads to $v^2 = 3.89$, so $v = 1.97$ m/s
(e) $mgd \sin\theta - 1/2\ kd^2 - \mu mg \cos\theta\ d = 0$; thus $d = 1.37$ m

6. (a) $P = F_D v = av + bv^3$. Substitute given values and solve to obtain a = 182 N, b = 0.517 kg/m;
(b) $P = 182(30) + 0.517(30)^3 = 19.4$ kW = 26 hp

7. We need the change in length $\ell_2 - \ell_1$.
From the law of sines:
$2/\sin 15^o = \ell_1/\sin 30^o = \ell_2/\sin 45^o$;
thus
$\ell_1 = 3.86$ m and $\ell_2 = 5.46$ m, and
so $\Delta\ell = 1.6$ m. $W = mg\Delta\ell = 392$ J

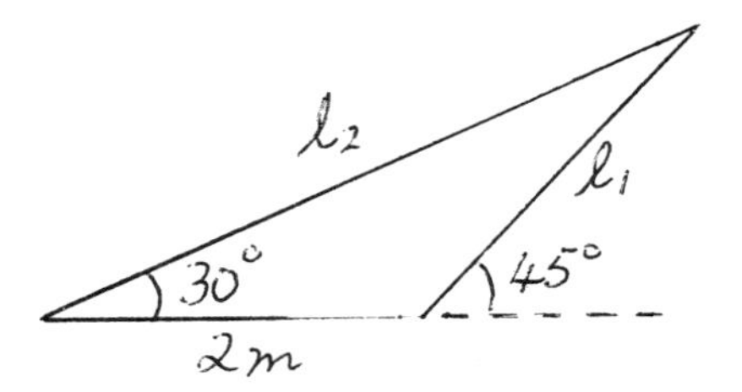

8. $W_{sp} = 1/2\ k(\ell - \ell_o)^2 = 1/2\ mv^2$, so
$v = (\ell - \ell_o)(k/m)^{1/2}$

9. (a) $P = Fv = (200 + 0.8v^2)v = 10.4$ kW = 13.9 hp
(b) $P = (200 + 0.8v^2 + mg\sin\theta + ma)v = 40.2$ kW = 53.9 hp
(c) FIP = P/Efficiency = 268 kW = 359 hp.

10. $\Delta K = 0.5(1200)(20^2 - 30^2) = -0.3$ MJ, and $\Delta t = \Delta x/v_{av} = 20$ s.
$W_g = -mgd\sin\theta = -1.02$ MJ; $W_f = -fd = -0.25$ MJ
$W_{net} = W_g + W_f + W_{eng} = \Delta K$, thus
$W_{eng} = \Delta K + mgd\sin\theta + fd = 0.97$ MJ.
$P_{eng} = W_{eng}/\Delta t = 48.5$ kW = 65 hp

11. $dW = m\ (d\underline{v}/dt).d\underline{s} = m\ \underline{v}.d\underline{v} = 1/2\ m\ d(v^2)$, so
$W = m(v_f^2 - v_i^2)/2$.

CHAPTER 8

Exercises

1. $1/2\ (m_1 + m_2)v^2 - m_1gh + m_2gh = 0$. Thus, $v = 1.83$ m/s

2. $1/2\ (m_1 + m_2)v^2 - m_1gh = 0$, so $v = 1.71$ m/s

3. $1/2\ (m_1 + m_2)v^2 - m_2gd \sin\theta + m_1gd \sin\theta = 0$; so $v = 1.18$ m/s

4. $\Delta U = mgh = 2.97 \times 10^5$ J. One gram of fat provides 5.64 kJ of mechanical energy. Thus, ΔU requires 52.7 g of fat.

5. With the given values $E = 1/2\ mv^2 + mgL(1 - \cos\theta) = 1.79$ J
   (a) $1/2\ mv^2 = 1.79$ J, so $v = 2.44$ m/s
   (b) $mgL(1 - \cos\theta_0) = 1.79$, so $\theta_0 = 53.6^o$

6. $E_1 = mgL_1(1 - \cos\theta_1) = 1.47$ J; $E_2 = mgL_2(1 - \cos\theta_2) = 1.47$ J thus $\theta_2 = 50^o$

7. (a) $1/2\ kA^2 = 0.8$ J $= 1/2\ mv_{max}^2$, so $v_{max} = 2.53$ m/s;
   (b) $1/2\ kx^2 + 1/2\ mv^2 = 0.8$; so $v = 2.19$ m/s
   (c) $U = E/2$, i.e., $1/2\ kx^2 = 1/4\ kA^2$; thus $x = 28.3$ cm

8. (a) 60 J; (b) -60J; (c) $\Delta K = -\Delta U$, so $v_f = 8.72$ m/s

9. $mg(x + 0.6) = 1/2\ kx^2$, $60x^2 - 4.9x - 2.94 = 0$, so $x = 26.6$ cm

10. $E_1 = mgL = 6.125$ J, $E_f = mgL(1 - \cos\theta) + 1/2\ mv^2$
    $\Sigma F_y = T - mg\cos\theta = mv^2/L$. Using $T = 6$ N,
    $mv^2 = L(6 - mg\cos\theta)$.
    Substitute mv^2 into $E_f = 6.125$ J to find $\theta = 65.9^o$

11. (a) $-mgD + 1/2\ kD^2 = 0$, so $D = 1.96$ m
    (b) $-mg(0.5) + 0.5(6)v^2 + 0.5(40)(0.25) = 0$, so $v = 2.21$ m/s

12. (a) $k = 4.9$ N/m, $E_i = 1/2\ kA^2 = 1/2\ (4.9)(0.3)^2 = 0.22$ J
    $E_f = mgh = 0.22$, so $h = 44.9$ cm.
    (b) $E_i = 1/2\ kA^2$, $E_f = 1/2\ kx^2 + mg(A + x) = 0.22$ J, Thus, $2.45x^2 + 0.49x - 0.073 = 0$, so $x = 0.1$ m

13. (a) $1/2\ kD^2 - m_1gD + m_2gD = 0$; $D = 1.23$ m

(b) $0.5(m_1 + m_2)v^2 + (m_2 - m_1)gd + 0.5kd^2 = 0$; $v = 0.95$ m/s

14. (a) $1/2\ mv^2 = mgd\sin\theta$, so $v = 6.26$ m/s;
(b) $E_i = 1/2\ mv^2 = 1.96$ J, $E_f = 1/2\ kA^2 - mgA\sin\theta$. Thus, $2.5A^2 - 0.49A - 1.96 = 0$, so $A = 0.99$ m

15. (a) $E_1 = 0.5mv^2 + mgy_1 = 165$ J; $U_2 = 157$ J. It gets over hill.
(b) $E_3 = 1/2\ kA^2 + mgy_3 = 165$, so $A = 1.31$ m

16. (a) $mgL = 1/2\ mv_{max}^2$. $\Sigma F_y = T - mg = mv_{max}^2/L$, so $T = 3mg = 23.5$ N
(b) $mgL = 1/2\ mv^2 + mgL(1 - \cos\theta)$, $\Sigma F_y = T - mg\cos\theta = mv^2/L$, thus $T = 2mg\cos\theta = 18.8$ N.

17. $E = mgH = 3U/4 + U = 7mgy/4$; thus $y = 4H/7$.

18. (a) $U = E/2$, thus $mgy = 1/2(1/2\ mv_i^2)$, so $y = 40.8$ m;
(b) $U = 2E/3$, thus $mgy = 2/3(1/2\ mv_i^2)$, so $y = 54.4$ m

19. $E_A = 1/2\ mv_A^2 + mgy_A = 2.19\text{x}10^5$ J.
(a) $E_B = 1/2\ (600)v_B^2 + (600g)(12) = 2.19\text{x}10^5$ J, so $v_B = 22.3$ m/s
(b) $E_C = 300v^2 + (600g)(25) = 2.19\text{x}10^5$, so $v_C = 15.6$ m/s.

20. (a) $ky_o = mg$, so $y_o = mg/k = 11.2$ cm;
(b) $\Delta E = 1/2\ ky^2 - mgy = 0$, so $y = 2mg/k = 22.4$ cm

21. (a) $1/2\ m(25)^2 + mg(40) = 1/2\ mv^2$; so $v = 37.5$ m/s.
(b) $1/2\ m(25)^2 + mg(40) = 1/2\ m(15)^2 + mgy$; thus $y = 60.4$ m.

22. (a) From Ch. 3: $\Delta y = (v_o \sin\theta)^2/2g = 1$ m, so $v_o = 4.58$ m/s
(b) $1/2\ mv_o^2 = 1/2\ mv^2 + mgy$, so $v = 1.18$ m/s ($= v_o\cos\theta$)

23. (a) $Mgh = (5\text{x}10^6)(9.8)(50) = 2.45\text{x}10^9$ J
(b) $2.33\text{x}10^7$ bulbs

24. $\Delta E/\Delta t = (\Delta m/\Delta t)(v^2/2 + gy) = 1080$ W $= 1.45$ hp

25. $\Delta E = \Delta K + \Delta U = m(v^2/2 + gy)$, where $v = 1.39$ km/s.
$\Delta E = m[(1.39\text{x}10^3)^2/2 + 9.8(25\text{x}10^3)] = 6.055\text{x}10^10$ J
$P = \Delta E/\Delta t = 1.01\text{x}10^9$ W $= 1.35\text{x}10^6$ hp

26. $\Delta E/\Delta t = (\Delta m/\Delta t)(v^2/2 + gd\sin\theta) = 549$ W

27. The cable lifts 100 persons, m = 7000 kg, at 1 m/s, so $P = Fv = (mg\sin\theta)v = 23.4$ kW.

28. $E_i = 1/2\ m(9)^2 = 2835$ J, $E_f = 1/2\ m(0.5)^2 + mgy$, so y = 4.1 m

29. $K_i = 1/2\ (83\ \text{kg})(38.9\ \text{m/s})^2 = 6.28\text{x}10^4$ J; $U_i = mgy = 81.3\text{x}10^4$ J, so $E_i = 876$ kJ. $E_f = 1/2\ (83)(49) = 2$ kJ, so W = 874 kJ.

30. $\Delta E = W_{NC}$: $1/2\ mv^2 - mgd\sin\theta = -\mu(mg\cos\theta)d$, so $\mu = 0.147$.

31. In each case use $E_f - E_i = -fd$.
(a) $-1/2\ mv^2 = -\mu mgd$, thus d = 1.36 m
(b) $mgd\sin\theta - 1/2\ mv^2 = -\mu mgd\cos\theta$, thus d = 0.80 m.
(c) $-mgd\sin\theta - 1/2\ mv^2 = -\mu mgd\cos\theta$, thus d = 41.6 m.

32. $E_1 = 1/2\ mv_1{}^2 + mgy = 41.2$ J.
Left of horizontal section: $E_2 = E_1 - \mu(mg\cos\theta_1)d_1 = 29.44$ J
Right of horizontal section: $E_3 = E_2 - \mu mgL = 17.68$ J
Up incline: $mgd_2\sin\theta_2 = E_3 - \mu(mg\cos\theta_2)d_2$, so $d_2 = 1.96$ m

33. $E_i = 1/2\ mv^2 = 17.3$ J, Ef = mgy = 15.3 J; thus $W_{NC} = -2.0$ J

34. $1/2\ kx^2 - mg(4 + x)\sin\theta = -\mu mg\cos\theta(4 + x)$,
$30x^2 - 3.96x - 15.85 = 0$, thus x = 0.796 m

35. $\Delta U = -\int C\ x^3\ dx$ leads to $U = -Cx^4/4$.

36. $\Delta U = -\int b/x^2\ dx = b/x$

37. $\Delta U = -\int ax\ dx/(b^2 + x^2)^{1/2}$ leads to $U = a/(b^2 + x^2)^{1/2}$

38. (a) $\underline{F}(x, y) = (Ax\underline{i} + Ay\underline{j})/(x^2 + y^2)^{3/2}$;
(b) $F_r = -dU/dr = Ae^{-Br}(B/r + 1/r^2)$

39. $\Delta U = -\underline{F}.\Delta\underline{s} = -(2\underline{i} - 5\underline{j}).(-5\underline{i} + 6\underline{j}) = +40$ J

42. Yes for all

43. (a) -70 J for both. No, must apply to any path;
(b) -140 J. No, should be zero.

44. (a) $W = \underline{F}.\underline{s} = 2$ J for both. Yes; (b) Zero, since $s = 0$.

45. (a) 0.13 nm and 0.33 nm;
(b) 0.8 nm and 4.1 nm, or 0.47 nm and 0.57 nm;
(c) $1.6\text{x}10^{-20}$ J

46. $E_i = 1/2\ m(v_{esc}/N)^2 - GmM/R_E$; and $E_f = -GmM/(R_E + h)$
Know that $v_{esc}{}^2 = 2GM/R_E$. Set $E_f = E_i$ to find $h = R_E/(N^2 - 1)$

47. $E_i = 1/2\ mv^2 - GmM/R_E$; $E_f = -\ GmM/5R_E$, thus
$v = (8GM/5R_E)^{1/2} = 10$ km/s.

48. $v_{esc} = (2GM/R)^{1/2}$. (a) 2.38 km/s;
(b) 5.04 km/s. (Note $R = 3.37\text{x}10^6$ m.)
(c) 60.4 km/s. (Note $R = 6.99\text{x}10^7$ m.)

49. $v_{esc} = (2GM/R)^{1/2} = 5.16\text{x}10^{-4}$ m/s

50. $GmM/R = mc^2$, so $R = GM/c^2 = 4.4$ mm

51. $E_i = 1/2\ mv_o{}^2 - GmM/R$; $E_f = -GmM/(R + h)$, and $g_o = GM/R^2$.
$E_i = E_f$: $v_o{}^2 - 2g_oR = -\ 2g_oR^2/(R + h)$, so
$h = v_o{}^2/(2g_o - v_o{}^2/R)$

52. To escape need $E = 1/2\ mv^2 - GmM/r = 0$, thus $v_{esc} =$ $(2GM/r)^{1/2}$. Since $v_{orb} = (GM/r)^{1/2}$, we see that $v_{esc} =$ $(2)^{1/2}v_{orb}$.

53. Note that $g_E = GM_E/R_E{}^2$.
(a) $\Delta E = Gm(-M_M/R_M + M_E/R_E) = m(-0.16g_ER_M + g_ER_E) = 60$ MJ.
(b) $E_1 = -GmM_E/R_E$; $E_2 = -GmM_E/2R_E$, so $\Delta E = mg_ER_E/2 = 31$ MJ.

54. Let $g_o = GM/R_E{}^2$.
$E_i = -GmM/(R_E + h) = -mg_oR_e{}^2/(R_E + h)$
$E_f = 1/2\ mv^2 - GmM/R_E = 1/2\ mv^2 - mg_oR_E$
Thus, $v^2 = 2g_oR_E - g_oR_E{}^2/(R_E + h) = 2g_oR_Eh/(R_E + h)$

55. $E_i = -GmM/R_E$; $E_f = -GmM/2r$, so $\Delta E = GmM(1/R_E - 1/2r) =$ $1.22\text{x}10^{11}$ J

56. (a) $0.5kA^2 = 0.5mv^2 + 0.5kx^2 = 0.115$ J gives $v = 0.657$ m/s.

(b) $v_m = (k/m)^{1/2}A = 0.759$ m/s.
$0.5kx^2 + 0.5m(v_m/2)^2 = 0.115$ J. Thus, x = 10.4 cm.

57. $\Delta E = W_{nc}$ is $-0.5kA^2 = -\mu(mg)d$ so $\mu = 0.15$.

58. (a) $E_i = mgL(1 - \cos\theta_m)$, $E_f = 0.5mv^2$ lead to v = 2.01 m/s.
(b) $E_i = mgL(1 - \cos\theta_m)$, $E_f = mgL(1 - \cos\theta)$ give
v = 1.83 m/s

59. (a) $E_i = mgd\sin\theta = 1.08d$, $E_f = 0.5kd^2$, lead to d = 0.6 m.
(b) $E_i = mgd\sin\theta = 0.432$ J,
$E_f = 0.5mv^2 + 0.5kd^2$, so v = 1.14 m/s.

60. mg - f = 0, so P = mgv = 44.8 hp

61. $\Delta E = W_{nc}$: $0.5(m_1 + m_2)v^2 + m_1gd\sin\theta - m_2gd + 0.5kd^2 =$
$-\mu(m_1g\cos\theta)d$. Find v = 1.51 m/s.

62. $\Delta E = W_{nc}$: $-mgy = -\mu(mg)d$, so $\mu = y/d = 0.482$.

63. $0.5mv^2 + mgA\sin\theta - 0.5kA^2 = -\mu(mg\cos\theta)A$, gives v = 1.46 m/s.

64. From $v_e = (2GM/R)^{1/2}$ find $M = 5.68 \times 10^{26}$ kg.

65. $E_i = -GmM/R$, $E_f = -GmM/2r$, $\Delta E = GmM(1/R - 1/2r) = 1.98 \times 10^9$ J

66. (a) $K = 2.7 \times 10^9$ J; (b) $U = -2K = -5.4 \times 109$ J;
(c) $E = -2.7 \times 10^9$ J; (d) $|E| = +2.7 \times 10^9$ J.

67. (a) $-mgA\sin\theta + 0.5kA^2 = 0$ leads to A = 0.894 m.
(b) $0.5mv^2 - mgd\sin\theta + 0.5kd^2 = 0$, thus v = 0.672 m/s.

68. $mgd\sin\theta - 0.5mv^2 = -fd$, so f = 3.53 N.

69. $\Delta K = -fd$, leads to f = 0.93 N

70. $E_i = 0.5mv^2 = 65.3$ J, $E_f = mgy = 53.3$ J, so y = 37.5 m.

71. $0.5(m_1 + m_2)v^2 - m_1gd = -\mu(m_2g)d$, thus v = 1.75 m/s.

72. $0.5mv^2 - mgh = -\mu(mg\cos\theta)(h/\sin\theta)$ leads to v = 6.49 m/s.

73. $E_i = 0.5mv_i^2 - GmM/R$, $E_f = 0.5mv_f^2 - GmM/5R$; thus
$v_i = 11.7$ km/s.

74. $E_i = -GmM/4R$, $E_f = -GmM/5R$; so $\Delta E = GmM/20R = 7.2 \times 10^8$ J.

75. $\Delta K = W_{nc}$: $0.5m(v_f^2 - v_i^2) = -\mu mg(2\pi r)$ gives $\mu = 0.094$.

76. $E_i = 0.5kA^2 = 0.691$ J, $E_f = mgd\sin\theta = 0.67d$,
$W_{nc} = -\mu(mg\cos\theta)d$. From $\Delta E = -W_{nc}$ find $d = 0.776$ m.

77. $0.5kA^2 = mgh$ leads to $h = 7.5$ m.

Problems

1. (a) $U(x) = C(a^2 + x^2)^{-1/2}$; $F_x = -dU/dx = Cx/(a^2 + x^2)^{3/2}$
(b) $dF/dx = C/(a^2 + x^2)^{3/2} - 3Cx^2/(a^2 + x^2)^{5/2} = 0$, so
$x = \pm a/(2)^{1/2}$

2. $F_r = -dU/dr = -C/r$

3. $U = U_o[(r_o/r)^{12} - 2(r_o/r)^6]$
(a) $(r_o/r) = 2^{1/6}$, thus $r = r_o/2^{1/6} = 0.891r_o$.
(b) $dU/dr = 0$ leads to $U_{min} = -U_o$ at $r = r_o$.
(c) From (b), $F_r = 0$ at $r = r_o$
(d) Sketch resembles Fig. 8.18.

4. (a) $F_r = -(U_o/r)(1 + r_o/r)e^{-r/r_o}$;
(b) $-2U_oe^{-1}/r_o = -2.45$ kN; $-4U_oe^{-3}/9r_o = -73.8$ N

5. TOP: $T_1 + mg = mv_1^2/R$; BOTTOM: $T_2 - mg = mv_2^2/R$, so
From $E_1 = E_2$, we find: $v_2^2 = v_1^2 + 4gR$, thus $T_2 - T_1 = 6mg$.

6. (a) $E_i = mgL$, $E_f = 1/2\ mv^2 + mgL/2$.
$\Sigma F_y = T + mg = mv^2/(L/4)$, thus $T = 3mg$;
(b) If $T = 0$, then $v^2 = gL/4$ at the top of the circle.
$E_i = mgL(1 - \cos\theta)$; $E_f = 1/2\ mv^2 + mgL/2 = 5mgL/8$.
Set $E_i = E_f$ to find $\cos\theta = 3/8$.

7. $E_i = mgL$; $E_f = 1/2\ mv^2 + 2mg(L - y)$,
At highest point: $\Sigma F_y = mg = mv^2/(L - y)$, thus $v^2 = g(L - y)$
Set $E_i = E_f$ to find $y = 3L/5$.

8. $E_i = mgH$, $E_f = 1/2\ mv^2 + mgr$.
At the highest point $\Sigma F_y = mg = v^2/r$, so $v^2 = gr$.
Set $E_f = E_i$ to find $H = 3r/2$.

9. (a) $f = T_2 - T_1$; (b) $P = fv = (T_2 - T_1)(2\pi R)(1/N)$
$= 2\pi RN(T_2 - T_1)$

10. (a) $E_i = mgH$, $E_f = 1/2\ mv^2 + mg(2R)$
At the highest point $\Sigma F_y = mg = mv^2/R$, so $v^2 = Rg$.
Set $E_i = E_f$ to find $H = 5R/2$;
(b) $E_i = mg(5R)$, $E_f = 1/2\ mv^2 + mg(2R)$, thus $v^2 = 6Rg$.
$\Sigma F_y = mg + N = mv^2/R$, thus $N = 5mg$.

11. (a) $\Sigma F_r = mg\cos\theta = mv^2/R$; so $v^2 = gR\cos\theta$.
$E = 1/2\ mv^2 = mgR(1 - \cos\theta)$, leads to $\cos\theta = 2/3$.
(b) Lower point, greater θ

12. $\underline{F}.d\underline{s} = F\ r\ d\theta$. $\int F\ r\ d\theta = 2\pi rF$ which is not zero.

13. (a) $\int_0^L F_x\,dx = L^2y^2/2 = 0$; $\int_0^L F_y\,dy = 0$

(b) $\int_0^L x\,L^2\,dx = L^4/4$;

(c) No

14. (a) $3L^2$; (b) $2L^3$; (c) No

15. $k = 9.8/0.4 = 24.5$ N/m
(a) $U = 1/2\ kx^2 - mgx$;
(b) $E_i = -1.8375$ J, $E_f = v^2/2 - 1.96$ J, thus $v = 0.495$ m/s
(c) $E = 1/2\ kx^2 - mgx$: $12.25x^2 - 9.8x + 1.8375 = 0$, thus
$x = 0.5$ m and 0.3 m, so $H = 0.1$ m
(d) $2mg/k = 0.8$ m

16. (a) $v = (GM/r)^{1/2} = 7.8$ km/s;
(b) $E = -GmM/2r = -2.31\times10^9$ J;
(c) $\Delta E/\Delta t = (1.73\times10^7\ \text{J})/(3.156\times10^7\ \text{s}) = 0.55$ W;
(d) $f = P/v = (0.55\ \text{W})/(7.8\ \text{km/s}) = 71\ \mu$N

CHAPTER 9

Exercises

1. $p = 700$ kg.m/s. (a) $v = p/m = 3.5\text{x}10^4$ m/s; (b) 0.47 m/s

2. (a) $p = 3\text{x}10^5$ kg.m/s, so $v = 250$ m/s;
(b) $K = 4.5\text{x}10^4$ J, so $v = 86.6$ m/s

3. (a) $K = p^2/2m$, thus $K_B/K_R = m_R/m_B = 3000$;
(b) $p = (2mK)^{1/2}$, so $p_B/p_R = (m_B/m_R)^{1/2} = 0.0183$.

4. Σp_x: $0 = m(20) - m(15\cos45^o) + mv_{3x}$, so $v_{3x} = -9.39$ m/s
Σp_y: $0 = m(15\cos45^o) + mv_{3y}$; so $v_{3y} = -10.6$ m/s.
$v_3 = 14.2$ m/s at 48.5^o S of W

5. $60\underline{i} = 5(2\underline{i} - \underline{j}) + 5\underline{v}$, so $\underline{v} = 10\underline{i} + \underline{j}$ m/s

6. Σp_x: $23.96 = 6\cos53^o - 6 + v_{3x}$, so $v_{3x} = 26.35$ m/s
Σp_y: $-18.05 = 6\sin53^o + v_{3y}$, so $v_{3y} = -22.84$ m/s
Thus $\underline{v}_3$ is 34.9 m/s at 40.9^o S of E

7. Σp_x: $0 = -3v_1\cos60^o + 2v_2\sin25^o$;
Σp_y: $-18 = -3v_1\sin60^o - 2v_2\cos25^o$
Solve to find $v_1 = 3.1$ m/s, $v_2 = 5.5$ m/s

8. $K_f = 0.75K_i = 6.75$ J $= 1/2\ (0.5 + m)v^2$.
Σp: $3 = (0.5 + m)v$. Substitute for v into K_f to find
$v = 4.5$ m/s, then $m = 167$ g

9. $4.5 = 2.7v$, so $v = 1.85$ m/s

10. Σp: $2u_1 = 5V$. Also $K_i - K_f = u_1^2 - 2.5V^2 = 60$ J.
Use $V = 0.4u_1$ in the second equation to find $u_1 = 10$ m/s

11. Σp_x: $0 = 3\cos37^o - 4\cos53^o + p_{3x}$, so $p_{3x} = 0.01\text{x}10^{-24}$ kg.m/s
Σp_y: $0 = 3\sin37^o - 4\sin53^o + p_{3y}$, so $p_{3y} = 1.39\text{x}10^{-24}$ kg.m/s

12. (a) $(1.2)(-\underline{i} + 3\underline{j}) + (1.8)(3\underline{i} + 4\underline{j}) = (1.2)(2\underline{i} + 1.5\underline{j}) +$
$1.8\underline{v}_B$. Thus $\underline{v}_B = \underline{i} + 5\underline{j}$ m/s;
(b) $K_i = 28.5$ J, $K_f = 27.15$ J, thus $\Delta K = -1.35$ J

13. (a) $4v = 9$, so $v = 2.25$ m/s;

(b) $84v = 9$, so $v = 0.107$ m/s;

(c) 0.107 m/s

14. Σp_x: $u_1 = v_1\cos 30^o + 2v_2\cos 45^o$; where $v_2 = 10$ m/s.
Σp_y: $0 = v_1\sin 30^o - 2v_2\sin 45^o$, so $v_1 = 2.83v_2 = 28.3$ m/s
Substitute v_1 into Σp_x to find $u_1 = 38.6$ m/s.

15. (a) $v = 2$ m/s, $\Delta K/K = -2/3$;
(b) $v = 4$ m/s, $\Delta K/K = -1/3$.

16. Σp: $u = (1 + m)v$. Also, $(0.4)(1/2\ u^2) = 1/2\ (1 + m)v^2$
Solve to find $m = 1.5$ kg

17. (a) $F = -\mu(m_1 + m_2)g = (m_1 + m_2)a$, thus $a = \mu g = -5.88$ m/s^2.
Using $\Delta x = 4$ m in $v^2 = 2a\Delta x$, we find the speed just after the collision is $v = 6.86$ m/s.
(b) Σp: $1400u = 2400(6.86)$, thus $u = 11.8$ m/s.

18. $(1500)(20) = (2500)v$, so $v = 12$ m/s just after the collision.
ΣF: $-\mu(m_1 + m_2)g = (m_1 + m_2)a$, so $a = -4.9$ m/s^2.
Use $v^2 = 2a\Delta x$ to find $\Delta x = 14.7$ m

19. (a) $\Sigma \underline{p} = 825\underline{i} + 720\underline{j} = 200\underline{v}$, thus $\underline{v} = 4.13\underline{i} + 3.60\underline{j}$ m/s.
(b) $\Delta K = 3000 - 5270 = -2970$ J

20. Σp: $64\text{x}10^3 = (1.5\text{x}10^6)v$, so $v = 4.27$ cm/s

21. $K_\alpha = 1/2\ mv_\alpha^2 = 6.72\text{x}10^{-13}$ J, thua $v_\alpha = 1.42\text{x}10^4$ m/s.
(a) $(222)v_R = 4v_\alpha$, $v_R = 2.56\text{x}10^5$ m/s
(b) $K_R = 1.21\text{x}10^{-14}$ J

22. (a) $(5\text{x}10^8)(10^4) = (6.0\text{x}10^{24})v$, so $v = 8.3\text{x}10^{-13}$ m/s;
(b) $\Delta K \approx K_i = 2.5\text{x}10^{16}$ J = 6.0 megatons

23. (a) Σp_x: $-164\text{x}10^7 + 17.4\text{x}10^7 = 5.8\text{x}10^7 v_x$, so $v_x = -25.3$ km/h;
Σp_y: $47.9\text{x}10^7 = 5.8\text{x}10^7 v_y$, so $v_y = 8.26$ km/h
$v = 26.6$ km/h (7.39 m/s) at 72° W of N
(b) $K_i = 1/2\ (4.1\text{x}10^7)(11.1)^2 + 1/2\ (1.7\text{x}10^7)(8.33)^2 =$
$3.12\text{x}10^9$ J.
$K_f = 0.5(5.8\text{x}10^7)(7.39)^2 = 1.54\text{x}10^9$ J, so $\Delta K = -1.54\text{x}10^9$ J

24. Σp: $75v_1 = (0.5)(24)$, so $v_1 = 0.16$ m/s,
Σp: $60v_2 = 12$, thus $v_2 = 0.2$ m/s

25. (a) $2.4\underline{i}$, $K_i = 150$ kJ, $K_f = 144$ kJ, $\Delta K = -6$ kJ
(b) $-0.9\underline{i}$; $K_i = 274$ kJ, $K_f = 20.3$ kJ, $\Delta K = -254$ kJ

26. (a) Σp_x: $15 = 6\cos 30^o + 3v_{2x}$; so $v_{2x} = 3.27$ m/s
Σp_y: $-8 = -6\sin 30^o + 3v_{2y}$; thus $v_{2y} = -1.67$ m/s
$v_2 = 3.67$ m/s at 27.1^o S of E;
(b) $K_i = 53.5$ J, $K_f = 29.2$ J, thus $\Delta K = -24.3$ J, inelastic.

27. (a) From Example 9.5: $u = (M + m)(2gH)^{1/2}/m = 160$ m/s.
(b) $K_i = 192$ J, $K_f = 1.4$ J, so $\Delta K = -190.6$ J, loss is 99.3%.

28. (a) $mu = (M + m)v$ and $v^2 = 2gH$, so $H = [mu/(M + m)]^2/2g =$ 81.6 cm; (b) $K_i = 90$ J, $K_f = 12$ J, thus $\Delta K = -78$ J

29. Σp: $4 = 1 + 2.5v$, so $v = 1.2$ m/s.
(a) $H = v^2/2g = 7.3$ cm; (b) $W = -\Delta K = +750$ J

30. Σp: $mu = (M + m)v$, so $v = 3$ m/s.
$\Delta E = W_{NC}$: $1/2\ kA^2 - 1/2\ (m + M)v^2 = -fA$, thus $f = 8$ N

31. $v_2 - v_1 = -(u_2 - u_1)$, so $v_2 - v_1 = u_1$ (i).
Σp: $4u_1 = 197v_2 + 4v_1$ (ii).
Thus, $v_2 = 8u_1/201 = 5.97\text{x}10^5$ m/s

32. (a) From Eq. 9.10: $v_2 - v_1 = u_1$, and Σp: $u_1 = v_1 + v_2$.
Thus, $v_1 = 0$ and $v_2 = u_1 = 160$ km/h.
(b) From Eq. 9.10: $v_2 - v_1 = u_1$, and Σp: $2u_1 = 2v_1 + v_2$
Thus $v_1 = u_1/3$, $v_2 = 4u_1/3 = 213$ km/h.

33. (a) Cons. of E: $mgH = 1/2\ mu^2$, thus $u = (2gH)^{1/2}$.
Σp: $mu = 3mv$, so $v = u/3$.
From $E_i = E_f$: $1/2\ (3m)(u/3)^2 = 3mgh_2$.
Thus, $2gH/18 = gh_2$, so $h_2 = H/9$.
(b) $v_2 - v_1 = u$, and Σp: $2v_2 + v_1 = u$.
Thus, $v_2 = 2u/3$, $v_1 = -u/3$.
Cons. of E leads to $h_1 = H/9$ and $h_2 = 4H/9$.

34. (a) $v_2 - v_1 = u_1$ and Σp: $3u_1 = 3v_1 + v_2$.
Thus, $v_1 = u_1/2$, $v_2 = 3u_1/2$.
(b) $v_2 - v_1 = u_1$ and Σp: $u_1 = v_1 + 3v_2$
Thus, $v_1 = -u/2$, $v_2 = u/2$.

35. See In-Chapter Exercise 6: $f_2 = 1 - f_1 = 4m_1m_2/(m_1 + m_2)^2$.
(a) $8/9 = 0.89$; (b) $48/(13)^2 = 0.28$;
(c) $4(208)/(209)^2 = 0.019$.

36. Let $r = m_2/m_1$. From Eq. 9.11: $v_1 = u(1 - r)/(1 + r)$
(a) $(1 - r)/(1 + r) = -1/3$, so $m_2 = 2m_1$;
(b) $(1 - r)/(1 + r) = 1/2$, so $m_2 = m_1/3$

37. Use Eq. 9.12. Proton $v_2^P = 2m_1u_1/(m_1 + m_2) = 2mu/(m + 1)$;
Nitrogen: $v_2^N = 2mu/(m + 14)$.
Thus, $m + 14 = 7.5m + 7.5$, which leads to $m = 1$ u.

38. (a) $v_2 - v_1 = -(u_2 - u_1)$ leads to $v_1 = -u/2$.
Σp: $2u = 2v_1 + m_2u/2$ leads to $m_2 = 6$ kg;
(b) In-chapter Exercise 6: $f_2 = 4m_1m_2/(m_1 + m_2)^2 = 1/3$ gives
$24m_2 = (2 + m_2)^2$. Thus, $m_2 = 0.2$ kg, 19.8 kg

39. $F = (46\text{x}10^{-3})(61.1)/(5\text{x}10^{-4}) = 5620$ N

40. $F = (1)(35)/(0.1) = 350$ N

41. $F = (0.01)(300)/(0.01) = 300$ N

42. (a) $(0.15)(60)/(10^{-3}) = -9\mathbf{i}$ kN;
(b) $(0.15)(40\mathbf{j} - 30\mathbf{i})/(10^{-3}) = -4.5\mathbf{i} + 6\mathbf{j}$ kN

43. $(0.5)(4)/(10^{-3}) = 2000$ N

44. (a) 10 m/s; (b) $\Delta p/\Delta t = (10^4)/0.2 = 5\text{x}10^4$ m/s;
(c) A: $(70)(5)/0.2 = 1.75$ kN, B: $(70)(10)/0.2 = 3.5$ kN

45. $F = dp/dt = v\, dm/dt = 15$ N

46. $\Delta v = 12(2)^{1/2}$, $F = \Delta v.dm/dt = 8.5$ N

47. $(15\text{x}10^{-3})(10)(450) = 67.5$ N

48. (a) $\Delta p_x = (0.4)16\cos53^o - (0.4)20\cos37^o = -2.54$
$\Delta p_y = -(0.4)16\sin53^o - (0.4)20\sin37^o = -9.93$
$\underline{I} = -2.54\underline{i} - 9.93\underline{j}$ kg.m/s;
(b) Gravity and air resistance; (c) No

49. From $v^2 = 2gh$, $v_i = 8.85$ m/s, $v_2 = 7.67$ m/s
$\Delta p = m(8.85 + 7.67) = 3.30$ kg.m/s in 0.01 s, thus F = 330 N

50. F = (0.06)(30)/(0.04) = 45 N

51. $\Delta p_x = m(v_2 \cos\theta_2 - v_1 \cos\theta_1) = -0.11$,
$\Delta p_y = m(v_2 \sin\theta_2 + v_1 \sin\theta_1) = +1.56$
(a) $\underline{I} = -0.11\underline{i} + 1.56\underline{j}$ kg.m/s; (b) $\underline{F} = \underline{I}/\Delta t = -22\underline{i} + 312\underline{j}$ N

52. (a) Area = 350 kg.m/s; (b) $F = I/\Delta t = 70$ N

53. (a) $I_F = 435$ kg.m/s; (b) $I_G = 260$ kg.m/s;
(c) Net I = 175 kg.m/s = mv, so v = 2.64 m/s;
(d) $H = v^2/2g = 0.355$ m;
(e) P = fv = (1350)(2.64) W = 4.78 hp.

54. (a) Area = 275 kg.m/s; (b) 275 = 50v, so v = 5.5 m/s

55. (a) $\underline{I} = 1200(-16\underline{i} - 12\underline{j})$ kg.m/s; 24,000 kg.m/s at 37° S of W;
(b) $\underline{F} = \underline{I}/\Delta t = -2400\underline{i} -1800\underline{j}$ N; 3000 N at 37° S of W

56. ΣP_x: $16 = 2v_1\cos30^o + v_2\cos\theta_2$;
Σp_y: $0 = 2v_1\sin30^o - v_2\sin\theta_2$
K: $64 = v_1^2 + 0.5v_2^2$
Find $3v_1^2 - 16(3)^{1/2}v_1 + 64 = 0$, so
$v_1 = 8/(3)^{1/2}$ m/s, $v_2 = 16/(3)^{1/2}$ m/s, $\theta_2 = 30^o$ below the x axis

57. (a) Σp_x: $Mu = 2Mv_2 \cos\theta$ (i);
Σp_y: $0 = Mv_1 - 2Mv_2 \sin\theta$ (ii);
ΣK: $u^2 = v_1^2 + 2v_2^2$ (iii).
Square and add (i) and (ii):
$$u^2 = v_1^2 + 4v_2^2 \quad \text{(iv)}.$$
From (iii) and (iv): $v_1 = u/(3)^{1/2}$, $v_2 = u/(3)^{1/2}$.

Then $\sin\theta = 0.5$ or $\theta = 30^o$.

(c) $K_\alpha = 1/2\ (2M)(u^2/3) = 2K_D/3$.

58. From In-chapter Exercise 8, $\theta_2 = 60^o$.
Σp_x: $20 = v_1\cos 30^o + v_2\cos 60^o$
Σp_y: $0 = v_1\sin 30^o - v_2\sin 60^o$
Thus, $v_1 = 17.3$ m/s, $v_2 = 10$ m/s.

59. Σp_x: $3 = 2\cos\theta_1 + v_2\ \cos\theta_2$ (i)
Σp_y: $0 = 2\sin\theta_1 - v_2\ \sin\theta_2$ (ii)
K: $3^2 = 2^2 + v_2{}^2$, thus $v_2 = (5)^{1/2}$
Find $\cos\theta_1 = 2/3$, so $\theta_1 = 48.1^o$ and $\theta_2 = 41.8^o$

60. $v_f - v_i = v_{ex}\ \ln(M_i/M_f)$, where $v_i = 0$.
(a) Since $v_f/v_{ex} = 2.5$, find $M_i = 2.718M_f$.
Thus $M_F/M_i = 1.718/2.718 = 0.632$;
(b) $M_i = 12.2M_f$, so $M_F/M_i = 11.2/12.2 = 0.92$

61. From $\Delta v = v_{ex}\ \ln(M_i/M_f)$, we have 2 km/s = $v_{ex}\ \ln(1/0.4)$,
thus, $v_{ex} = 2.18$ km/s.

62. (a) $m_1u_1 + m_2u_2 = (m_1 + m_2)V$; V = 6.85 m/s due N. $\Delta K = -67.5$ J
(b) $m_1u_1 - m_2u_2 = (m_1 + m_2)V$, so V = 3 m/s due N. $\Delta K = -1.3$ kJ

63. (a) $m_2u_2\sin 30^o = (m_1 + m_2)V_x$; $-m_1u_1 + m_2u_2\cos 30^o =$
$(m_1 + m_2)V_y$, so $V_x = 1$ m/s and $V_y = 0.668$ m/s, or
V = 1.2 m/s at 33.7^o N of E.
(b) $K_i = 0.5m_1u_1{}^2 + 0.5m_2u_2{}^2$; $K_f = 0.5(m_1 + m_2)V^2$,
$\Delta K = -45.4$ J.

64. (a) $m_1u_1 = m_1v_1 + m_2v_2$, so $v_2 = 1.38$ m/s.
(b) $K_i = 735$ J, $K_f = 88.3$ J, $\Delta K/K = -647/735 = -88\%$.

65. Σp_x: $-\ 6u_1 = -2v_1\sin 50^o + 4v_{2x}$, so $v_{2x} = -8.68$ m/s
Σp_y: $0 = -2_{v2}\cos 50^o + 4v_{2y}$, so $v_{2y} = 2.25$ m/s.
Thus, $v_2 = 8.97$ m/s at 14.5^o N of W.

66. mu = (m + M)V, so V = 2.98 m/s.
Then, $-0.5(1.612)V^2 = -\mu(1.612)gd$, so d = 2.06 m.

67. (a) Σp_x: $-m_1u_1 + m_2u_2 = -m_1v_1\sin 53.1^o + m_2v_{2x}$, so
$v_{2x} = 1.10$ m/s.
Σp_y: $0 = m_1v_1\cos 53.1^o + m_2v_{2y}$, so $v_{2y} = -0.45$ m/s.
Thus, $\underline{v}_2 = 1.19$ m/s at 22.2^o S of E.
(b) $K_i = 10.2$ J, $K_f = 3.05$ J, so $\Delta K = -7.15$ J

68. Σp: $mu = (m + M)V$, so $V = 1$ m/s. Then, $E_i = 0.5(m + M)V^2$;
$E_f = (m + M)g(L - L\cos\theta)$. Find $\theta = 21.3^o$.

69. Σp_x: $-m_1u_1\cos 20^o - m_2u_2\sin 30^o = (m_1 + m_2)V_x$,
so $V_x = -1.20$ m/s.
Σp_y: $-m1u_1\sin 20^o + m_2u_2\cos 30^o = (m_1 + m_2)V_y$, so $V_y = 0.55$ m/s.
So $\underline{V} = 1.32$ m/s at 24.6^o N of W.
(b) $K_i = 9.82$ J, $K_f = 5.05$ J, so $\Delta K = -4.77$ J

70. (a) $(m_1 + m_2)u = 50v_1 + 16$, so $v_1 = 3.84$ m/s.
(b) $(m_1 + m_2)u = 50v_1 - 16$, so $v_1 = 4.48$ m/s.

71. (a) $m_1u_1 = m_1v_1 + m_2v_2$ and $(v_2 - v_1) = -(u_2 - u_1) = u_1$,
so $v_1 = -u_1/3$, $v_2 = 2u_1/3$. Then, $K_2/K_i = 88.9\%$.
(b) Find $v_1 = u_1/3$, $v_2 = 4u_1/3$, and $K_2/Ki = 88.9\%$.

72. (a) $m_1u_1 + m_2u_2 = m_1v_1 + m_2v_2$ and $(v_2 - v_1) = -(u_2 - u_1)$.
Find $v_1 = -1.57$ m/s, $v_2 = 3.93$ m/s.
(b) $\Delta K_1 = -3.84$ J, $\Delta K_2 = +7.45$ J

73. $m_1u_1 = m_1v_1 + m_2v_2$ and $(v_2 - v_1) = -(u_2 - u_1) = u_1$.
Find $v_1 = 0.162$ m/s, $v_2 = 0.762$ m/s.

74. From $v_f - v_i = v_e \ln(M_i/M_f)$, find $v_f = 3.15$ km/s.

75. $K_f = 0.81K_i$, so $v_f = 0.9v_i = 27$ m/s.
$\Delta p = mv_f - mv_i = 3.42$ kg.m/s

76. $m_1u_1 = m_1v_1 + m_2v_2$ and $(v_2 - v_1) = -(u_2 - u_1) = u_1$.
Find: $v_1 = -0.174$ m/s, $v_2 = 0.426$ m/s.

77. $m_1u_1 = m_1v_1 + m_2v_2$; $(v_2 - v_1) = -(u_2 - u_1) = u_1$.
Find $m_2 = 1.82$ kg

78. $v_f = v_e \ln(M_i/M_f) = v_e \ln(1/0.3) = 3.37$ km/s.

79. From $2v_e = v_e \ln(M_i/M_f)$, $M_i - M_f = M_i(1 - e^{-2}) = 3.46 \times 10^5$ kg.

Problems

1. From $v_2 - v_1 = -(u_2 - u_1)$ we see that $v_2 - v_1 = -(-u_o - u_1)$. But $v_1 = u_1$ since racket is held, thus $v_2 = u_o + 2u_1$.

2. (a) $-60v + (6 - v) = 0$, $v = 6/61$ m/s $= 9.84$ cm/s;
   (b) $-60.5v_1 + (0.5)(6 - v_1) = 0$, $v_1 = 3/61$
   $-60.5v_1 = -60v_2 + (0.5)(6 - v_2)$, $v_2 = 9.88$ cm/s;
   (c) $v_f = v_{ex} \ln(M_i/M_f) = 6 \ln(61/60) = 9.92$ cm/s

3. Σp: $m_1u_1 - m_2u_2 = (m_1 + m_2)V$;
   $|\Delta K| = 1/2\ (m_1u_1^2 + m_2u_2^2) - 1/2\ (m_1 + m_2)V^2$
   $= m_1m_2(u_1 + u_2)^2/2(m_1 + m_2)$.

4. Let u′ be the velocity of the block relative to the wedge.
   Σp_x: $0 = -MV + m(u'\cos\theta - V)$, thus $u'\cos\theta = (M + m)V/m$
   Cons. of E: $mgH = 1/2\ MV^2 + 1/2\ m[(u'\cos\theta - V)^2 + (u'\sin\theta)^2]$
   Substitute for u′ to find V.

5. Σp_x: $mV \cos\alpha - mu = Mu$; thus
   $u = mV \cos\alpha/(M + m)$.
   $\tan\theta = V \sin\alpha/(V \cos\alpha - u)$
   $= (M + m)\tan\alpha/M$

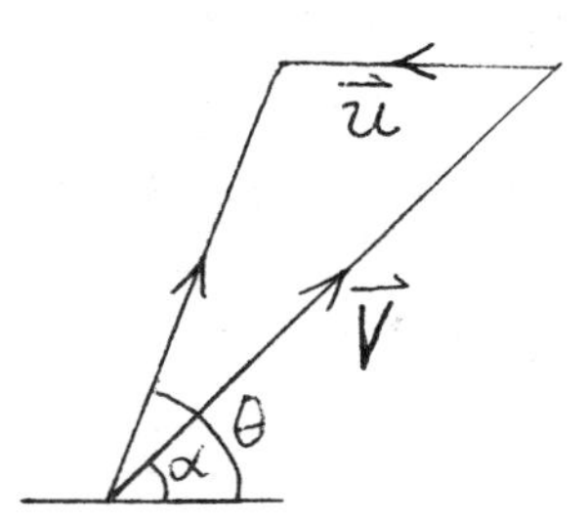

6. $v_1 = (m_1 - m_2)u/(m_1 + m_2)$, where
   $u = (2gH)^{1/2} = 4.43$ m/s. Then $h' = v_1^2/2g$.
   (a) $v_1 = -2u/3$, $h' = 4H/9 = 0.444$ m;
   (b) $v_1 = 3u/7$, $h' = 9H/49 = 0.184$ m

7. $m_1u_1 + m_2u_2 = m_1v_1 + m_2v_2$ and $v_2 - v_1 = -u_2 + u_1$.
   Substituting for v_2: $(m_1 - m_2)u_1 + 2m_2u_2 = (m_1 + m_2)v_1$,
   which leads to the given v_1 and then v_2.

8. Σp_x: $200 = 80 + 2v_{3x}$; thus $v_{3x} = 60$ m/s. (i)
   Σp_y: $0 = 60 - 3v_2 + 2v_{3y}$, thus $v_2 = 20 + 2v_{3y}/3$ (ii)
   ΣK: $12\text{x}10^3 = 10^3 + 1.5v_2^2 + (v_{3x}^2 + v_{3y}^2)$ (iii)
   Substitute for v_{3x} and v_2 from (i) and (ii) into (iii):
   $v_{3y}^2 + 24v_{3y} - 4080 = 0$, thus $v_{3y} = 53.0, -77$

From (ii) $v_2 = 55.3, -31.3$ (reject)
Thus, $v_3 = 60i + 53j$ m/s.

9. From 9.11: $v_1 = (u - 2u)/3 = -u/3$; $v_2 = (4u + u)/3 = 5u/3$.
Then $H = v^2/2g$: $H_1 = H/9$; $H_2 = 25H/9$.

10. Initial speed of water: $v_o^2 = 2g\Delta y$ thus $v_o = 8.85$ m/s.
$Mg = v\, dm/dt$ leads to $v = 1.96/0.7 = 2.8$ m/s.
But $v^2 = v_o^2 + 2gH$, so $H = 3.6$ m

11. $v_2 - v_1 = -e(u_2 - u_1)$; Σp: $m_1u_1 + m_2u_2 = m_1v_1 + m_2v_2$.
Substitute for v_2 into Σp:
$$(m_1 - em_2)u_1 + (m_2 + em_2)u_2 = (m_1 + m_2)v_1.$$
$$(m_1 + m_2)v_1 = (m_1 - em_2)u_1 + m_2(1 + e)u_2$$
Similarly,
$$(m1 + m_2)v_2 = m_1(1 + e)u_1 + (m_2 - em_1)u_2$$

12. $K_i = 1/2\ (m_1u_1^2 + m_2v_2^2)$; $K_f = 1/2\ (m_1v_1^2 + m_2v_2^2)$.
Substitute for v_1 and v_2 from Pb. 11 into K_f.

13. First M collides with 4M on right: $v_M = -3u/5$, $v_{4M} = 2u/5$
Second M collides with 4M on left: $v_M = 4u/25$, $v_{4M} = -6u/25$
Since $4u/25 < 2u/5$, M cannot catch the 4M on the right.

14. First collision: $v_1 = u/5$, $v_2 = 6u/5$
Second collision: $v_{2'} = -v_2/5 = -6u/25$, $v_3' = 4v_2/5 = 24u/25$

15. From Example 9.7: $f_1 = [(m_1 - m_2)/(m_1 + m_2)]^2 = 1/4$;
If $r = m_2/m_1$, then $2(1 - r) = \pm(1 + r)$, so $m_2/m_1 = 3$ or $1/3$.
Thus $m_2 = 3m$. (Since m_1 rebounds, $m_2 > m_1$.)

16. Σp: $30 = -v_1 + 2v_2$, so $v_1 = 2v_2 - 30$.
K: $250 = v_1^2 + v_2^2$
Find $6v_2^2 - 120v_2 + 400$, so $v_2 = 15.8$ m/s.
Find $v_1 = 1.54$ m/s

17. (a) $16 = -3 + 6v_2$ so $v_2 = 19/6 = 3.17$ m/s.
(b) $v_2 - v_1 = -e(u_2 - u_1)$ leads to
$19/6 + 1.5 = 8e$, so $e = 7/12 = 0.58$.

18. (a) Σp: $16 = 6v$, $v = 8/3$ m/s.
$E_i = K_i = 64$ J, $E_f = K_f + U_f = 1/2\ 6v^2 + 1/2\ kA^2 =$
$= 64/3 + 200A^2$. Set $E_f = E_i$ to find $A = 0.462$ m;
(b) $v_2 - v_1 = -(u_2 - u_1) = 8$, Σp: $16 = 2v_1 + 4v_2$.
Find $v_1 = -8/3$ m/s, $v_2 = 16/3$ m/s

19. (a) $F(y) = \lambda yg + v\ dm/dt = \lambda gy + \lambda v\ dy/dt = \lambda gy + \lambda v^2$
But $v^2 = 2gy$, thus $F(y) = 3\lambda gy$.
(b) When $y = L$, $F = 3Mg$.

20. Σp: $5\text{x}8 = 8v$, so $v = 5$ m/s. $K_i = 1/2\ (8)(5^2) = 100$ J
$\Delta E = W_{NC}$: $mgd\sin\theta - 1/2\ mv^2 = -\mu(mg\cos\theta)d$. Thus $d = 1.59$ m

21. $m_1u_1 = m_1v_1\ \cos\theta_1 + m_2v_2\ \cos\theta_2$; (i)
$0 = m_1v_1\ \sin\theta_1 - m_2v_2\ \sin\theta_2$ (ii)
$m_1u_1^2 = m_1v_1^2 + m_2v_2^2$ (ii)
Square (i) and (ii) and add:
$m_1^2(u_1 - v_1\ \cos\theta_1)^2 + (m_1v_1\ \sin\theta_1)^2 = (m_2v_2)^2 = m_2m_1(u_1^2 - v_1^2)$
$$(m_1 + m_2)v_1^2 - (2m_1u_1\ \cos\theta_1)v_1 + (m_1 - m_2)u_1^2 = 0$$
For real solution need $\cos^2\theta_1 = 1 - (m_2/m_1)^2$ or $\sin\theta_1 = m_2/m_1$

CHAPTER 10

Exercises

1. (a) x_{CM} = 35(0.13 nm)/36 = 0.126 nm from H;

   (b) x_{CM} = (0.2 nm)(cos52.5^o)/18 = 6.76 pm from O along the line of symmetry.

2. x_{CM} = (-4 - 9 + 15)/10 = 0.2 m;
   y_{CM} = (6 + 12 - 5)/10 = 1.3 m

3. x_{CM} = [30(40) + 20(120) + 10(200)]/60 = 93.3 cm

4. x_{CM} = [4M(0) - M(R/2)]/3M = -R/6 (left of center)

5. Mass of plate = $\sigma(4R^2)$, mass of hole = $-\sigma\pi(R/2)^2 = -\sigma\pi R^2/4$.
   $x_{CM} = [4R^2(0) - (\pi R^2/4)(R/2)]/(4 - \pi/4)\sigma R^2 = -0.122R = y_{CM}$

6. $x_{CM} = [0 - \rho 4\pi r^3 d/3]/(R^3 - r^3)4\pi\rho = -r^3 d/(R^3 - r^3)$

7. Square $m_1 = \sigma L^2$ at x_1 = L/2, y_1 = L/2;
   Triangle $m_2 = \sigma L^2(3)^{1/2}/4$ at x_2 = L/2, $y_2 = L + L/2(3)^{1/2}$
   x_{CM} = L/2; y_{CM} = 0.738L

8. (a) x = [4(d/2) + 1(1.5d) + 1(2.5d)]/6 = d = 2 cm,
   y = [3(d/2) + 1.5d + 2.5d + 3.5d]/6 = 3d/2 = 3 cm;
   (b) x = [4(d/2) - 2(d/2) - 1.5 d + 1.5d]/8 = d/8 = 0.25 cm,
   y = [4(d/2) + 1.5d + 2.5d + 2(3.5d)]/8 = 13d/8 = 3.25 cm

9. $x_{CM} = [M_E(0) + M_m r]/(M_E + M_m)$ = 4680 km from earth's center

10. Canceling $4\pi\rho/3$: $x = [0 + R^3(3R)]/(8R^3 + R^3) = R/3$

11. (a) $m1v_1\Delta t = m_2u_2\Delta t$, so (60 kg)(1.5 m) = 75d, so d = 1.2 m.
    Separation = 5 - (1.5 + 1.2) = 2.3 m;
    (b) x_{CM} = 60(5)/135 = 2.22 m.

12. $x_{CM} = 0 = (20)(-3x10^{-4}) + m(0.18)$, thus m = 33.3 g

13. (a) x_{CM} = [(1)(L/2) + (1)(L/4) + (1)(3L/4)]/3 = L/2.
    y_{CM} = [0 + 2(1/2)(0.866L)]/3 = 0.29L
    (b) $x_{CM} = [(1)(L/2) + 0 + (2)^{1/2}(L/2)]/[(2 + 2^{1/2}] = 0.354L$

14. (a) $8\mathbf{v}_{CM} = (10\mathbf{i} - 6\mathbf{j} + 8\mathbf{k}) + (-18\mathbf{i} + 12\mathbf{j} - 6\mathbf{k})$, thus $\mathbf{v}_{CM} = -\mathbf{i} + 0.75\mathbf{j} + 0.25\mathbf{k}$ m/s;
(b) $\mathbf{P} = M\mathbf{v}_{CM} = -8\mathbf{i} + 6\mathbf{j} + 2\mathbf{k}$ kg.m/s

15. m collides with the box. Σp: $-mu = 4mv_1$, so $v_1 = -u/4$. In the time $t_1 = L/u$ it took m to collide with box, 2m moves L/2. The relative speed of the box and 2m is $(u/2 + u/4) = 3u/4$. The box moves for a time $(L/2)/(3u/4) = 2L/3u$, so its displacement $\Delta x = (2L/3u)(-u/4) = -L/6$.

16. (a) $\mathbf{r}_{CM} = (5\mathbf{i} + 10\mathbf{j})/10 = 0.5\mathbf{i} + \mathbf{j}$ m;
(b) $\mathbf{v}_{CM} = -0.21\mathbf{i} - 0.37\mathbf{j}$ m/s;
(c) $\mathbf{r}_2 = \mathbf{r}_1 + \mathbf{v}_{CM}\,\Delta t = -0.13\mathbf{i} - 0.11\mathbf{j}$ m

17. (a) $\mathbf{r}_{CM} = 6.67\mathbf{i}$ m; $\mathbf{v}_{CM} = 0.67\mathbf{i} + 2.67\mathbf{j}$ m/s.
(b) $\mathbf{a}_1 = \mathbf{F}_1/m_1 = 10\mathbf{j}$ m/s^2; $\mathbf{a}_2 = \mathbf{F}_2/m_2 = 4\mathbf{i}$ m/s^2;
(c) $\mathbf{a}_{CM} = 2.67\mathbf{i} + 3.33\mathbf{j}$ m/s^2
(d) $8\mathbf{i} + 10\mathbf{j} = 3\mathbf{a}_{CM}$;
(e) $\mathbf{r}_2 = \mathbf{r}_1 + \mathbf{v}_{CM}\,t + 1/2\,\mathbf{a}_{CM}\,t^2 = 13.4\mathbf{i} + 12\mathbf{j}$ m

18. (a) $\mathbf{r}_{CM} = (-21\mathbf{i} + 11\mathbf{j})/7 = -3\mathbf{i} + 1.57\mathbf{j}$ m;
(b) $\mathbf{v}_{CM} = (13\mathbf{i} - 10\mathbf{j})/7 = 1.86\mathbf{i} - 1.43\mathbf{j}$ m/s;
(c) $\mathbf{P} = 2(-\mathbf{i} + 5\mathbf{j}) + 5(3\mathbf{i} - 4\mathbf{j}) = 13\mathbf{i} - 10\mathbf{j}$ kg.m/s;
(d) $\mathbf{r}_2 = \mathbf{r}_1 + \mathbf{v}_{CM}\,\Delta t = 0.72\mathbf{i} - 1.29\mathbf{j}$ m

19. (a) $0 = (60)(0.3) + 75v$, so $v = 0.24$ m/s
(b) $x_{CM} = (60)(10)/135 = 4.44$ m.

20. (a) $x_{CM} = [25(8) + 60(10)]/85 = 9.41$ m;
(b) $9.41 = [60d + 25(d + 2)]/85$, thus $d = 8.82$ m

21. $y = 0 = 100 + 40t - 5t^2$, so $t = 10$ s, and range $R = 300$ m. $x_{CM} = 300 = [(4)(200) + 2x]/6$, leads to $x = 500$ m.

22. (a) $\Sigma\mathbf{p}$: $61(2)\mathbf{i} = 5\mathbf{j} + 60\mathbf{v}$, so $\mathbf{v} = 2.03\mathbf{i} - 0.083\mathbf{j}$ m/s;
(b) $\Delta\mathbf{r}_{CM} = \mathbf{v}_{CM}\,\Delta t = 8\mathbf{i}$ m (east)

23. (a) $\mathbf{v}_{CM} = [(1000)(15\mathbf{i}) + (1800)(10\mathbf{j})]/2800 = 5.36\mathbf{i} + 6.43\mathbf{j}$ m/s;
(b) $16.1\mathbf{i} + 19.3\mathbf{j}$ m

24. $\underline{v}_{CM} = (7\underline{i} + 4\underline{j})/7$, and $\underline{r}_2 = 0 = \underline{r}_1 + \underline{v}_{CM}\,\Delta t$, so
$\underline{r}_1 = -3\underline{i} - 1.71\underline{j}$ m

25. (a) $\underline{v}_{CM} = (2.4\underline{i} - 6\underline{i})/2 = -1.8\underline{i}$;
(b) $\underline{u}_1 = 3 - (-1.8) = 4.8\underline{i}$; $\underline{u}_2 = -5 - (-1.8) = -3.2\underline{i}$;
(c) $K = 3.6 + 15 = 18.6$ J;
(d) $K_{CM} = 1/2\ (2)(1.8)^2 = 3.24$ J;
(e) $K = 1/2[(0.8)(4.8)^2 + (1.2)(3.2)^2] = 9.22 + 6.14 = 15.36$J

26. (a) $v_{CM} = 10$ m/s, $K_{CM} = 1/2\ (6)(10^2) = 300$ J,
$K_{rel} = 1/2\ (5)(2^2) + 1/2\ (1)(10^2) = 60$ J;
(b) $v_{CM} = 2$ m/s, $K_{CM} = 12$ J,
$K_{rel} = 1/2\ (1)(10^2) + 1/2\ (5)(2^2) = 60$ J

27. $\underline{v}_{CM} = 7\underline{i}$ m/s. (a) $70\underline{i} = 40\underline{i} + 6\underline{v}_2$, so $\underline{v}_2 = 5\underline{i}$;
(b) $K_1 = 1/2\ (10)(49) = 245$ J; (c) $K_f = 200 + 75 = 275$ J;
(d) $K_{CM} = 1/2\ (10)(49) = 245$ J; (e) $K_{CM} = 245$ J;
(f) $K_{Rel} = 0$; (g) $K_{Rel} = 30$ J

28. (a) $v_{CM} = v/21$, $K_{rel} = 0.5(1\ u)(20v/21)^2 + 0.5(20\ u)(v/21)^2$
Thus $E = (10v^2/21)(1.66\text{x}10^{-27}$ kg)
$K_i = 1/2\ (1\ u)v^2 = 1/2\ (21E/10) = 8.4\text{x}10^{-19}$ J;
(b) $v_{CM} = 20v/21$, $K_{rel} = 0.5(1\ u)(20v/21)^2 +$
$0.5(20\ u)(v/21)^2$. Thus $E = (10v^2/21)(1.66\text{x}10^{-27}$ kg)
$K_i = 1/2\ (20\ u)v^2 = 21E = 1.68\text{x}10^{-17}$ J

29. (a) $\underline{v}_{CM} = (24\underline{i} + 6\underline{i})/6 = 5\underline{i}$ m/s;
(b) $\underline{v}_1' = \underline{i}$ m/s, $\underline{v}_2' = -2\underline{i}$ m/s;
(c) $v_2 - v_1 = -(u_2 - u_1) = +3$; and $4v_1 + 2v_2 = 30$.
Thus, $v_1 = 4$ m/s, $v_2 = 7$ m/s; $\underline{v}_1' = -\underline{i}$ m/s; $\underline{v}_2' = 2\underline{i}$ m/s

30. (a) $\underline{v}_{CM} = (12\underline{i} - 15\underline{i})/5 = -0.6\underline{i}$ m/s;
(b) $\underline{v}_1' = 6.6\underline{i}$ m/s, $\underline{v}_2' = -4.4\underline{i}$ m/s;
(c) $\underline{v}_1' = -6.6\underline{i}$ m/s, $\underline{v}_2' = 4.4\underline{i}$ m/s (See E29(c) above.)

31. $\underline{v}_{CM} = [4\text{x}5\underline{j} + 0]/8 = 2.5\underline{j}$ m/s.
So, $\underline{v}_1' = 1.5\underline{j}$ m/s, and $\underline{v}_2' = -2.5\underline{j}$ m/s.
(a) $K_{Rel} = 1/2\ (3)(2.5)^2 + 1/2\ (5)(1.5)^2 = 15$ J;
(b) KCM $= 1/2\ (8)(2.5)^2 = 25$ J.

32. (a) $T = v_{ex}\ dm/dt = (2.6x10^3)(1.5x10^4) = 3.9x10^7$ N;
(b) $T - mg = ma$, so $a = 5.8$ m/s^2

33. Thrust = $v_{ex}\ dm/dt$, so $dm/dt = 900$ kg/s. SInce $v_i = 0$,
$v_f = v_{ex}\ \ln(M_i/M_f) - gt = 2000\ln(135/126) - 98 = 40$ m/s.

34. $M\ dv/dt = F_{EXT} - v\ dm/dt$, thus
$(1.6x10^4)\ dv/dt = 10^3$ N - (0.1 m/s)(10^2 kg/s) = 990 N
Thus $dv/dt = 6.19$ cm/s^2

35. From $(m_1 + m_2)x_{CM} = m_1x_1 + m_2x_2$, find $m_2 = 3.5$ kg.

36. $(m_1 + m_2 + m_3)x_{CM} = m_1x_1 + m_2x_2 + m_3x_3$, so $x_{CM} = 1.00$ m
$(m_1 + m_2 + m_3)y_{CM} = m_1y_1 + m_2y_2 + m_3y_3$, so $y_{CM} = 1.08$ m

37. (a) $\underline{P} = m_1\underline{v}_1 + m_2\underline{v}_2 = 6.70\underline{i} - 8.08\underline{j}$ kg.m/s
(b) $\underline{P} = (m_1 + m_2)\underline{v}_{CM}$, thus $\underline{v}_{CM} = 1.20\underline{i} - 1.44\underline{j}$ m/s

38. (a) $(m_1 + m_2)\underline{v}_{CM} = m_1\underline{v}_1$, so $\underline{v}_{CM} = 2\underline{i}$ m/s.
(b) Relative to CM: $\underline{v}_1 = 6\underline{i}$, $v_2 = -2\underline{i}$ m/s.
(c) Relative to CM: $\underline{p}_1 = m_1\underline{v}_1 = 12\underline{i}$ kg.m/s;
$\underline{p}_2 = -12\underline{i}$ kg.m/s.

39. (a) $(m_1 + m_2)\underline{v}_{CM} = m_1\underline{v}_1 + m_2\underline{v}_2$, thus $\underline{v}_{CM} = 16.5\underline{i}$ m/s
(b) Relative to CM: $\underline{v}_1 = -2.5\underline{i}$ m/s, $\underline{v}_2 = 5.5\underline{i}$ m/s
(c) Relative to CM: $\underline{p}_1 = -6875$ kg.m/s; $\underline{p}_2 = 6875$ kg.m/s

40. Replace the two halves by point masses at their centers.
$d = 15\cos 30^o = 13$ cm.

Problems

1. $y = hx/b$, $dm = \sigma y\ dx = (\sigma hx/b)\ dx$
and $M = \sigma bh/2$.
$\int x\ dm = (\sigma h/b) \int x^2\ dx = \sigma hb^2/3$.
$\int y\ dm = \int \sigma\ y(b - x)dy$
$= \int \sigma(by - by^2/h)\ dy = \sigma bh^2/6$
Thus, $x_{CM} = 2b/3$, $y_{CM} = h/3$.

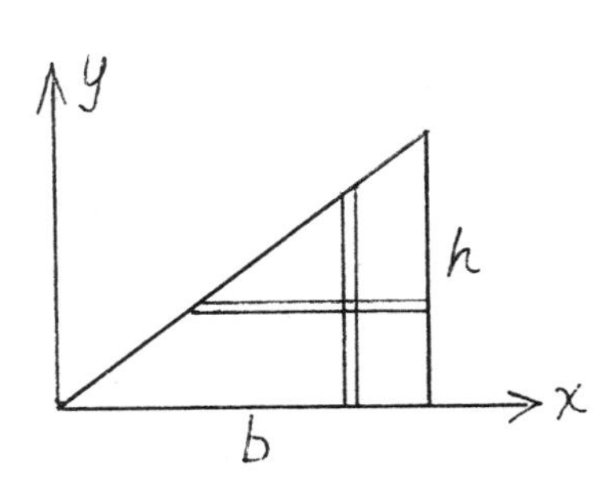

2. From Example 10.3, $y_{CM} = 2r/\pi$ for a semi-circular ring.
$y_{CM} = \int y\ dm/M = \int (2r/\pi)\sigma\pi r\ dr/(\sigma\pi R^2/2) = 4R/3\pi = 0.424R$

3. ARC: The mass of the arc is $\lambda\pi R/2$.
For an element on the arc $dm = \lambda R\, d\theta$, and $y = R \sin\theta$.

$$y_{CM} = \lambda R^2 \int_0^{\pi} \frac{\sin\theta\, d\theta}{\pi\lambda R/2} = 2R/\pi = x_{CM} \text{ (by symmetry)}$$

WHOLE FIGURE: $m_1 = m_2 = \lambda R$, $m_3 = \lambda\pi R/2$

$$x_{CM} = \frac{(\lambda R)(R/2) + (\pi R\lambda/2)(2R/\pi)}{(2 + \pi/2)\lambda R}$$

$= 3R/(\pi + 4) = 0.420R = y_{CM}$

4. $x_{CM} = \int x(\lambda dx)/\int \lambda dx = [ax^2/2 + bx^3/3]/[ax + bx^2/2] =$
$= L(3a + 2bL)/(6a + 3bL)$

5. $dm = 2\pi\sigma\ell \sin\alpha\, d\ell$; $M = \pi\sigma H^2 \sin\alpha/\cos^2\alpha$.
$\int y\, dm = 2\pi\sigma \sin\alpha \cos\alpha\, L^3/3$. Thus, $y_{CM} = 2H/3$.

6. The CM is 4680 km from earth's center. The CM moves in a circular orbit, thus in one half-period of a lunar orbit, the earth moves 2x4680 = 9360 km

7. (a) $x_{CM} = [(75)(0) + (25)(2)]/100 = 0.5$ m;
(b) Σp: $0 = -25v + 75(2 - v)$, so $v = 1.5$ m/s
(c) $d = vt = 3$ m.

8. (a) Σp: $130\text{x}2 = 125v + 5(4 + v)$, so $v = 1.85$ m/s;
(b) $\Delta t = 1$ s, so $\Delta x = 1.85$ m;
(c) $v_{CM} = 2$ m/s, so $\Delta x = 2$ m

9. (a) Thrust $= v_{ex}\, dm/dt = 2\text{x}10^5$ N;
(b) $M_i = 50{,}000$ kg, $M_f = 47{,}000$ kg.
$v_f = v_i + v_{ex} \ln(M_i/M_f) = 10^3 + (2000) \ln(50/47) = 1.12$ km/s.

10. $ma = v_{ex}\, dm/dt$, thus $dm/dt = (1200)(2)/(2000) = 1.2$ kg/s

11. $x_{CM} = 2m(L/2)/(3m) = L/3 = y_{CM} = 2$ cm.

12. $mu = (M + m)V$, $K_{rel} = E = 1/2\, m(u - V)^2 + 1/2\, MV^2$
$= mMV^2/2(M + m)$. Thus, $1/2\, mu^2 = (M + m)E/M$

CHAPTER 11

Exercises

1. 33 1/3 rpm = 3.49 rad/s. (a) $\alpha = 1.75$ rad/s^2; (1.16)
   (b) $\theta_1 = 1/2\ \alpha t^2 = 3.49$ rad, $\theta_2 = 3\text{x}3.49$ rad, $\theta = 2.22$ rev.
   (c) $\Delta\theta_2 = 9.08$ rad, $\Delta t_2 = 2.6$ s, $\Delta t = 4.6$ s;
   (d) $a_r = \omega^2 r = 0.46$ m/s^2; $a_t = \alpha r = 0.26$ m/s^2.
   (e) $a_r = 1.83$ m/s^2; $a_t = 0$

2. (a) $v = \omega r = (45\pi/30)(0.145) = 0.683$ m/s;
   (b) $v = \omega r = 0.292$ m/s;
   (c) $\Delta r/\Delta t = (-8.3$ cm/20 min$) = -6.92\text{x}10^{-5}$ m/s
   (d) In 20 min there are 900 revs and the needle moves radially by 8.3 cm, thus it moves (8.3 cm/900 rev) = $9.22\text{x}10^{-5}$ m, which would be the average groove width.

3. (a) $\omega = 2\pi/1$ d $= 7.27\text{x}10^{-5}$ rad/s;
   (b) $\omega = 2\pi/1$ y $= 1.99\text{x}10^{-7}$ rad/s;
   (c) $v_s = \omega r = 463$ m/s; $v_{orb} = 29.7$ km/s.
   Thus $v_C = 29.2$ km/s, $v_F = 30.2$ km/s.

4. (a) $\omega = 2\pi/1$ d $= 7.27\text{x}10^{-5}$ rad/s, $v = \omega R = 463$ m/s;
   (b) $v = \omega R\cos 41^o = 349$ m/s

5. (a) $\theta_1 = 31$ rad, $\theta_2 = 9$ rad, $\omega_{av} = 11$ rad/s;
   (b) $v = \omega R = (-5 + 8t)(0.06) = 0.66$ m/s;
   (c) $a_r = \omega^2 r = 7.26$ m/s^2, $a_t = \alpha R = 0.48$ m/s^2.

6. $a = (a_r^2 + a_t^2)^{1/2} = (\omega^4 r^2 + \alpha^2 r^2)^{1/2}$

7. $\omega_o = 5.24$ rad/s, $\Delta t = \Delta\omega/\alpha = 10.5$ s.
   $\Delta\theta_1 = \omega_o t + 1/2\ \alpha t^2 = 82.6$ rad,
   When $\alpha = 0$, $\Delta\theta_2 = (10.47)(9.5) = 99.5$ rad.
   $\Delta\theta = 182$ rad, or 29 rev.

8. (a) $\omega = (30$ m/s$)/(0.3$ cm$) = 100$ rad/s; $\alpha = 10$ rad/s^2;
   (b) $\theta = 1/2\ \alpha t^2 = 500$ rad or 79.6 rev;
   (c) $a_r = \omega^2 r = (10^4)(0.3) = 3\text{x}10^3$ m/s^2

9. Bullet makes 60/25 = 2.4 rev as it passes through barrel.
   $\Delta t = 0.6/850 = 7.06\text{x}10^{-4}$ s,

Rotational frequency = 2.4 rev/Δt = $2.04\text{x}10^5$ rpm.

10. (a) $\omega = 2\pi/60 = \pi/30$ rad/s; (b) $v = \omega r = 0.838$ cm/s

11. (a) $\omega = v/r = (4.8 \text{ cm/s})/0.75 \text{ cm}) = 6.4$ rad/s.
(b) $\omega = 4.8/1.75 = 2.74$ rad/s.
(c) $\Delta t = 86.4 \text{ m}/(4.8\text{x}10^{-2} \text{ m/s}) = 1.8\text{x}10^3$ s,
$\alpha = \Delta\omega/\Delta t -2.03\text{x}10^{-3}$ rad/s^2.

12. (a) $\omega = 2 - 10t + 8t^3$, $\alpha = -10 + 24t^2 = 14$ rad/s^2;
(b) $\Delta\omega/\Delta t = (46 - 0)/1 = 46$ rad/s^2;
(c) $\omega_{av} = \Delta\theta/\Delta t = [16 - (-1)]/1 = 17$ rad/s

13. (a) $\Delta\theta = \omega_o t + 1/2\ \alpha t^2$, thus
$(90\text{x}2\pi) = 4\pi(60) + 1/2\ \alpha(60)^2$,
so $\alpha = -\pi/30 = -0.105$ rad/s^2.
(b) $\omega = \omega_o + \alpha t = 4\pi - (\pi/30)(60) = 2\pi$ rad/s. For last leg
$v_o = \omega r = 0.4\pi$ m/s, $a = -0.021$ m/s^2:
$0 = v_o^2 + 2a\Delta x$, $\Delta x = 37.7$ m.

14. $\alpha = (-750\pi/30 \text{ rad/s})/25 \text{ s} = -\pi$ rad/s^2, $a = \alpha r = -0.7865$ m/s^2
At 500 rpm, $v = (500\pi/30)(0.25) = 13.1$ m/s.
$\Delta x = (13.1)^2/2(0.785) = 109$ m

15. (a) $v_o = 27.8$ m/s, $a = -v_o^2/2\Delta x = -7.73$ m/s^2
and $\alpha = a/r = -30.9$ rad/s^2.
(b) 50 m = $(2\pi r)N$ thus N = 31.8 rev.

16. (a) $a = 3$ m/s^2, $\alpha = a/r = 12$ rad/s^2, so $a_t = \alpha r = 3$ m/s^2.
$a_r = v^2/r = 16$ m/s^2 (down), thus $\underline{a} = 3\underline{i} - 16\underline{j}$ m/s^2;
(b) $a_t = \alpha(2r) = 6$ m/s^2, $a_r = (2v)^2/2r = 32$ m/s^2
So $\underline{a} = 6\underline{i} - 32\underline{j}$ m/s^2

17. $\theta = \omega t - 1/2\ \alpha t^2$: $80\pi = 50\pi/3 - 12.5\alpha$, so $\alpha = -15.9$ rad/s^2.

18. (a) $\omega = (5300\pi/30)$ rad/s, $\alpha = \Delta\omega/\Delta t = 370$ rad/s^2;
(b) $a_r = \omega^2 r = 1.23\text{x}10^4$ m/s^2, $a_t = \alpha r = 33.3$ m/s^2

19. (a) $a_r = 21.6$ m/s^2, $a_t = 12$ m/s^2; $a = 24.7$ m/s^2.
(b) $a_r = 2.74$ m/s^2

20. (a) $\omega^2 = \omega_o^2 + 2\alpha\Delta\theta$, where $\Delta\theta = 80\pi$ rad, so $\alpha = 4.58\text{x}10^{-2}$ rad/s^2;
(b) $\Delta\theta = \omega^2/2\alpha$, $N = \Delta\theta/2\pi = 7.62$ rev

21. $v = \omega_1 r_1 = \omega_2 r_2$; where $r_2 = 100/60 = 1.67$ cm

22. (a) $I_x = \Sigma\ my^2 = 16$ kg.m^2;
(b) $I_y = \Sigma\ mx^2 = 36$ kg.m^2;
(c) $I_z = \Sigma\ mr^2 = I_x + I_y = 52$ kg.m^2

23. (a) $I = 2M(2d^2) + M(4d^2) = 8Md^2$.
(b) $I = 2M\ell^2 + M(2\ell^2) = 4M\ell^2$.

24. $I = \int r^2\ dm = R^2 \int dm = MR^2$

25. (a) $I = 7$ kg.m^2;
(b) $I_{CM} = 2(10/7)^2 + 5(4/7)^2 = 5.71$ kg.m^2.

26. $I = 1/2\ mr_1^2 + 4[m_2L^2/12 + m(2^2)] + M_3r_3^2 =$
$= 1/2\ (2)(4) + 4(16/12 + 16) + 2(6^2) = 145$ kg.m^2

27. (a) $ma^2 = 2\pi\lambda a^3 = 6.28\lambda a^3$;
(b) $4m(L^2/12 + a^2) = 32\lambda a^3/3 = 10.7\lambda a^3$.
(c) $3(2\lambda a)(4a^2)(1/12 + 1/12) = 4\lambda a^3$

28. (a) Ma^2;
(b) $M(4a^2/12 + a^2) = 4Ma^2/3$;
(c) $M(4a^2/12 + a^2/3) = 2Ma^2/3$

29. $2m[2R^2/5 + (2.5R)^2] + m(3R)^2/12 = 14.1mR^2$

30. (a) $I = m_H(d\ \sin 75^o)^2 = 1.26\text{x}10^{-47}$ kg.m^2;
(b) $I = 2m_H(d\ \sin 37.5^o)^2 = 1.00\text{x}10^{-47}$ kg.m^2

31. $I = (2\pi Rh\sigma)R^2 + 2(\pi\sigma R^2/2)R^2 = \pi\sigma R^3(2h + R)$

32. Distances of elements to axis reduced from x to x $\sin\alpha$, so ML/3 becomes $ML^2\ \sin^2\alpha/3$

33. (a) $\int \lambda x^2\ dx$ from $-L/4$ to $+3L/4$ leads to $7ML^2/48$.
(b) $I = ML^2/12 + M(L/4)^2 = 7ML^2/48$

34. (a) $I_x = my^2$, $I_y = mx^2$

(b) $I_z = mr^2 = m(x^2 + y^2) = I_x + I_y$

35. $I_z = 2I = MR^2$, thus $I = 1/2\ MR^2$.

36. $I = \int \lambda x^2\, dx$ from $(h - L/2)$ to $(h + L/2)$ leads to

$I = \lambda[(h + L/2)^3/3 - (h - L/2)^3/3] = M(h^2 + L^2/12)$

37. $E_i = MgL/2$, $E_f = 1/2(ML^2/3)\omega^2$, thus $\omega^2 = 3g/L$.

$v_{end} = \omega L = (3gL)^{1/2}$.

38. $E = 1/2\ (1/2\ MR^2)\omega^2 = 1/4\ (15)(0.2\ m)^2\ (1.4x10^3\ \pi/30)^2 =$

$= 3.22x10^5$ J

39. $\Delta E = 1/2\ mv^2 + 1/2\ I(v/R)^2 + 1/2\ kx^2 - mgx = 0$.

(a) With $v = 0$, $x = 2mg/k = 0.98$ m;

(b) $1/2\ (5)v^2 + 40(0.2)^2 - 4g(0.2) = 0$, thus $v = 1.58$ m/s.

40. $I = 0.33MR^2 = 8.01x10^{37}$ kg.m^2

$\omega = 2\pi/1\ d = 7.27x10^{-5}$ rad/s

$K = 1/2\ I\omega^2 = 2.12x10^{29}$ J

41. $E_i = 1/2\ mv^2 + 1/2\ I\omega^2$, $E_f = mgH$.

Sphere: $E_i = 7mv^2/10$; Disk: $E_i = 3mv^2/4$.

(a) One; (b) $H_S/H_D = 14/15$;

42. $K = 1/2\ (4m)v^2 + 1/2\ (4I)\omega^2 = 2(mv^2 + I\omega^2) =$

$= 2(10)(55.5)^2 + 2(ML^2/3)(100\pi)^2 = 7.2x10^5$ J

43. $E_i = mgR$, $E_f = 1/2\ I\omega^2$ where $I = 3MR^2/2$. Thus $\omega = (4g/3R)^{1/2}$

$v = \omega(2R) = (16gR/3)^{1/2}$.

44. (a) $I = M_rL^2/3 + [1/2\ M_dR^2 + M_d(R + L)^2] = 0.314$ kg.m^2

(b) $E_i = [M_rgL/2 + M_dg\ell](1 - \cos\theta) = 0.814$ J

$E_f = 1/2\ I\omega^2 = 1/2\ (0.314)\omega^2$

Find $\omega = 2.28$ rad/s, so $v = 1.59$ m/s

45. (a) $K = 1/2\ Mv^2 + 4(I\omega^2/2) + 4(1/2\ mv^2) = 518$ kJ;

(b) $518x10^3 = Mgd\sin 10^o$, so $d = 276$ m

46. $\tau_1 = (F_1\sin 60^o)(L/2) = 34.6$ N.m,

$\tau_2 = (F_2\cos 45^o)(L/4) = 21.2$ N.m; $\tau_3 = F_3L/2 = 32$ N.m

47. $(0.5\ m)(10^5\ N)\cos 5^o = 4.98x10^4$ N.m

48. In each case τ = Fdcosθ

(a) 24 N.m; (b) 20.8 N.m; (c) 17 N.m; (d) 12 N.m

49. (a) $\alpha = -1/3$ rad/s^2, $\tau_f = I\alpha = 0.01$ N.m;

(b) $\tau_E - \tau_f = I(4\ rad/s^2)$, thus $\tau_E = 0.13$ N.m

50. (a) $\alpha = \tau/I = 4$ rad/s^2;

(b) $\Delta\theta = 1/2\ \alpha t_1^2 + \omega t_2 = 150$ rad or 23.9 rev

51. (a) $P = fv = \mu NR\omega = 62.8$ W

(b) $\alpha = fR/I = -15$ rad/s^2, thus $t = (20\pi/3)/15 = 1.4$ s

(c) $\theta = \omega_o t + 1/2\ \alpha t^2 = 14.6$ rad or 2.32 rev.

52. (a) $\omega_o = 40\pi$ rad/s. $\alpha_1 = -\pi/6$ rad/s^2, $a_2 = -4\pi/6$ rad/s^2

$\alpha_1 = \tau_f/I$ and $\alpha_2 = (\tau_f + 300)/I$. Solve to find I = 191 kg.m^2;

(b) $\tau_f = I\alpha_1 = -100$ N.m

53. (a) $\alpha = 12$ rad/s^2, then $I = \tau/\alpha = 48/12 = 4$ kg.m^2.

Since $\omega = 60$ rad/s, $\alpha = -60/12 = -5$ rad/s^2, so $\tau_F = -20$ N.m

(b) $\tau = \tau_M + \tau_{f,}$ thus $\tau_M = 68$ N.m

54. (a) ΣF: mg - T = ma, $\Sigma\tau$: TR = Iα, so T = Ma/2.

Find $a = mg/(M/2 + m) = 4.9$ m/s^2

(b) T = Ma/2 = 9.8 N

(c) $v^2 = 2g\Delta y$, thus v = 1.98 m/s

55. ΣF: mg sinθ - T = ma; $\Sigma\tau$: TR = Iα, so T = Ma/2.

Find $a = mg\sin\theta/(M/2 + m) = 3.92$ m/s^2,

$\alpha = a/R = 7.84$ rad/s^2.

(b) $v^2 = 2a\Delta x$ leads to v = 2.8 m/s.

56. (a) $Mgh = 1/2\ (3/2\ MR^2)\omega^2$ leads to $v = (4gh/3)^{1/2}$;

(b) $v^2 = 2ah$, thus a = 2g/3;

(c) Take τ about the point of contact with the string.

$\tau = I\alpha$: mgR = $(3/2\ MR^2)(a/R)$, thus a = 2g/3;

(d) mg - T = ma, thus T = mg/3;

(e) Need T = mg. Take τ about the center:

$\tau = I\alpha$: TR = $(1/2\ MR^2)\alpha$, leads to $\alpha = 2g/R$

57. ΣF: $T_1 - m_1g = m_1a_1$, $m_2g - T_2 = m_2a_2$
$\Sigma\tau$: $T_2r_2 - T_1r_1 = I\alpha = I(a_1/r_1) = 4a_1$, or $I(a_2/r_2) = 2a_2$.
Solving the three equations: $a_1 = 0.527$ m/s^2;
$a_2 = 1.05$ m/s^2, $T_1 = 10.3$ N and $T_2 = 26.2$ N.

58. (a) ΣF: $F - f = Ma$, $\Sigma\tau$: $fR = I\alpha$ yields $f = Ma/2$.
From ΣF we find $a = 2F/3M$;
(b) $f = Ma/2 = F/3$

59. (a) $mg \sin\theta - f = ma$; $fR = I\alpha$ which leads to $f = Ma/2$.
Solve to find $a = 2g \sin\theta/3$
(b) $f = Ma/2 = mg \sin\theta/3 = \mu(mg \cos\theta)$. Thus $\mu = \tan\theta/3$

60. $I = ML^2/12 + M(0.15)^2 = 4.23\text{x}10^{-3}$ kg.m^2
$\tau = mgd\cos\theta = I\alpha$, thus $\alpha = 13.1$ rad/s^2

61. $I = 1/2\ mR^2 = 2.25\text{x}10^{-2}$ kg.m^2, $\omega = 10\pi/9$ rad/s.
$\alpha = \Delta\omega/\Delta t = -\pi/18$ rad/s^2, thus the frictional torque is
$\tau_f = I\alpha = -3.93\text{x}10^{-3}$ N.m. Require equal and opposite torque to maintain ω, so $P = \tau\omega = 13.7$ mW

62. $\Delta K = 1/2\ I(\omega_f^2 - \omega_i^2) = 2370$ J. Thus $P = 237$ W

63. (a) $P = mgv = 29.4$ W;
(b) $F(0.4) = mg(0.03)$, thus $F = 11$ N.

64. $\tau = P/\omega = (4\text{x}10^4)(746)/(225\pi/30) = 1.27\text{x}10^6$ N.m for each engine

65. (a) $K = 1/2\ (1/2\ MR^2)\omega^2 = 49.1$ J;
(b) From $\tau = I\alpha$; $\alpha = 7.14\text{x}10^{-3}$ rad/s^2;
Then, $t = \omega/\alpha = 29.3$ s

66. (a) $W = \Delta K = 0.5(0.5MR^2)\omega^2 = 1/4\ (7600)(2.15\text{ m})^2(2\pi/30)^2 =$ 385 J;

(b) $F = W/d = 321$ N (Unlikely)

67. $\tau_1 = r_1F_1\sin30^o = 5.00$ N.m; $\tau_2 = r_2F_2\sin60^o = -2.60$ N.m;
$\tau_3 = r_3F_3\sin50^o = -5.09$ N.m

68. $\tau_1 = (3L/4)F_1\sin\alpha = 5.40$ N.m; $\tau_2 = (L)F_2\cos\theta = -5.41$ N.m.

69. $\tau_1 = (2L/3)F_1\sin\alpha = 3.45$ N.m; $\tau_2 = (L)F_2\sin\theta = -2.05$ N.m; $\tau_3 = (L/2)F_3\cos\theta = -1.41$ N.m.

70. $\Delta E = W_{nc}$: $0.5mv^2 - mgd\sin\theta + 0.5kd^2 + 0.5I\omega^2 = -\mu(mg\cos\theta)d$. Find v = 1.17 m/s.

71. $0.5(m_1 + m_2)v^2 + 0.5I\omega^2 + (m_1 - m_2)gd = 0$, thus v = 1.11 m/s.

72. (a) $m_2g - T_2 = m_2a$; $T_1 - m_1g = m_1a$; $(T_2 - T_1)R = (0.5MR^2)(a/R)$. Add all three to find $a = 1.55$ m/s^2; (b) $v^2 = 2a\Delta y$ leads to v = 1.11 m/s.

Problems

1. $dI = \sigma(2\pi r\ dr)\ r^2$; $I = \sigma\pi(B^4 - A^4)/2$ but $M = \sigma\pi(B^2 - A^2)$ thus $I = M(B^2 + A^2)/2$.

2. Cons. of E: $mgH = 1/2\ mv^2 + 1/2\ (2mr^2/5)(v/r)^2 + mg(2R)$
ΣF: $mg = mv^2/R$, thus $v^2 = Rg$. Find H = 2.7R

3. (a) $E_i = mgL/2$, $E_f = 1/2\ mv_{CM}^2 + 1/2\ I_{CM}\ \omega^2$.
On landing, $v_L = 0$ and $v_R = 2v_{CM}$, thus $\omega_f = v_{CM}/(L/2)$.
Substituting into $E_i = E_f$: $\omega = (3g/L)^{1/2}$;
(b) $v_L = 0$, $v_R = \omega L = (3gL/4)^{1/2}$.

4. $t = (2h/g)^{1/2}$. Relative to earth's surface
$v_{rel} = \omega(h + R) - \omega R = \omega h$, thus $d = v_{rel}.t = (2\omega^2h^3/g)^{1/2}$, East

5. $I = (2/5)(M_AA^2 - M_BB^2) = (2/5)(4\pi\rho/3)(A^5 - B^5)$.
$M = (4\pi\rho/3)(A^3 - B^3)$ thus $I = 2M(A^5 - B^5)/5(A^3 - B^3)$

6. (a) $dI = r^2\ dm = (R\sin\theta)^2\ [2\pi\sigma R\sin\theta.\ Rd\theta]$
$I = -2\pi\sigma R^4 \int (1 - c^2)\ dc = -2\pi\sigma R^4[c - c^3/3] = 8\pi\sigma R^4/3$
$M = 4\pi R^2\sigma$, so $I = 2MR^2/3$
(b) $dI = 2\ dm\ r^2/3 = 2(4\pi r^2\rho\ dr)r^2/3$ and $M = 4\pi\rho R^3/3$.
Integrate to find $I = 2MR^2/5$.

7. $I = 1/2\ (\sigma\pi R^2)R^2 - \sigma\pi a^2(a^2/2 + b^2)$, and $M = \sigma\pi(R^2 - a^2)$
Thus $I = M(R^4 - a^4 - 2a^2b^2)/2(R^2 - a^2)$

8. $\omega = 6t^2 - t^3 + 5$ rad/s, $\theta = 2t^3 - t^4/4 + 5t - 17$ rad

9. $dm = 2\sigma\pi\ell \sin\alpha\, d\ell$, so $M = 2\sigma\pi\sin\alpha(h^2/2 \cos^2\alpha)$,

or $M = \pi RL$.

$dI = dm\, r^2 = [2p\sigma\ell\sin\alpha\, d\ell](\ell \sin\alpha)^2$

$= 2\pi\sigma\sin^3\alpha\, \ell^3\, d\ell$, thus

$I = 1/2\, Mh^2 \tan^2\alpha$

10. $\Delta t = 60\theta/360N = \theta/6N$. Then $v = D/\Delta t = 6DN/\theta$.

11. Head: $2MR^2/5 = 2(5)(0.08)^2/5 = 1.28x10^{-2}$ kg.m^2

Arms: $2ML^2/12 + 2M(0.5)^2 = 9[(0.7)^2/12 + 0.5^2] =$

$261.8x10^{-2}$ kg.m^2

Legs: $MR^2 + 2M(0.09)^2 = 23.7x10^{-2}$ kg.m^2

Torso: $32(0.3^2 + 0.15^2)/12 = 30x10^{-2}$ kg.m^2

Total $I = 3.17$ kg.m^2

12. (a) $\tau = I\alpha$: $mgR\sin\theta = (3/2\, MR^2)(a/R)$, thus $\alpha = 2g \sin\theta/3$;

(b) ΣF: $mg\sin\theta - f = ma$, $\Sigma\tau$: $fR = I\alpha$.

$f = Ma/2 = mg\sin\theta/3 = \mu(mg\cos\theta)$, thus $\mu = \tan\theta/3$

13. $I_x = Ma^2/12$, $I_y = Mb^2/12$, thus $I_z = M(a^2 + b^2)/12$

14. (a) $dm = \rho dV = \rho_0(1 - r/2R)(4\pi r^2\, dr)$,

Thus $M = \int dm = 5\rho_0\pi R^3/6$

(b) For a shell, $I = 2mr^2/3$, thus

$dI = 2(\rho\, 4\pi r^2\, dr)r^2/3 = 8\pi\rho r^4\, dr/3$, thus

$I = 8\pi\rho_0/3\, [r^5/5 - r^6/12R] = 14\pi\rho_0 R^5/45$.

Finally, $I = 0.37MR^2$

15. (a) $\Delta t = 1/8N = 1/f_0$, so $N = f_0/8$.

(b) $\Delta N = +0.05N$, means N/20 rev/s forward

(c) $\Delta N = -0.125N$, means N/8 rev/s backward

16. (a) $n = 10cm/4\, cm = 2.5$;

(b) $(100)(0.16) = 16$ N.m;

(c) $16 = T(0.1)$, so $T = 160$ N;

(d) $\tau = 160(0.04) = 6.4$ N.m;

(e) $P = 16x4\pi = 201$ W;

(f) $\omega = 10\pi$ rad/s , $v = \omega\theta = (0.35)(10\pi) = 11$ m/s;

(g) $6.4 = 0.35f$, so $f = 18.3$ N

CHAPTER 12

Exercises

1. $\underline{\ell}_1 = (3\underline{i} - 2\underline{j})x(6\underline{j}) = 18\underline{k}$ kg.m^2/s;
   $\underline{\ell}_2 = (2\underline{i} + 3\underline{j})x(1.5\underline{i} - 2.6\underline{j}) = -9.70\underline{k}$ kg.m^2/s
   $\underline{\ell}_3 = (-3\underline{i} + \underline{j})x(5.14\underline{i} + 6.13\underline{j}) = -23.5\underline{k}$ kg.m^2/s;
   $\underline{\ell}_4 = (-4\underline{i} - 2\underline{j})x(7.5\underline{i}) = 15\underline{k}$ kg.m^2/s
   $\underline{L} = -0.2\underline{k}$ kg.m^2/s

2. $(2\underline{i} - 3\underline{j})x(20\underline{i} + 28\underline{j}) = 116\underline{k}$ kg.m^2/s^2

3. $\ell = mv(a + d) - mva = mvd$

4. $K = 1/2\ I\omega^2$, $L = I\omega$, so $K = L^2/2I$

5. $\Sigma\ell_x = 0$, $\Sigma\ell_y = 2\ell$

6. $\underline{a}_r = \underline{\omega}$ x $\underline{v}$; $\underline{a}_t = \underline{\alpha}$ x $\underline{r}$

7. $\underline{v} = \underline{\omega}$ x r, $\underline{a}_r = \underline{\omega}$ x $\underline{v}$

8. ΣF: $ke^2/r^2 = mv^2/r$, so $v = (ke^2/mr)^{1/2}$.
   Use v in $mvr = nh/2\pi$, to find $r = (nh/2\pi)^2/kme^2$.

9. (a) $\tau = (m_2 - m_1)gR = 1.57$ N.m;
   (b) $L = (m_1 + m_2)vR + I(v/R) = 0.64v + 0.16v = 0.8v$;
   (c) $\tau = dL/dt$: $1.57 = 0.8a$, thus $a = 1.96$ m/s^2.

10. $L = mvR + I\omega = (m + M/2)vR$, while $\tau = mgR$
    From $\tau = dL/dt$ we find $a = mg/(m + M/2) = 4.9$ m/s^2

11. $\underline{v} = 3At^2\underline{i} + (2Bt - C)\underline{j}$; $\underline{a} = 6At\underline{i} + 2B\underline{j}$.
    (a) $\underline{L} = \underline{r}$ x $\underline{p} = MAt^3(2C - Bt)\underline{k}$; (b) $\underline{F} = m\underline{a} = M(6At\underline{i} + 2B\underline{j})$

12. (a) To the left;
    (c) $L = I\omega$, $\Delta L = L\Delta\theta = L(v\Delta t/r) = \omega L\Delta tR/r$
    Thus, $\Delta L/\Delta t = \omega LR/r = I\omega^2R/r = 2KR/r$

13. (a) $L_1 = I_1\omega_1 = 0.024$, $L_2 = I_2\omega_2 = 0.0143\omega_2$.
    From $L_2 = L_1$, we find $\omega_2 = 1.68$ rad/s.
    (b) $K_i = 0.024$ J, $K_f = 0.020$ J, thus $\Delta K = -4\text{x}10^{-3}$ J

14. (a) $L = [3m(d^2) + m(2d)^2 + 2m(3d)^2]\omega = 25md^2\omega$;

(b) $L = [Md^{/}3 + M(d^2/12 + (1.5d)^2) + M(d^2/12 + (2.5d)^2)]\omega +$
$+ 25md^2\omega = (9M + 25m)d^2\omega$

15. (a) $I_i\omega_i = I_f\omega_f$ leads to $\omega_f = 6.4/1.8 = 3.555$ rad/s.
(b) $K_i = 0.144$ J, $K_f = 0.128$ J, thus $\Delta K = -0.016$ J
(c) $\Delta\omega = 0.445$ rad/s, $\alpha = 0.223$ rad/s^2,
so $\tau = I\alpha = 4.5\text{x}10^{-3}$ N.m

16. (a) $L_1 = 1/2\ (1/2\ MR^2)\omega_1{}^2$; $L_2 = [I_1 + m(0.1)^2]\omega_2$
Set $L_2 = L_1$ to find $\omega_2 = 1.64$ rad/s
(b) $K_1 = 45$ mJ, $K_2 = 37$ mJ, thus $\Delta K = -8$ mJ

17. (a) $L = muR = 900$ kg.m^2/s; $L_f = (1/2\ MR^2 + mR^2)\omega = 990\omega$,
thus $\omega = 0.909$ rad/s.
(b) $1/2\ mv^2 = 750$ J, $K_f = 1/2(1/2\ MR^2 + mR^2)\omega^2 = 409$ J
Thus $\Delta K = -341$ J

18. (a) $L_i = [ML^2/12 + 2M(L/4)^2](20\text{ rad/s}) = 50ML^2/12$
$L_f = [ML^2/12 + 2M(L/2)^2]\omega = 7ML^2\omega_2/12$
Thus $\omega_2 = 50/7 = 7.14$ rad/s
(b) 7.14 rad/s; (c) No centripetal force

19. (a) $L_1 = (1/2\ MR^2)\omega_1 = 320$ kg.m^2/s, $L_2 = (160 + 40\text{x}4)\omega_2$
Thus $\omega_2 = 1$ rad/s.
(b) $L_3 = (1/2\ MR^2)\omega_3$, thus $\omega_3 = 2$ rad/s
(c) $K_i = 1/2\ (320)(1^2) = 160$ J, $K_f = 320$ K, thus $\Delta K = +160$ J.

20. (a) $L_1 = 0$; $L_2 = (mR^2)\omega_G - (1/2\ MR^2)\omega_P$, thus $\omega_P = 1.2$ rad/s
(b) Zero

21. $L_M = mR(1 - v)$, $L_C = (1/2\ MR^2)(v/R)$: $160(1 - v) = 100v$,
so $v = 0.615$ m/s, and $\omega = 0.308$ rad/s.

22. (a) $L = I\omega = 2\pi I/T$, thus $T = (2\pi/L)I$, so $dT = (2\pi/L)dI$
and $dT/T = dI/I$.
(b) $dI/I = mR^2/(0.33MR^2) = m/0.33M = 5.05\text{x}10^{-13}$
Thus $dT = (dI/I)T = (5.05\text{x}10^{-13})(3.156\text{x}10^7\text{ s}) = 16\ \mu$s.

23. $L_1 = (1/2\ MR^2)\omega_1$; $L_2 = [1/2\ MR^2 + m(R\sin 45^o)^2]\omega_2$,
so $\omega_2 = M\omega_1/(M + m) = 4.16$ rad/s.

24. (a) $L_1 = 2muR = 240$; $L_2 = 2mR^2\omega_1$, thus $\omega_1 = 2.0$ rad/s
(b) $L_3 = 2m(R/2)^2\omega_2$, so $\omega_2 = 8.0$ rad/s
(c) $K_2 = 1/2\ (2mR^2\omega_1^2)$; $K_3 = 1/2\ (mR^2/2)\omega_2^2$, so $\Delta K = 720$ J

25. $mv_1^2/r_1 = 2Mg$; $mv_2^2/r_2 = Mg$, thus $v_1^2r_2 = 2v_2^2r_1$ (i)
$L_1 = m_1v_1r_1$, $L_2 = m_2v_2r_2$, thus $v_1r_1 = v_2r_2$ (ii)
From (i) and (ii) find $r_2^3 = 2r_1^3$ or $r_2 = 1.26r_1$.

26. $F_1d_1 = F_2d_2$, thus $F_2 = 160/2 = 80$ N

27. (a) $w_1\ell_1 + w_2\ell_2 - w_3\ell_3 = 0$;
$(7\text{x}10^{-3}g)(0.2\text{ m}) + (12\text{x}10^{-3}g)(0.07\text{ m}) - (60\text{x}10^{-3}g)\ell_3 = 0$
Thus $\ell_3 = (1.4 + 0.84)/60 = 3.73$ cm
(b) Weight of cartridge is 0.0686 N. Since LP exerts a 0.02 N force upward, net downward force at end of arm is 0.0486 N
$\Sigma\tau$: $(0.0486)(0.2) + (0.1176)(0.07) - (0.588)\ell_3' = 0$
Thus $\ell_3' = 3.05$ cm. Thus $\Delta\ell_3 = 0.68$ cm

28. $1.5d + 2(0.5) - 2.4(1.5) = 0$, so $d = 1.73$ m on the 2 kg side.

29. (a) $(8\text{x}0.8) + (5\text{x}1.8) + (60\text{x}2.3) = 3.6T_1$, so $T_1 = 417$ N
(b) $(60\text{x}1.3) + (5\text{x}1.8) + (2.8\text{x}8) = 3.6T_2$, so $T_2 = 298$ N

30. (a) Center of sign is 20 + 36 = 56 cm from right end, or (120 - 56) = 64 cm from the pivot.
$(2\text{ kg})(0.6\text{ m}) + (3\text{ kg})(0.64\text{ m}) - (T\sin 30^o)(1.2\text{ m}) = 0$,
Thus, T = 51 N
(b) ΣF_x: $H - T\cos 30^o = 0$, so $H = 44.2$ N
ΣF_y: $V + T\sin 30^2 - 5g = 0$, thus $V = 23.5$ N

31. $\Sigma\tau$: $3T_1 - (20g)(2)\cos 30^o - (4W)\cos 30^o = 0$, thus $W = 768$ N
ΣF_x: $H - T\sin 30^o = 0$, so $H = 500$ N;
ΣF_y: $V + T\cos 30^o - 196 - 768 = 0$, so $V = 98$ N

32. (a) From the law of cosines, the length of the rope = 2.395 m. From the law of sines, the angle of rope to vertical is $\beta = 51.7^o$.
$\Sigma\tau$: $-W(1)\cos 20^o + T(2)\sin 18.3^o = 0$, so $T = 44.0$ N

(b) ΣF_x: $H - T\sin\beta = 0$, so $H = 34.5$ N
ΣF_y: $V + T\cos\beta - 3g = 0$, thus $V = 2.13$ N

33. (a) $(4g\times1.2 + 2g\times2)\cos45^o - (2.4)T\sin75^o = 0$, so $T = 26.3$ N
(b) ΣF_x: $H = T\sin60^o = 22.8$ N;
ΣF_y: $V = 6g - T\cos60^o = 45.7$ N

34. $5g(0.3)\cos30^o + 2g(0.15)\cos30^o - (0.04)T\sin55^o = 0$,
so $T = 466$ N

35. $T(1.2\text{ cm})\cos30^o + 2g(13.8\text{ cm})\cos30^o - (50\text{ N})(28.8\text{ cm})\cos30^o$
$= 0$. Thus $T = 974$ N

36. $(30g)(16) = T(4)$, thus $T = 1180$ N

37. (a) $T = 200g = 1960$ N;
(b) ΣF_x: $H - T\cos30^o = 0$, so $H = 1700$ N;
ΣF_y: $V - T - T\sin30^o = 0$, so $V = 1.5T = 300g = 2940$ N

38. $4T = W$, so $T = W/4$.

39. (a) $\Sigma\tau$: $(100)(2d/3) - (T\sin12^o)(2d/3) + 450(d/2) = 0$;
so $T = 2100$ N.
(b) ΣF_x: $H = T\cos12^o = 2054$ N; ΣF_y: $V = 550 - T\sin12^o =$
113 N. $F = (H^2 + V^2)^{1/2} = 2060$ N

40. (Left support) $\Sigma\tau$: $60g\times2.5 - V_1(0.5) = 0$, so $V_1 = 2940$ N
down.
(Right support) ΣF_y: $V_2 - (360\,g) = 0$, thus $V_2 = 3530$ N, up.

41. $Fh\cos\theta = WH\cos(\alpha + \theta) = WH\cos51.8^o$,
$H = (D^2 + L^2)^{1/2}/2 = 0.539$ m; and $\tan\alpha = 0.4$. Thus
$F = 200\times0.539\times\cos51.8^o/(1.1\cos30^o) = 70$ N.

42. $130x_{CM} = 20\times15 + 40\times40 + 50\times85 + 20\times50$, thus $x_{CM} = 55$ cm.

43. $300x = 350(1.6 - x)$, thus $x = 0.862$ m from feet.

Problems

1. (a) Σp: $mu = (2M + m)v_{CM}$, so $v_{CM} = 0.8$ m/s.
(b) Position of CM after collision: $d = MR/(M + m) = 0.8$ m
ΣL: $mud = (M + m)d^2\omega + M(R - d)^2\omega$, thus $\omega = 2/3$ rad/s.

2. $mu_1 = -mu_2 + MV$, so $V = m(u_1 + u_2)/M$ to the right.
$mu_1 d = -mu_2 d + I_{CM}\,\omega$, so $\omega = m(u_1 + u_2)d/I_{CM}$. This tends to move the end of the rod to the left. For these tendencies to cancel we need $V = \omega(L/2)$. This leads to $d = L/6$.

3. $L_o = mv_o r_o$.
(a) $F = mv^2/r = L_o^2/mr^3$.
(b) $W = \int F_r\, dr = (L_o^2/2m)(1/r_2^2 - 1/r_1^2)$
(c) $K = L_o^2/2mr^2$ thus $\Delta K = (L_o^2/2m)(1/r_2^2 - 1/r_1^2)$; (d) Yes.

4. $K = 1/2\ m(R - a)^2\omega^2 + 1/2\ m(R + a)^2\omega^2 = m(R^2 + a^2)\omega^2$
OR: $K = K_{CM} + K_{rel} = m(\omega R)^2 + m(a\omega)^2$

5. (a) $f = Ma$, so $a = \mu g$; $fR = I\alpha$, so $\alpha = 2\mu g/R$
(b) $v = \omega R$ becomes $at = (\omega_o - \alpha t)R$, thus $t = \omega_o R/3\mu_k g$.
(c) $x = 1/2\ at^2 = (\omega_o R)^2/18\mu g$

6. (a) $L_i = L_{spin} = I\omega_o = 1/2\ MR^2\omega_o$
(b) $\tau = 0$
(c) $L_f = L_{spin} + L_{orb} = I\omega + MvR$
(d) When rolling starts $\omega = v/\theta$, thus $L_f = 3MRv/2$.
Set $L_f = L_i$ to find $v = \omega_o R/3$.

7. $f = Ma$, $a = \mu g$; $fR = I\alpha = (2MR^2/5)\alpha$, thus $\alpha = 5\mu g/2R$.
For rolling $v = \omega R$, thus $v_o - at = \alpha tR$ or $v_o = (a + \alpha R)t$
$= 7\mu gt/2$ and $t = 2v_o/7\mu g$.
(a) $v = v_o - at = 5v_o/7$
(b) $v^2 = v_o^2 - 2a\Delta x$; $\Delta x = 24v_o^2/98\mu g$.

8. (a) ΣF: $F + f = Ma$; $\Sigma\tau$: $Fr - fR = I\alpha$
For no slipping $\alpha = a/R$, $a = \alpha R$.
Find $a = F(1 + r/R)/(M + I/R^2)$
(b) $f = Ma - F = F(r - I/MR)/(R + I/MR)$
(c) The direction of f changes.

9. (a) Yes, central force so $\tau = 0$
(b) $L_i = L_f$: $mu_o b = mvr$;
$E_i = E_f$: $1/2\ mu_o^2 = C/r + 1/2\ mv^2 = C/r + 1/2\ m(u_o b/r)2$

$$(mu_o^2)r^2 - 2Cr - mu_o^2b^2 = 0,$$

thus

$$r = [C + (C^2 + (mbu_o^2)^2)^{1/2}]/mu_o^2$$

10. ΣF: $T - f = Ma$; $\Sigma\tau$: $fR - Tr = (1/2\ MR^2)\alpha$
For no slipping $\alpha = a/R$. Note also that $f = \mu mg$.
Find $T = 3\mu MgR/(R + 2r)$

11. ΣF: $Mg - N = Ma$; $\Sigma\tau$: $MgL/2 = (ML^2/3)\alpha$
Thus, $\alpha = 3g/2L$, and $a = \alpha L/2 = 3g/4$. Finally, $N = Mg/4$.

12. ΣF: $F - f = Ma$, $\Sigma\tau$: $fR = I\alpha = (1/2\ MR^2)(a/R)$
Thus $f = Ma/2$ and then $F = 3Ma/2 = 3f = 3\mu Mg$.
Thus $\mu = F/3Mg$.

13. $DW\cos(\theta + \alpha) = N_1(3)$ where
$D = (1.8^2 + 0.8^2)^{1/2} = 1.97$ m.
$\tan\alpha = 0.8/1.8$ thus $\alpha = 24^o$;
$\tan\beta = 0.8/1.2$ thus $\beta = 33.7^o$
$N_1 = 1.97W\cos 44^o/3 = 0.472W = 5550$ N
$dW\cos(\beta - \theta) = N_2(3)$ where
$d = (1.2^2 + 0.8^2)^{1/2} = 1.44$.
$N_2 = 1.44W\cos(13.7^o)/3 = 0.466W = 5480$ N

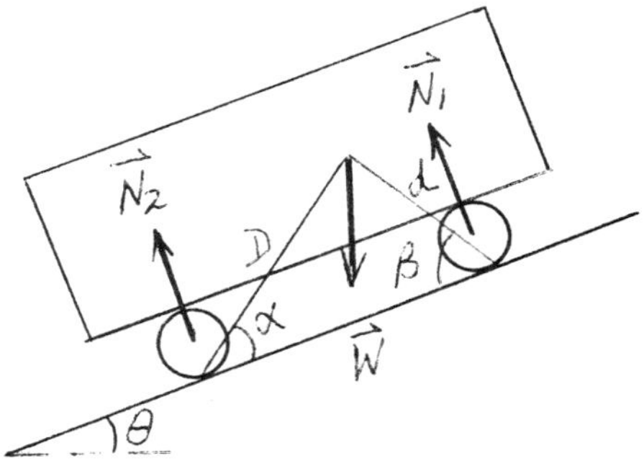

14. (a) $N = mg/2 = 343$ N
$NL\cos\theta = T(L/2)\sin\theta$, thus $T = mg\cot\theta = 277$ N
(b) $\Sigma\tau$: $mg(1)\cos\theta + 4N_2\cos\theta - 4N_1\cos\theta = 0$, so
$mg = 4(N_2 - N_1)$.
ΣF: $mg = N_1 + N_2$, thus $N_1 = 5mg/8 = 429$ N, $N_2 = 257$ N
$\Sigma\tau$: $N_2L\cos\theta = T(L/2)\sin\theta$, so $T = 208$ N

15. (a) Clearly $d_1 = L/2$;
(b) $2m\ x_{CM} = 2mL = m(d_2 + L/2) + m(d_2 + L)$, so $d_2 = L/4$;
(c) $(d_3 + L/2) + (d_3 + d_2 + L/2) + (d_3 + d_2 + L) = 3L$,
so $d_3 = L/6$

16. (a) When W passes through lower corner: $\tan\theta = b/h$
(b) ΣF: $mg\sin\theta - \mu_k(mg\cos\theta) = 0$, then $\tan\theta = \mu_k$
(c) Slide; (d) Topple

17. (a) About the right corner $Fd = Wb/2$; so $d = Wb/2F = 0.613$ m
(b) Torque about the CG: $fh/2 - Nx = 0$. Also $f = F$ and $N = W$
Thus, $x = Fh/2W = 0.224$ m from the central line.

18. (a) $F_1(R - h) = Wx$, where $x = [R^2 - (R - h)^2]^{1/2} = 0.125$ m
Thus, $F_1 = 32.2$ N
(b) $F_2(2R - h) = Wx$, leads to $F_2 = 15.7$ N.

19. $[W_2(1.5) + W_1(1)]\cos\theta = 3N_2 \sin\theta$;
$N_2 = (15 + 50)g/3\tan\theta = 77.3$ N
ΣF_y: $N_1 = W = 588$ N, and ΣF_x: $f_1 = N_2 = 77.3$ N.

20. ΣF_y: $N_1 = 13W$, ΣF_x: $f_1 = \mu(13W) = N_2$
$\Sigma\tau$: $(wL/2 + 12Wd) \cos\theta = N_2 \, L\sin\theta = 13WL\sin\theta$
Thus, $d = L[13\mu\tan\theta - 1/2]/12 = 0.733L$

21. $N_1 = W$; $f_1 = N_2 = \mu N_1$.
$2WL\cos\theta/3 = N_2L \sin\theta$, thus $\tan\theta = 2/3\mu$.

22. $W(b/2) = H(h)$, $H_1 = H_2 = 24.5$ N

CHAPTER 13

Exercises

1. (a) $a = Gm/r^2 = 6.67\text{x}10^{-9}$ m/s^2
   (b) $E_i = -Gm^2/1$, $E_f = mv^2 - Gm^2/(0.5)^2$.
   $E_i = E_f$ leads to $v = 8.17\text{x}10^{-5}$ m/s

2. $F = GmM/r^2$. (a) $2.0\text{x}10^{20}$ N; (b) $4.42\text{x}10^{20}$ N

3. $F = GmM/r^2$. (a) $2.33\text{x}10^{-3}$ N; (b) $4.19\text{x}10^{-1}$ N

4. Since $F_E = F_M$ we find $M_E/x^2 = M_M/(d - x)^2$ which leads to $9(d - x) = +x$, thus $x = 0.9d$.

5. $F_{31} = 50G$, $F_{32} = 160G$.
   $F_{3x} = -(50 + 320/(5)^{1/2})G = -1.29\text{x}10^{-8}$ N
   $F_{3y} = 160G/(5)^{1/2} = 4.77\text{x}10^{-9}$ N

6. From $4/x^2 = 9/(1 - x)^2$, we find $x = 0.4$ m

7. (a) $\Sigma F_x = -\ GM^2/L^2\ (2 + 2(2)^{1/2}) = -4.84GM^2/L^2$
   $\Sigma F_y = GM^2/L^2\ (6 + 2(2)^{1/2}) = +\ 8.82GM^2/L^2$
   (b) $\Sigma F_x = -GM^2/L^2\ (12 + 1.5/(2)^{1/2}) = -13.1\ GM^2/L^2$
   $\Sigma F_y = -GM^2/L^2\ (6 + 1.5/(2)^{1/2}) = -7.06\ GM^2/L^2$

8. (a) ΣF_x: $-(6 + 7.5)GM^2/L^2 = -13.5GM^2/L^2$;
   ΣF_y: $15\sin 60^o GM^2/L^2 = 13.0GM^2/L^2$
   (b) ΣF_x: $(15 - 10)\cos 60^o GM^2/L^2 = 2.5GM^2/L^2$
   ΣF_y: $-25\sin 60^o GM^2/L^2 = -21.7GM^2/L^2$

9. $\Delta g = GM(1/R_1{}^2 - 1/R_2{}^2) = 6.48$ cm/s^2

10. If M is mass of original sphere, M/8 is mass of removed sphere. $F = GmM/d^2 - GmM/8(d - R/2)^2$.

11. (a) $g' = g_o(T'/T_o)^2 = (9.810)(1440/1441)^2 = 9.796$ m/s^2
    (b) $(R + h)/R^2 = g'/g_o$ thus $h = 4.55$ km.

12. $g_o = GM/R^2$, $g = GM/(R + h)^2$. The equation $g = g_o/N$ leads to $h = [(N)^{1/2} - 1]R$

13. $T_2 = T_1\ (g_1/g_2)^{1/2} = 2(6)^{1/2} = 4.92$ s.

14. $F = GmM[1/(r - a)^2 - 1/(r + a)^2] = 4GmMra/(r^2 - a^2)^2$
If $r >> a$, then $F \approx 4GmMa/r^3$

15. $F = GmM/d^2$. (a) 0.43 N; (b) $7.5\text{x}10^{17}$ N; (c) $2.8\text{x}10^{-7}$ N

16. $g = GM/R^2$. (a) 11.6 N/kg; (b) 24.6 N/kg; (c) $1.67\text{x}10^{11}$ N/kg

17. (a) $g_o = 4\pi\rho RG/3 \; \alpha \; \rho R$;
(b) $g = 4g_o$; (c) $g = g_o$
(d) $V = kR^3$, $R' = 1.26R_o$, thus $g = 1.26g_o$

18. (a) $T = kg^{-1/2}$, so $dT/dg = -0.5 \; g^{-3/2}$ and $dT/T = -dg/2g$.
(b) +2.5min/1440 min = $-dg/2g$, thus $dg = -5g/1440 = g' - g_o$.
$g' = g_o(1 - 5/1440) = 9.77$ N/kg.
(c) Astronomical observations

19. $mg - W_1 = m(v + \omega R)^2/R$; $mg - W_2 = m(-v + \omega R)^2/R$,
$W_2 - W_1 = 4m\omega v$

20. (a) $v = (GM/R)^{1/2} = 7.9$ km/s
(b) $T = 2\pi R/v = 84.4$ min
(c) $\Delta E = \Delta K = 1/2 \; mv^2 = GmM/2R = 3.14\text{x}10^7$ J
(d) $E = -GmM/2R = -3.14\text{x}10^7$ J

21. (a) $v = (GM/r)^{1/2} \; \alpha \; r^{-1/2}$;
(b) $T = 2\pi(r^3/GM)^{1/2} \; \alpha \; r^{3/2}$;
(c) $p = mv = m(GM/r)^{1/2} \; \alpha \; r^{-1/2}$;
(d) $L = mvr = m(GMr)^{1/2} \; \alpha \; r^{1/2}$.

22. $T^2 = 4\pi^2 r^3/GM$, where $M = 4\pi\rho R^2/3$, thus
$\rho = 3\pi(R + 10^5)^3/GT^2R^3 = 3330$ kg/m^3

23. $v_A = r_P v_P/r_A = 29.3$ km/s

24. $v_P = r_A v_A/r_P = 913$ m/s

25. (a) $E = -GmM/2a$; where $2a = r_A + r_P$. Thus $E = -2.39\text{x}10^9$ J
(b) $T^2 = \kappa a^3$; thus $T = 96.5$ min
(c) From Eq. 13.7a: $v_P{}^2 = (GM/a)(r_A/r_P)$, thus
$v_P = 7.97$ km/s.

26. (a) $E = -GmM/2a$; where $2a = r_A + r_P$. Thus $E = -3.38\times10^8$ J
(b) $T^2 = \kappa a^3$; thus T = 125 min
(c) From Eq. 13.7a: $v_P^2 = (GM/a)(r_A/r_P)$, thus $v_P = 8.49$ km/s

27. Since $L = (3)^{1/2}r$; $F = (2Gm^2/L^2)((3)^{1/2}/2) = Gm^2/r^2(3)^{1/2}$;
Also $F = m\omega^2 r$, thus $\omega^2 = Gm/r^3(3)^{1/2}$

28. (a) $r_A = R + h_A = 1.865\times10^6$ m, $r_P = R + h_P = 1.84\times10^6$ m
From Eq. 13.7: $v_P^2 = (GM/a)(r_A/r_P)$, thus $v_P = 1.64$ km/s.
Similarly, $v_A = 1.62$ km/s.
(b) $T^2 = \kappa a^3 = (4\pi^2/GM)a^3$, where $a = (r_A + r_P)/2 =$
1.853×10^6 m. Thus T = 119 min.

29. $r_A = 6820$ km, $r_P = 6685$ km. (a) $T^2 = \kappa a^3$, so T = 92 min.
(b) $E = -GmM/2a = -2.25\times10^{12}$ J
(c) From Eq. 13.7: $v_P^2 = (GM/a)(r_A/r_P) = 7.76$ km/s.
Similarly, $v_A = 7.61$ km/s.

30. (a) 2a = 15800 km. $T^2 = \kappa a^3$, so T = 116 min
(b) $E = -GmM/2a = -1.89\times10^9$ J
(c) From Eq. 13.7, $v_A^2 = (GM/a)(r_P/r_A)$, so $v_A = 7.05$ km/s.
Similarly, $v_P = 7.16$ km/s.

31. $r_A v_A = r_P v_P$; $2GM(1/r_P - 1/r_A) = v_p^2 - v_A^2 =$
$v_P^2(r_A^2 - r_P^2)/r_A^2$. Thus,

$$v_P^2 = (GM/a)(r_A/r_P)$$

Use this expression in E:
$E = 1/2\, mv_P^2 - GmM/r_P = (GmM/2r_P)(r_A/a - 2) = -GmM/2a$.

Problems

1. $T^2 = 4\pi^2 R^3/GM = 3\pi/G\rho$

2. $\Delta g = GM[1/(R + h)^2 - 1/R^2] = -GM(2Rh + h^2)/R^2(R + h)^2$
When R >> h, ignore h^2 and $(R + h) \approx h$. Then using
$g = GM/R^2$, find $\Delta g = -2gh/R$
(a) 1.33×10^{-3} N/kg; (b) 2.72×10^{-2} N/kg; (c) 3.08×10^{-1} N/kg

3. $U(x) = 2GmM/(a^2 + x^2)^{1/2}$;
$F_x = -dU/dx = -2GmMx/(a^2 + x^2)^{3/2}$
$dF_x/dx = 0$ leads to $x = \pm a/(2)^{1/2}$

4. (a) $F_r = -GmMr/R^3$

(b) $U(r) - U(R) = (GmM/R^3) \int r\,dr = (GmM/2R)(r^2/R^2 - 1)$

But $U(R) = -GmM/R$, thus

$$U(r) = (GmM/2R)(r^2/R^2 - 3)$$

5. (a) $E_i = 1/2\ m(0.85v_{esc})^2 - GmM/R = -0.278GmM/R$.

$E_f = -GmM/r$, thus $r = 3.6R$

(b) $E_i = -0.278GmM/R$,

$E_f = -GmM/r + 1/2\ mv^2$;

$L = m(0.85v_{esc})R = mvr$, where $v_{esc} = (2GM/R)^{1/2}$.

Obtain an expression for v and substitiute it into $E_f = E_i$:

$0.278r^2 - Rr + 0.723R^2 = 0$; thus $r = 2.6R$.

6. $E_i = 1/2\ mv_o^2 - GmM/R = -GmM/2R$

$E_f = 1/2\ mv^2 - GmM/d$

$L = mv_oR(3)^{1/2}/2 = mvd$

Thus, $1/2\ mv^2 = 3m(v_oR/d)^2/8$. Use this in $E_f = E_i$:

$4d^2 - 8Rd + 3R^2 = 0$, thus $d = 3R/2$, $R/2$ (reject).

7. (a) $E_i = -GmM/R$; $E_f = - GmM/(R + h)$;

Thus, $\Delta E_1 = GmM(1/R - 1/(R + h))$

(b) $E_i = -GmM/R$; $E_{orb} = -GmM/2(R + h)$

Thus, $\Delta E_2 = GmM[1/R - 1/2(R + h)]$

(c) $[1/R - 1/2(R + h)] = 2[1/R - 1/(R + h)]$, leads to

$h = R/2$

8. $E_i = -Gm_1m_2/r$; $E_f = p_1^2/2m_1 + p_2^2/2m_2 - Gm_1m_2/(R_1 + R_2)$

But $p_1 = p_2$, so

$$-Gm_1m_2/r = m_2^2v_2^2[1/m_1 + 1/m_2]/2 - Gm_1m_2/(R_1 + R_2)$$

Solve to find $v_2 = 4.85$ km/s, then $v_1 = 8.08$ km/s.

9. (a) $dF = [Gm\lambda d\ell/(R^2 + b^2)] \cos\theta$ where

$\cos\theta = b/(R^2 + b^2)^{1/2}$.

$\int d\ell = 2\pi R$, thus $F = GmMb/(R^2 + b^2)^{3/2}$.

(b) When $b >> R$, $F \approx GmM/b^2$

10. $E_P = 1/2\ mv_C^2(1.13) - GmM/r_C = 0.87E_C$, where $r_P = r_C$.

$$-0.87GmM/2r_P = -GmM/(r_P + r_A)$$

Thus $r_A = (1.13/0.87)r_P = 1.3r_P$.

11. (a) $GmM(r)/r^2 = [GmM/8]/(R/2)^2 = GmM/2R^2$
(b) $GmM/(1.5R)^2 = 0.44GmM/R^2$
(c) $2GmM/(2.5R)^2 = 0.32GmM/R^2$

12. (a) $E = 1/2\ m(v_r^2 + v_\theta^2) - GmM/r$
$L = mv_\theta r$, so $v_\theta^2 = (L/mr)^2$
(b) $v_r = 0$ at these points

13. $dM = \rho\ dV = [\rho_0(1 - r/2R)](4\pi r^2\ dr)$
Thus $M(r) = \int dM = 4\pi\rho_0 r^3(1/3 - r/8R)$
$g(r) = GM(r)/r^2 = 4\pi\rho_0 G(r/3 - r^2/8R)$

14. $1/2\ mv^2 = m(GM/R_E^2)H$, thus $v^2 = 2GMH/R_E^2$.
Know $v_{esc}^2 = 2GM/R_a$, where $M = 4\pi\rho R_a^3/3$
Equate $v^2 = v_{esc}^2$, to find $R_a = (HR_E)^{1/2}$

15. $m_1\omega^2 r_1 = m_2\omega^2 r_2 = Gm_1m_2/(r_1 + r_2)^2$ thus
CM : $r_2 = m_1(r_1 + r_2)/(m_1 + m_2)$ and $r_1 = m_2(r_1 + r_2)/(m_1 + m_2)$
$\omega^2 = Gm_1/r_2(r_1 + r_2) = G(m_1 + m_2)/(r_1 + r_2)^3$
$$T^2 = 4\pi^2(r_1 + r_2)^3/G(m_1 + m_2)$$

16. Kepler's 3 rd law: $\omega^2 = GM/r^3$, thus
$2\omega\ d\omega = -3GMr^{-4}\ dr$, so $d\omega/\omega = -1.5dr/r$
$d\omega = -1.5\omega(dr/r)$ where $r = 4.23\text{x}10^7$ m and $\omega = 2\pi$ rad/day.
Find $d\omega = 2.58\text{x}10^{-9}$ rad/s or 4.66^o per year.

17. From the law of sines:
$\sin\theta/m\omega^2 r = \sin(\lambda + \theta)/mg$
$r = R\cos\lambda$, and if $\lambda >> \theta$,
then $\sin(\lambda + \theta) \approx \sin\lambda$
Thus, $\sin\theta = \omega^2 R \sin 2\lambda/2g$

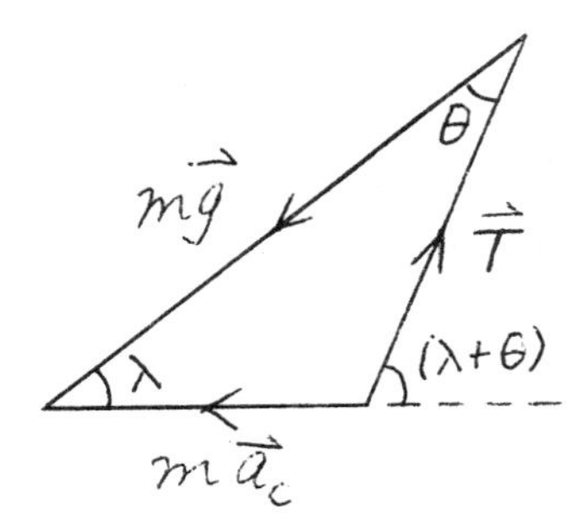

CHAPTER 14

Exercises

1. (a) $m = (0.7V)(0.8) + (0.3V)(1) = 0.86V$, thus $\rho = 0.86\ g/cm^3$

   (b) $V = (0.7m)/0.8 + (0.3m)/1$. Then $\rho = m/V = 0.851\ g/cm^3$

2. $V = (100\ g)/(1\ g/cm^3) = 100\ cm^3$. Mass of liquid = 115 g, thus $\rho = 115/100 = 1.15\ g/cm^3$

3. (a) $\rho = m/V$ where $V = 4\pi r^3/3$. $\rho = 8.95 \times 10^{14}\ kg/m^3$

   (b) $V = 4\pi r^3/3 = M/\rho$, find $r = 1.17$ km

4. $Y = FL/(A.\Delta L) = 6.67 \times 10^{10}\ N/m^2$

5. $A = (Y/F)(\Delta L/L) = \pi D^2/4$ leads to $D = 1.75$ mm

6. $B = -\Delta P.V/\Delta V = +\ 1.05 \times 10^9\ N/m^2$

7. $A = (Y/F)(\Delta L/L)$. $F/A = 1.2 \times 10^8$ N/m2 where $F = m(g + a) = 9040$ N. Then $A = 9040/1.2 \times 10^8 = \pi D^2/4$ leads to $D = 9.8$ mm.

8. $\Delta V/V = -\Delta P/B = -5.14 \times 10^{-2}$

9. $F = AY\Delta L/L$, thus $F = (3 \times 10^{-4})(10^{10})(10^{-2}) = 3 \times 10^4$ N

10. $F = (\pi d^2/4)S = 3.96 \times 10^4$ N

11. $F = (\pi dh)S = 1.06 \times 10^4$ N

12. (a) $mg = PA$, so $m/A = P/g = 10.3 \times 10^3\ kg/m^2$.
    Since $A = 4\pi R^2$, $M = 5.25 \times 10^{18}$ kg.

    (b) $V = 4\pi/3\ [(R + h)^3 - R^3] \approx 4\pi R^2 h$
    $m = \rho V$, thus $h = 8$ km.

13. $F = \Delta P\ A = (0.6 \times 101.3\ kPa)(1.68\ m^2) = 1.02 \times 10^5$ N

14. $F = \Delta P\ A = 600$ N

15. $F = PA = (20/760)(1.013 \times 10^5\ N/m^2)(\pi)(5 \times 10^{-3}\ m)^2 = 0.21$ N

16. $P = P_o + \rho gh = 1.013 \times 10^5 + 9800h$
    (a) 131 kPa; (b) 1.08 MPa; (c) 107 MPa

17. Let $y = 0$ at the mercury-water interface in one arm. The pressure is the same at this level in the other arm containing just Hg.
$V_w = Ah_w = 25$ mL, so $h_w = 25/\pi(0.4)^2 = 49.74$ cm.
If h_m is the height of Hg above the $y = 0$ level:
$A\rho_m h_m = A\rho_w h_w$, thus $h_m = (49.74)(1)/13.6 = 3.66$ cm
The difference in heights is $\Delta h = 49.74 - 3.66 = 46.1$ cm.

18. $P = (15\text{x}10^{-3}\ \text{N})/\pi(4\text{x}10^{-10}\ \text{m}^2) = 1.19\text{x}10^7 = \rho gh$
Thus $h = 1.21$ km.

19. $\Delta P = \rho gh = 1.85\text{x}10^4$ Pa.

20. $\Delta P = \rho gh = 1.26\text{x}10^4$ Pa
$F = \Delta P.A = 0.63$ N

21. $\rho_1 gh_1 = \rho_2 gh_2$. thus $\rho_2 = 4.8/4.4 = 1.09\ \text{g/cm}^3$

22. $P = P_o + \rho(g + a)h$

23. $\Delta P = \rho gh + 5\text{x}10^5 = 1.96\text{x}10^6\ \text{Pa} + 0.5\text{x}10^6\ \text{Pa} = 2.46$ MPa.

24. $h = \Delta P/\rho g = 81.6$ cm

25. $P = P_o + \rho gh = 101.3\ \text{kPa} + 3\ \text{kPa} = 104$ kPa

26. $P = (13.6\text{x}10^3)g(20\ \text{mm}) = (1025)gh$, so $h = 26.5$ cm

27. $Mg = 4AP$, $M = 4AP/g = 980$ kg

28. $F = \Delta P.A = (0.9\ \text{atm})(\pi)(0.3)^2 = 2.58\text{x}10^4$ N

29. $\Delta F_B = \rho(\pi r^2 h)g = mg$, thus $m = 6.03$ g

30. $W = (600)(3\text{x}3\text{x}0.16)g = 864g = 8.47\text{x}10^3$ N
$F_B = \rho_f gV = (10^3)(9.8)(0.8\text{x}0\text{x}0.16) = 11.3\text{x}10^3$ N
Thus load is $2.82\text{x}10^3$ N or 288 kg.

31. $F_{B1} = \rho_1 g(0.6V)$; $F_{B2} = \rho_2 g(0.7V)$. Thus,
$\rho_2 = (0.6/0.7)(1000) = 857\ \text{kg/m}^3$

32. $F_B = \rho_f gV = mg - T = 2.6$ N. Also, $V = m/\rho = 2/9000\ \text{m}^3$.
Thus, $\rho_f = 2.6/(9.8)V = 1190\ \text{kg/m}^3$

33. $(0.4V)(1) = 400g$, so $V = 10^3\ cm^3$
$V(1) = 10^3 g$, thus $\Delta m = 600\ g$

34. $mg = 12\ N$, $mg - T = F_B = 4\ N = \rho_f gV$, thus $V = 4.08 \times 10^{-4}\ m^3$.
$\rho_s gV = 12$, thus $\rho_s = 12/(4/\rho_f) = 3000\ kg/m^3$

35. $F_B{}^1 = 30 = \rho_1 gV_1$, $F_B{}^2 = 40 = \rho_2 gV_2$. Thus, $V_1 = 3V/4$.
$W = 30 = \rho_B gV = \rho_f gV(0.75)$, thus $\rho_B = 0.75\rho_f = 750\ kg/m^3$.

36. $mg = (950)g(0.12h)$, thus $h = 12.6\ cm$

37. $W - F_B = 4.2 = (0.5)g - (800)gV$, thus $V = 8.93 \times 10^{-5}\ m^3$
$\rho = 0.5/V = 5600\ kg/m^3$

38. $F_B = \rho_f gV = mg$, thus $m = 1.54 \times 10^7\ kg$

39. $F_B = \rho_f gV_s = \rho_f g(V - 2.5) = \rho_p gV$, thus
$\rho_p = 1000(V - 2.5)/V$; $60 = \rho_p V$ thus $\rho_p = 960\ kg/m^3$.

40. $\rho_f g(V - 10^6) = \rho_i gV$, thus $V = 9.76 \times 10^6\ m^3$, so
$m = 8.98 \times 10^9\ kg$.

41. $W = 3(1025)A = 1000Ah$, $h = 3.075$, so $\Delta h = 7.5\ cm$.

42. $mg = (V_B + Ah)\rho_w g = [V_B + A(h + 1.5\ cm)]\rho_L g$
Use $(V_B + Ah) = m/\rho_w = 5\ cm^3$ to find $\rho_L = 0.944\ g/cm^3$.

43. $V = 200/9 = 22.2\ cm^3$; $F_B = \rho_A gV = 0.28\ mN$.
So $\Delta m = F_B/g = 0.029\ g$ and $m = 200.029\ g$

44. $F_B = \rho_A gV = 3.97 \times 10^4\ N$.
$m_{He} = \rho_{He} V = 180\pi\ kg$, thus
$\Delta m = (1.29\pi \times 10^3 - 180\pi) = 3490\ kg$.

45. $V = \pi(2^2)(1.25) = 15.7\ m^3 = Av\Delta t$, so $\Delta t = V/Av = 1100\ min$.

46. $A = 0.25\ m^2$, $V = 36\ m^3$, thus $V/\Delta t = 0.03\ m^3/s = Av$,
so $v = (0.03)/(0.25) = 0.12\ m/s$.

47. $A_1 v_1 = A_2 v_2$, $v_2 = (2.4)(1.59/1.28)^2 = 3.7\ m/s$
$H = v_2{}^2/2g = 0.7\ m$

48. $v_2 = (20g)^{1/2} = 14$ m/s.
$P_1 = 1/2\ \rho v_2^2 + 2\rho g = 118$ kPa (gauge) or 219 kPa (absolute)

49. $F = \Delta P.A = (1/2\ \rho\ v^2)(150\ m^2) = 155$ kN.

50. $1/2\ \rho(v_2^2 - v_1^2) = 4.90\ kN/m^2$, so $F = \Delta P.A = 392$ kN

51. (a) $A_1 v_1 = A_2 v_2$, so $v_2 = 3.6$ m/s
(b) $P_1 + 1/2\ \rho v_1^2 = P_2 + \rho gh + 1/2\ \rho v_2^2$
Find $P_1 = 370$ kPa.

52. $A_1 v_1 = A_2 v_2$, so $v_2 = 4v_1 = 9.6$ m/s.
$P_1 + 1/2\ \rho v_1^2 = P_2 + 1/2\ \rho v_2^2 = P_2 + 8\rho v_1^2$
$P_2 = P_1 - 7.5\rho v_1^2 = 117$ kPa.

53. $v_2 = (1.59/1.28)^2(2.4) = 3.7$ m/s.
$P_1 + 1/2\ \rho v_1^2 = 1/2\ \rho v_2^2$, so $P_1 = 3.79$ kPa.

Problems

1. $dF = W\ dy(\rho g)(H - y)$;

$$F = \rho g W \int_0^H (Hy - y^2)\ dy = \rho g W H^2/2$$

$F = 3.53 \times 10^9$ N

2. $d\tau = y\ dF$, $\tau = \rho g W \int (Hy - y^2)\ dy = \rho g W H^3/6 = F(H/3)$.

3. $d\rho/dV = -m/V^2 = -\rho/V$, thus $d\rho/\rho = -dV/V$
$B = -\Delta P/(\Delta V/V)$ leads to $B = \rho gh/(\Delta\rho/\rho)$,
so $\Delta\rho = \rho^2 gh/B$
$\Delta\rho = 53.4\ kg/m^3$ using $1025\ kg/m^3$ at the surface.

4. $dF = dm\omega^2 r = (2\pi r\rho h\ dr)\omega^2 r$
$= dP(2\pi rh)$, thus
$dP = \rho\omega^2 r\ dr$.
After integration:
$P = P_a + \rho\omega^2 r^2/2$

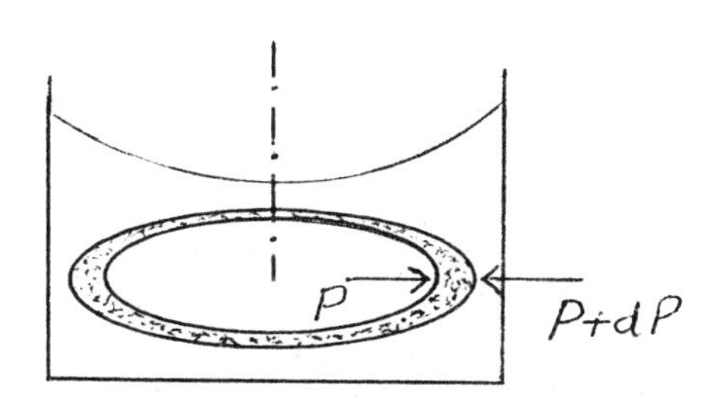

5. $m = \rho_s(2\pi Rt)(L + R)$, and
$F_B = mg = \rho_f(\pi R^2 h)g$
Fraction $= h/L = 2\rho_s t(R + L)/RL\rho_f$

6. $F_B = \rho_f g V_s = mg = \rho_A(V_s + 3/10^3)$
$\rho_f V_s = \rho_A(V_s + 3/10^3) = 90$, thus $V_s = 90/10^3\ m^3$
$F_B = \rho_f g(V_s + 3/10^3) = (90 + m)g$, thus
$m = \rho_f(V_s + 3/10^3) - 90 = 3$ kg.

7. $mg = \rho_f g(0.6V)$; $F = \rho_f gA\ x$, thus
$$W = \int F\ dx = \int_0^{0.4h} \rho_f\ gA\ x\ dx = \rho_f gA(0.4h)^2/2 = 0.67\ J$$
Note that $V = m/(\rho_f 0.6V) = 5x10^{-3}\ m^3$, $h = 0.171$ m

8. $A_1 v_1 = A_2 v_2$ gives $R^2 v_o = r^2 v$
$v^2 = v_o^2 + 2gy$ thus $(R^2 v_o^2/r^2)^2 = v_o^2 + 2gy$
$r^4 = R^4[1 + 2gy/v_o^2]^{-1}$

9. $A_1 v_1 = A_2 v_2$; $v_2^2 = v_1^2 + 2gh$, so $v_2^2 = 2gh/(1 - A_2^2/A_1^2)$

10. $P_B + 1/2\ \rho_f v_B^2 + \rho_f gh = P_A + \rho_f gh$, so $\Delta P = 1/2\ \rho_f v_B^2$
Also, $\Delta P = \rho_m gh$, thus $v_B = (2\rho_m gh/\rho_f)^{1/2}$

11. $P_1 + 1/2\ \rho\ v^2 + \rho\ gy + P_2 + 1/2\ \rho\ v^2$; $P_1 = P_2 = P_o$.
$v^2 = 2g\ \Delta y = 2g(1 - h)$
$\Delta y = 1/2\ gt^2$, thus $t = 0.09$ s; $R = vt$; $t = (2H/g)^{1/2}$, thus
$R = 2[h(H - h)^{1/2}$

CHAPTER 15

Exercises

1. $\cos(\theta + 5\pi/3)$, $\sin(\theta + \pi/6)$

2. (a) $(20\pi t + \pi/4) = \pi/2$, thus $t = 1/80$ s

   (b) $v = \omega A\cos(\omega t + \phi)$. Need $(20\pi t + \pi/4) = 2\pi$, so $t = 7/80$ s

   (c) $a = -\omega^2 A\sin(\omega t + \phi)$. Need $(20\pi t + \pi/4) = 3\pi/2$, so $t = 5/80$ s

3. (a) $k = \Delta F/\Delta x = 1.47\text{x}10^5$ N/m;

   (b) $T = 2\pi(m/k)^{1/2} = 0.656$ s.

4. (a) $a = -\omega^2 x = -11.5$ m/s^2

   (b) $\sin(12t + 0.2) = 0.5$, so $(12t + 0.2) = \pi/6$, $5\pi/6$
   But $v < 0$ means $\cos(12t + 0.2) < 0$, so use $5\pi/6$ which gives $t = 0.201$ s

5. $v = -1.26 \sin(3.6t - 0.5) = \underline{+}0.5(1.26)$
   Thus $(3.6t - 0.5) = (2n + 1)\pi/6$, so
   $t = 0.284$s, 0.866 s, 1.16 s, 1.74 s

6. $x = A\cos(\omega t)$, $v = -\omega A\sin(\omega t)$, $a = -\omega^2 A\cos(\omega t)$

   (a) $\sin(\omega t) = +0.5$, so $\omega t = (2n + 1)\pi/6$, $t = (2n + 1)T/12$
   Find $\cos(\omega t) = +0.866$, thus $x = +0.866A$

   (b) $x = +A/2$. Since $\cos(\omega t) = +0.5$, $\omega t = \pi(1, 2, 4, 5)/3$ or $t = T(1, 2, 4, 5)/6$

7. (a) $x = A \sin(10t + \phi)$, $v = 10A \cos(10t + \phi)$: $A = 0.206$ m.
   $(1 + \phi)$ is in the 4th quadrant: $(1 + \phi) = -1.33$,
   so $\phi = -2.33$ rad

   (b) $x = 0.206 \sin(10t - 2.33)$ m

   (c) $(10t + \phi)$ is in the 2nd quadrant: $\sin(10t + \phi) = 0.971$,
   $(10t - 2.33) = 1.81$ so $t = 0.414$ s.

8. $x = A\sin(4t + \phi)$, so $v = 4A\cos(4t + \phi)$, $a = -16A\sin(4t + \phi)$
   At $t = 0.15$ s,

   $$-0.174 = 4A\cos(0.6 + \phi), \qquad 0.877 = -16A\sin(0.6 + \phi)$$

   Find $\tan(0.6 + \phi) = 1.26$ and argument must be in 3rd quadrant. With $(0.6 + \phi) = 4.04$, $\phi = 3.44$ rad.

Then $A = 0.07$ m. $x = 0.07\sin(4t + 3.44)$ m

9. (a) $k = mg/x_o = 30.6$ N/m; $\omega = (k/m)^{1/2} = 7.83$ rad/s.
Since $x' = A$ at $t = 0$, $\phi = \pi/2$, so displacement from equilbrium
$x' = 0.08\sin(7.83t + \pi/2)$ m $= 0.08\cos(7.83t)$ m.
(b) If $x = 0.1$ m, then $x' = -0.06$ m, so $\cos(7.83t) = -0.75$
and $|\sin(7.83t)| = 0.661$, so $|v| = \omega A\sin(\omega t) = 0.414$ m/s.
$a = -\omega^2 x = +3.68$ m/s^2.

10. $f_1/f_2 = (m_2/m_1)^{1/2}$, so $1.2/0.9 = [(m + 50)/m]^{1/2}$
Thus $m = 64.3$ g. $f_1 = (k/m)^{1/2}/2\pi$, so $k = 3.66$ N/m.

11. $\omega = (k/m)^{1/2} = 6.83$ rad/s. $x = A\sin(\omega t + \phi)$;
$v = \omega A\cos(\omega t + \phi)$, and $a = -\omega^2 x$.
(a) $\sin(\omega t + \phi) = -1/3$, so $\cos(\omega t + \phi) = \pm 0.943$
Thus $v = \pm 0.773$ m/s and $a = +1.87$ m/s^2.
(b) $\sin(\omega t + \phi) = +2/3$. $\cos(\omega t + \phi) = 0.745$
Thus $v = \pm 0.611$ m/s; and $a = -3.73$ m/s^2

12. (a) $F = k_1x_1 = k_2x_2 = (x_1 + x_2)k_{eff}$; $F/k_{eff} = F/k_1 + F/k_2$, so
$T = 2\pi(m/k_{eff})^{1/2} = 2\pi[m(k_1 + k_2)/k_1k_2]^{1/2}$.
(b) & (c): $F = k_1(x + x_{o1}) - k_2(x_{o2} - x) = (k_1 + k_2)x = k_{eff}x$

13. $x = R\cos(\omega t)$ and $y = R\sin(\omega t)$. Both indicate SHM.

14. $\omega = (k/m)^{1/2} = 25.3$ rad/s, $A = 0.2$ m
(a) $K = 1/2\ mv^2 = 1/2\ m[-\omega A\sin(2\pi/5)]^2 = 0.579$ J
$U = 1/2\ kx^2 = 1/2\ k[A\cos(2\pi/5)]^2 = 6.11\text{x}10^{-2}$ J
(b) $\cos^2(\omega t) = 1/4$, $\sin^2(\omega t) = 3/4$
$K = 1/2\ m(\omega A)^2\sin^2(\omega t) = 480$ mJ
$U = 1/2\ kA^2\cos^2(\omega t) = 160$ mJ
(c) $K = U$ when $\omega t = (2n + 1)\pi/4$, thus $t = (2n + 1)T/8$
where $T = 2\pi/\omega = 248$ ms.

15. If $K = U/2$, then $U = 2E/3$ thus $x = \pm(2/3)^{1/2}A$.
$\text{Sin}(27.4t) = +(2/3)^{1/2}$, $t = 34.8$ ms, 79.8 ms,
$\text{Sin}(27.4t) = -(2/3)^{1/2}$, $t = 150$ ms, 194 ms

16. (a) $v_{max} = 2\pi fA = 314$ m/s. $E = K_{max} = 1/2\ mv_{max}^2 = 4.93\text{x}10^{-22}$ J
(b) 314 m/s; (c) $a = \omega^2 A = (2\pi f)^2 A = 1.97\text{x}10^{15}\ \text{m/s}^2$
(d) $k = (2\pi f)^2 m = 0.395$ N/m

17. (a) $m = k/\omega^2 = 0.75$ kg
(b) $E = 1/2\ kA^2 = 0.24$ J;
(c) $U = 0.5kx^2 = 2E/3 = 0.16$ J, thus $x = 0.163$ m,
so $(4t - 0.8) = \pm 0.618$, which gives $t = 45.5$ ms.
(d) $a = -\omega^2 x = -2.95\ \text{m/s}^2$

18. (a) A is unchanged; (b) E is unchanged; (d) ϕ is unchanged
(c) $T_2/T_1 = (m_2/m_1)^{1/2} = (2/3)^{1/2}$;

19. (a) $E_i = 0$; $E_f = 1/2\ kx^2 - mgx$, thus $x = 2mg/k = 0.245$ m
(b) $\Delta t = T/2 = \pi(m/k)^{1/2} = 0.351$ s

20. (a) $x = 0.08\cos(6.98t) = -0.04$ m
(b) $\cos(\omega t) = -5/8$, thus $\sin(\omega t) = +0.7806$
$v = -A\omega\sin(\omega t) = -0.436$ m/s
(c) $a = -\omega^2 x = +2.44\ \text{m/s}^2$
(d) $E = 1/2\ m(v_{max})^2 = 1/2\ m(\omega A)^2 = 9.35$ mJ

21. $1/2\ kx^2 + 1/2\ mv^2 = 1/2\ kA^2$; thus $v^2 = (k/m)(A^2 - x^2)$

22. (a) $\omega = (g/L)^{1/2} = 3.5$ rad/s. Given $\theta = 0.15$ rad and
$v = Ld\theta/dt = +60$ cm/s, thus $d\theta/dt = v/L = 0.75$.
$\theta = \theta_0\sin(\omega t + \phi)$; $d\theta/dt = \omega\theta_0\cos(\omega t + \phi)$;
At $t = 0$, $\tan\phi = (0.15)/(0.75/3.5) = 0.7$, so $\phi = 0.61$ rad.
At $t = 0$, $\sin\phi = \theta/\theta_0$, so $\theta_0 = 0.262$ rad.
(b) $E = 1/2\ mv^2 + mgL(1 - \cos\theta) = 10.7$ mJ
(c) $y = L(1 - \cos\theta_0) = 2.73$ cm

23. $T = 2\pi(I/mgd)^{1/2}$.
(a) $I = ML^2/12 + M(L/2)^2 = 0.333M$, thus $T = 1.64$ s;
(b) $I = ML^2/12 + M(0.1)^2 = 0.0933M$, thus $T = 1.94$ s.

24. $T = 2\pi(I/MgR)^{1/2} = 2\pi(3R/2g)^{1/2}$

25. $\omega = (\kappa/I)^{1/2} = 126$ rad/s, thus $f = \omega/2\pi = 20.1$ Hz.

26. $T = 2\pi(I/\kappa)1/2 = 0.181$ s

27. (a) $T = 2\pi(L/g)^{1/2} = 1.27$ s;
(b) $(d\theta/dt)_{max} = \omega\theta_o$ and $v_{max} = L(\omega\theta_o) = 0.69$ m/s
(c) $E = 1/2\ mv_{max}{}^2 = 11.9$ mJ

28. $T_2/T_1 = (I_2/I_1)^{1/2}$, thus $4/3 = [(0.5 + I)^{1/2}/0.5]$
Find $I = 0.389\ kg.m^2$

29. $T\ \alpha\ (I)^{1/2}\ \alpha\ L(m)^{1/2}$. Thus,
$T^2/T^1 = (L_2/L_1)(m_2/m_1)^{1/2} = 1/(2)^{1/2}$, thus $T_2 = 0.636$ s.

30. (a) $L = g/\pi^2 = 0.993$ m;
(b) $T_2/T_1 = (g_1/g_2)^{1/2}$, thus $T_2 = 2(6)^{1/2} = 4.90$ s

31. (a) $T = 2\pi(L/g)^{1/2} = 1.80$ s; $\omega = 3.5$ rad/s.
(b) $\theta = (\pi/6)\cos(3.5t)$
(c) $E = 1/2\ mv_{max}{}^2$ where $v_{max} = L(\omega\theta^o)$, so $E = 21$ mJ
(d) At $\theta = 15^o$, $\cos(\omega t) = 0.5$, $\sin(\omega t) = 0.866$.
Thus $v = L\ d\theta/dt = -L\omega\theta_o \sin(\omega t) = 1.27$ m/s

32. Say $\theta = \theta_o\sin(\omega t)$, where $\omega = \pi$ rad/s.
Given $+0.5 = \sin(\pi t)$, thus $t = +1/6$ s. The time to go from $-\theta_o/2$ to $+\theta_o/2$ is 1/3 s.

33. (a) $T = 0.8$ s, so $\omega = 7.85$ rad/s. (b) $F_{max} = ma_{max} = m\omega^2 A =$ 1.11 N; (c) $\omega A = 0.942$ m/s.

34. (a) $T = 2\pi/5\pi = 0.4$ s; (b) 0.25 m; (c) $\pi/4$ rad;
(d) $\omega A = 3.93$ m/s; (e) $\omega^2 A = 61.7\ m/s^2$

35. (a) -0.177 m; (b) $\omega A\cos(5\pi/4) = -2.78$ m/s;
(c) $a = -\omega^2 x = 43.7\ m/s^2$

36. (a) 7.67×10^{-2} m; (b) $v = 1.28\sin(0.5\ rad) = 1.12$ m/s;
(c) $a = -\omega^2 x = -4.91\ m/s^2$.

37. (a) $a_m = (2\pi f)^2 A$, thus $A = 7.04 \times 10^{-2}$ m. $d = 4A = 0.282$ m.
(b) $\omega A = 0.531$ m/s.

38. (a) $\omega A = 1.25$ and $\omega^2 A = 9$, so $\omega = 7.2$ rad/s and $A = 0.174$ m

(b) With $k = m\omega^2$, then $0.5kA^2 = 0.5kx^2 + 0.5mv^2$ gives $v = 0.907$ m/s

39. (a) A = 12 cm; $\omega = 2\pi/1.2 = 5.24$ rad/s.
(b) $v_m = \omega A = 0.628$ m/s; (c) $a_m = \omega^2 A = 3.29$ m/s^2

40. (a) $\omega A = 15$, $\omega^2 A = 90$, so $T = 2\pi/6 = 1.05$ s;
(b) A = 2.5 cm.

41. (a) $\omega A = (2\pi f)A = 0.314$ m/s; (b) $\omega^2 A = 98.7$ m/s^2.

42. (a) $\omega A = (2\pi f)A = 2.21$ m/s; (b) $\omega^2 A = 6110$ m/s^2.

43. (a) $\omega^2 A = 28$, so A = 0.113 m. At t = 0, $\omega^2 A = -\omega^2 A\sin(\phi)$, so $\sin\phi = -1$ and $\phi = 3\pi/2$ rad. (b) $x = 0.113\sin(15.7t + 3\pi/2)$ m

44. (a) $1 = \cos(5.15t)$, $5.15t = 2\pi$, t = 1.22 s.
(b) $-1 = \sin(5.15t)$, $5.15t = 3\pi/2$, t = 0.915 s.
(c) $-1 = \cos(5.15t)$, $5.15t = \pi$, t = 0.61 s.

45. (a) No change; (b) 3.54 s, (c) 1.77 s

46. Find k = 1.53 N/m, then $T = 2\pi(m/k)^{1/2} = 1.02$ s.

47. $v_{max} = \omega A$, so $\omega = 3.5$ rad/s. Then $m = k/\omega^2 = 0.2$ kg.

48. (a) $T = 0.7 = 2\pi(m/k)^{1/2}$, thus k = 15.3 N/m.
(b) $\omega A = 2.02$ m/s; (c) $\omega^2 A = 18.1$ m/s^2.

49. k = 35 N/m, then $m = k/\omega^2 = 0.307$ kg.

50. (a) $v_m = \omega A$, so A = 3.30 cm; (b) $k = m\omega^2 = 1.55$ N/m;
(c) 4A/T = 18.5 cm/s.

51. $k = m\omega^2 = 3.16$ N/m. $E = 0.5mv^2 + 0.5kx^2 = 3.62$ mJ and $E = 0.5kA^2$, thus A = 4.79 cm.

52. (a) $E = 0.5kA^2 = 0.4$ J, thus k = 20 N/m;
(b) $m = k/\omega^2 = 0.324$ kg.

53. (a) $E = 0.5m\omega^2 A^2 = 15.8$ mJ; (b) $\omega A = 0.628$ m/s;
(c) $E = 0.5mv^2 + 0.5kx^2 = 15.8$ mJ, so v = 0.544 m/s.

54. (a) $E = 0.5kA^2 = 0.2$, so A = 10 cm; (b) $E = 0.5mv_m{}^2 = 0.2$,

v_m = 1.83 m/s; (c) $E = 0.5mv^2 + 0.5kx^2 = 0.2$, so x = 7.02 cm; (d) $kA/m = 33.3\ m/s^2$.

55. (a) $E = 0.5m\omega^2A^2 = 0.344$ J, thus A = 0.21 m;
(b) $\omega A = 2.93$ m/s;
(c) $E = 0.5mv^2 + 0.5kx^2$, then v = 2.58 m/s.

56. (a) $E = 0.5(m\omega^2)A^2 = 0.249$ J; $E = 0.5mv^2 + 0.5kx^2$, so v = 2.71 m/s;
(b) $E = 0.5mv^2 + 0.5kx^2$ gives x = 0.205 m; (c) 0.249 J

57. (a) $0.5m\omega^2A^2 = 0.22$ J, so m = 93.0 g; (b) $\omega A = 2.18$ m/s;
(c) K = E - U = 0.185 J; (d) U = E - K = 0.153 J.

58. (a) $E = 0.5mv_m{}^2 = 0.18$, thus m = 0.230 kg;
(b) $E = 0.5kA^2$, so k = 18.4 N/m;
(c) $\omega = v_m/A = 8.93$ rad/s, so f = 1.42 Hz;
(d) $E = 0.5mv^2 + 0.5kx^2$, thus v = 1.08 m/s.

59. (a) $a = -\omega^2x$, so $\omega = 13.7$ rad/s;
(b) $k = m\omega^2$, so m = 0.192 kg;
(c) $E = 0.5mv^2 + 0.5kx^2 = 46.1$ mJ

60. (a) $E = 0.5mv^2 + 0.5kx^2 = 12.1$ mJ;
(b) $E = 0.5kA^2$, so A = 6.95 cm
(c) $E = 0.5mv_m{}^2$, so $v_m = 63.4$ cm/s.

61. (a) From $a_m = \omega^2A$, T = 0.640 s; (b) $E = 0.5m\omega^2A^2 = 23.1$ mJ.

62. $g = 4\pi^2\ell/T^2 = 9.80\ m/s^2$

63. $\ell = gT^2/4\pi^2 = 0.993$ m

64. From $T = 2\pi(I/mgd)^{1/2}$, find $I = 2.41 \times 10^{-3}\ kg.m^2$

65. $T = 2\pi(I/mgd)^{1/2} = 2\pi(2L/3g)^{1/2}$, so $\ell = 2L/3 = 0.8$ m.

66. Given A = 15 cm and $\omega = 2\pi/T = \pi$ rad/s. At t = 0, $x = A\sin\phi = 0$ so $\phi = 0, \pi$. But $v = \omega A\cos(\phi) < 0$, so $\phi = \pi$. So $x = 0.15\sin(\pi t + \pi)$ m

67. (a) A = 0.1 m and $\omega^2A = 3.6\ m/s^2$ so $\omega = 6$ rad/s. At t = 0,

$x = +A$, thus $\sin\phi = 1$ and $\phi = \pi/2$. So $x = 0.1\sin(6t + \pi/2)$ m
(b) $x = 0.1\sin(6t + \pi/2) = 0$, so $(6t + \pi/2) = \pi$, $t = 0.262$ s

68. (a) $\omega^2 A = 8.5\ m/s^2$, so $\omega = 5$ rad/s. Since $x = +A$ at $t = 0$, $\phi = \pi/2$. Finally $x = 0.34\sin(5t + \pi/2)$ m. (b) $\omega A = 1.70$ m/s (c) From $\cos(5t + \pi/2) = 1$, find $t = 0.942$ s.

69. From $0.15 = A\sin\phi$ and $1.3 = 6A\cos(\phi)$ find $A = 0.264$ m and $\phi = 0.606$ rad.

70. (a) $a = -\omega^2 x$, so $\omega = 4.5$ rad/s. (b) Note $x^2 + v^2/\omega^2 = A^2$, then $A = 2.68$ cm.

71. From $x = A$ at $t = 0$, find $x = A\cos(10.3t)$, and $v = -\omega A\sin(10.3t)$. Then v at $t = 0.05$ s leads to $A = 0.19$ m.

72. (a) $0.5kA^2 = 0.5mv^2 = 9$ mJ, thus $A = 4.9$ cm.
(b) Time in contact is $T/2 = \pi(m/k)^{1/2} = 0.257$ s.

73. Use $x = A\sin\phi$ and $v = \omega A\cos\phi$ to find (a) $\pi/2$; (b) $3\pi/2$; (c) π; (d) $\pi/6$; (e) $0.5 = \sin\phi$ and $\cos\phi < 0$, so $\phi = 5\pi/6$.

74. $\omega = (k/m)^{1/2} = 11.2$ rad/s. Since $x = A$ at $t = 0$, find $\phi = \pi/2$ and so $x = 0.05\sin(11.2t + \pi/2)$ m.

75. (a) $F_1 = kx_1$ and $(F_1 + 0.49) = k(x_1 + 0.38)$ or $k = \Delta F/\Delta x$ give $k = 1.29$ N/m. (b) $m_1 = kT^2/4\pi^2 = 0.021$ kg $= 21$ g.

76. Since $x = -A$ at $t = 0$, $\phi = 3\pi/2$. Know $\omega A = 90$ cm/s, so $\omega = 6$ rad/s. Then $x = 0.15\sin(6t + 3\pi/2)$ m.

77. (a) $a = -\omega^2 x = -1.97\ m/s^2$; (b) $0.05 = 0.08\sin(2\pi t)$ means $\cos(2\pi t) = \pm\ 0.7806$, so $v = \omega A\cos(2\pi t) = \pm\ 0.392$ m/s.

78. (a) $x = 15\sin(4.33t)$ cm. (b) $0.2 = \sin(\omega t_1)$, so $t_1 = 46.5$ ms and $0.8 = \sin(\omega t_2)$, so $t_2 = 214.2$ ms. Thus, $\Delta t = 168$ ms.

79. $T_2/T_1 = (\ell_2/\ell_1)^{1/2} = 1.125$. Then, $nT_2 = (n + 1)T_1$, so $n = 8$ and the time is $nT_2 = 14.5$ s.

80. $(2\pi f)^2 = Mgd/I$, where $I = ML^2/12 + Md^2$. Then $7.643d^2 - 9.8d + 0.6369 = 0$ leads to $d = 6.87$ cm.

81. $T = 2\pi(I/Mgd)$ where $I = 0.5MR^2 + Md^2$. Find $T = 1.15$ s.

82. $I = m(L/4)^2 + m(3L/4)^2 = 5mL^2/8$.
$T = 2\pi(I/2mgd)^{1/2} = 2\pi(5L/4g)^{1/2} = 2.03$ s.

83. From Energy Cons. $v_m{}^2 = 2gL(1 - \cos\theta_o)$ so $L = 0.389$ m, then $T = 2\pi(L/g)^{1/2} = 1.25$ s.

84. (a) From Energy Cons. $v_m{}^2 = 2gL(1 - \cos\theta_o)$, so $\theta_o = 0.353$ rad.
(b) $\theta = 0.353\sin(3.74t) = 0.2$, so $t = 0.161$ s.

85. (a) $\cos(4.7t - 0.23) = 1$, so $(4.7t - 0.23) = 0$, $t = 48.9$ ms.
(b) $\sin(4.7t - 0.23) = -1$, so $(4.7t - 0.23) = 3\pi/2$, $t = 1.05$ s.

86. (a) $x = A\cos(\omega t)$, so $v = -\omega A\sin(\omega t)$ where $\omega = 20$ rad/s. Use given v and t to find $A = 0.041$ m;
(b) $E = 0.5m\omega^2A^2 = 20.2$ mJ

87. (a) If $v = 0.5v_m$, then $K = 0.25K_m = 0.25E$ and $U = 0.75E$, thus $0.5kx^2 = 0.75(0.5kA^2)$ and $x = 17.3$ cm.
(b) If $K = U$, then $U = 0.5E$ or $x^2 = 0.5A^2$ and $x = 14.1$ cm.

Problems

1. $\omega = 8$ rad/s. $E = K_{max} = 1\ J = 1/2\ kA^2$, thus $A = 1/4$ m
$x = 0.25\sin(8t)$ m.

2. When $a = \omega^2A = g$, thus $f = (98)^{1/2}/2\pi = 1.58$ Hz

3. $\Delta E = 0 = 1/2\ mv^2 + 1/2\ I\omega^2 + 1/2\ (y + y_o) - mgy$,
$y_o = mg/k$. Set $dE/dt = 0$ to find $d^2y/dt^2 + 2ky/(M + 2m) = 0$

4. $\omega^2 = k/(M + m)$. From ΣF_x and ΣF_y find $a = \mu g$ for m.
$a_{max} = \omega^2A = \mu g$, thus $\mu = 0.136$

5. Identical to simple pendulum. Could also use $\tau = I\alpha$:
$-(mg\sin\theta)R = (MR^2)(d^2\theta/dt^2)$ leads to $\omega = (g/R)^{1/2}$.

6. (a) $F = ma$: $-2xA\rho g = \ell A\rho d^2x/dt^2$, thus
$d^2x/dt^2 + (2g/\ell)x = 0$, i.e. simple harmonic.
(b) $T = 2\pi/\omega = 2\pi(\ell/2g)^{1/2}$

7. $(m/F_o)dA/d\omega = -1/2[\ ]^{-3/2}\ [2(\omega_o^2 - \omega^2)(-2\omega) + 2\gamma^2\omega/m^2] = 0$
Find $\omega^2 = \omega_o^2 - \gamma^2/2m^2$

8. $F_B = \rho_f gAy$; Mass of block $= Ah\rho_B$. Let y be downward displacement, then F = ma: $-\rho_f gAy = (Ah\rho_B)\ d^2y/dt^2$
Thus, $\omega^2 = \rho_f g/\rho_B h$

9. (a) $dK = 1/2\ (m\ dx/L)(xv/L)^2$,
thus, $K = \int_0^L dK = mv^2/6.$

(b) $E = (M/2 + m/6)v^2 + 1/2\ kx^2$, set $dE/dt = 0$ to find $d^2x/dt^2 + (k/(M + m/3))\ x = 0$, thus $T = 2\pi[(M + m/3)/k]^{1/2}$

10. (a) N.m/rad; $[\kappa] = ML^2T^{-2}$
(b) $T = I^x\kappa^y$, thus $T = (ML^2)^x\ (ML^2T^{-2})^y$
$1 = -2y$; $0 = x + y$; $0 = 2x + 2y$. Thus $y = -1/2$ and $x = +1/2$
$T\ \alpha\ (I/\kappa)^{1/2}$

11. (a) 1.00430 s; (b) 1.01738 s; (c) 1.03963 s; (d) 1.07129 s

12. (a) $E = 1/2\ mv^2 + 1/2\ kx^2 - mgx$
(b) $dE/dt = 0$ leads to $md^2x/dt^2 + kx - mg = 0$
Let $x' = x - x_o = x - mg/k$, then
$d^2x'/dt^2 + (k/m)x' = 0$, where $\omega^2 = k/m$.

13. $F_x = -(mgr/R)\sin\theta$ where $\sin\theta = x/r$. From $F_x = ma$, find $\omega = (g/R)^{1/2}$ and $T = 2\pi/\omega \approx 84.4$ min.

14. $\tau = I\alpha$: $-(kx)L = I\ d^2\theta/dt^2$. Note that $x = L\theta$ and $I = ML^2/3$
Find $d^2\theta/dt^2 + 3k\theta/M = 0$, thus $T = 2\pi(M/3k)^{1/2}$

CHAPTER 16

Exercises

1. $\lambda = c/f$: (a) 188 m to 546 m; (b) 2.78 m to 3.41 m

2. $\omega = 10\pi/9$ rad/s, At the rim $v = \omega r = \pi/6$ m/s.
$f = v/\lambda = (\pi/6 \text{ m/s})/(1.2\text{x}10^{-3} \text{ m}) = 436$ Hz.

3. (a) $\lambda = 4$ cm, thus $f = v/\lambda = 10$ Hz;
(b) $\Delta\phi = (\Delta x/\Delta\lambda)2\pi = 5\pi/4$;
(c) $T = 1/10$ s, $\Delta t = (60/360)T = 1/60$ s;
(d) $-(\partial y/\partial t)_{max} = -\omega A = -1.26$ m/s

4. (a) $\omega = 10\pi/9$ rad/s. At the rim $v = \omega r = 10\pi(0.145)/9 = 0.506$ m/s. Signal $T = 10^{-4}$ s. In this time $\Delta s = vT = 5.06\text{x}10^{-5}$ m; (b) $\lambda = v_s T = 3.4$ cm

5. $\Delta t = \Delta x(1/v_s - 1/v_p)$, thus $\Delta x = 1440$ km

6. $F = v^2\mu = 13.3$ N

7. $m = FL/v^2 = 0.563$ kg

8. $F_2/F_1 = (v_2/v_1)^2$, thus $F_2 = 15(45/28)^2 = 38.7$ N

9. (a) $\lambda \propto (F)^{1/2}/f$, thus $f_2 = (2)^{1/2}f_1$;
(b) $v \propto (F)^{1/2}$, so $v_2 = (2)^{1/2}v_1$

10. (b) It takes $\Delta t = 0.5$ s to reach the peak, thus,
$\partial y/\partial t = (1 \text{ cm})/(0.5 \text{ s}) = 2$ cm/s, up

11. (b) It takes $\Delta t = 1$ s to go from the peak to zero, thus,
$\partial y/\partial t = 1$ cm/s, down

13. $f(x + vt) = (2 \text{ x } 10^{-3})/[4 - (x + 12t)^2]$ m

14. $v = 1.2 \text{ cm}/0.3 \text{ s} = 4$ cm/s, $\omega = 2\pi v/\lambda = 8\pi/7.5$, $\phi = \pi/2$.
$y = A \sin(kx - \omega t + \phi) = 2 \cos(2\pi x/7.5 - 8\pi t/7.5)$ cm

15. (a) $\partial y/\partial x = kA \cos(kx - \omega t)$;
(b) $(\partial y/\partial t)_{max} = v\ (\partial y/\partial x)_{max}$

16. (a) 2.5 m/s; (b) $(\Delta\phi/2\pi)\lambda = \lambda/3 = 10\pi/3 = 10.5$ cm

17. (a) $(\partial y/\partial t)_{max} = \omega A = (2\pi)(2.4) = 4.8\pi$ cm/s = 15.1 cm/s;
(b) $\partial y/\partial t = 4.8\pi \sin(0.75\pi/20 - \pi/2) = -15.0$ cm/s;
(c) $a_{max} = \omega^2 A = 94.7$ cm/s^2;
(d) $a = -\omega^2 A \cos(0.75\pi/20 - \pi/2) = -11.1$ cm/s^2

18. (a) $f = \omega/2\pi = 1.91$ Hz; (b) $v = \omega/k = 5$ cm/s;
(c) $A = 0.03$ cm;
(d) $\partial y/\partial t = +\omega A \sin[(2.4)(15) - (12)(0.2) + 0.1]$
$= 0.272$ cm/s;
(e) $a_{max} = \omega^2 A = 4.32$ cm/s^2

19. a, b, d, e

20. (a) $\Delta\phi = 2\pi(\Delta x/\lambda) = 2.51$ rad;
(b) $\Delta\phi = 2\pi(\Delta t/T) = 11$ rad (or $11 - 2\pi = 4.72$ rad)

21. (a) $\lambda = 2\pi/k = 15.7$ cm, $\phi = 0.8$ rad;
(b) $T = 2\pi/\omega = 0.126$ s;
(c) $A = 0.02$ cm;
(d) $v = \omega/k = -125$ cm/s;
(e) $\partial y/\partial t = \omega A\cos(0.4 + 25 + 0.8) = 0.483$ cm/s

22. (a) $\lambda = 2\pi/k = 10\pi = 31.4$ m
(b) $T = 2\pi/\omega = \pi = 3.14$ s
(c) $v = \omega/k = 10$ m/s

23. $y = 0.03 \sin(2\pi x/0.025 + 2\pi t/0.01 + \phi)$
$= 0.03 \sin(80\pi x + 200\pi t + \phi)$
$y(0, 0) = -0.02$ m $= 0.03\sin\phi$, so $\phi = -0.730$ rad, -2.301 rad.
Since $v > 0$, it means $\cos\phi > 0$, so $\phi = -0.730$ rad.
$y = 0.03 \sin[80\pi x + 200\pi t - 0.730]$ m

24. With given data: $y = 0.05\sin(0.1x + 5t + \phi)$ m.
$y(0, 2) = 0.0125$ leads to $\sin(10 + \phi) = 0.25$, thus
$(10 + \phi) = 0.253$ rad, 2.89 rad. Since $\partial y/\partial t < 0$,
$\cos(10 + \phi) < 0$, so $(10 + \phi) = 2.89$ and $\phi = -7.11$ rad.

25. (a) $A \sin[(2\pi/\lambda)(x - vt)]$; (b) $A \sin[2\pi f(x/v - t)]$
(c) $A \sin[k(x - vt)]$; (d) $A \sin[2\pi(x/\lambda - ft)]$

26. (a) $\lambda = 2\pi/k = 12.6$ cm, $f = \omega/2\pi = 4.77$ Hz,
$v = \omega/k = 60$ cm/s
(b) $\partial y/\partial t = -120\sin(0.5x)\sin(30t) = 101$ cm/s

27. (a) $\lambda = 40/8 = 5$ cm, $\omega = 2\pi f = 16\pi$ Hz
$y = 4\sin(0.4\pi x)\cos(16\pi t)$ cm;
(b) $\lambda/2 = 2.5$ cm;
(c) $y = 4\sin(0.2\pi) = 2.35$ cm

28. $f_2 = v/L = (F/\mu)^{1/2}/L$, thus $F = f_2{}^2L^2\mu = 109$ N

29. (a) $\lambda_n = 2L/n$ thus $2L = 36n = 32(n + 1)$, so $n = 8$.
$L = n\lambda_n/2 = 8(36)/2 = 144$ cm;
(b) $f_1 = (1/2L)(F/\mu)^{1/2} = 1/(2.88\text{ m})\ (10^4/4)^{1/2} = 17.4$ Hz

30. (a) $f_{n+1} - f_n = f_1 = 120$ Hz.
(b) $v = (F/\mu)^{1/2} = 67.9$ m/s. $L = v/2f_1 = 679/240 = 0.283$ m

31. (a) $L_1 = \lambda/2 = v/2f = (F/\mu)^{1/2}/2f = 2.30$ cm;
(b) $L_3 = 3L_1 = 6.89$ cm

32. $v = (F/\mu)^{1/2} = 70.71$ m/s, $\lambda = 0.24$ m, thus $f = 295$ Hz
$y = A\sin(2\pi x/\lambda)\cos(2\pi ft) = (2\text{x}10^{-3})\sin(26.2x)\cos(1850t)$ m

33. $f_2/f_1 = (\mu_1/\mu_2)^{1/2}$. For given length: $\mu L = \pi r^2 L\rho$, thus
$\mu \alpha r^2\rho$.
$f_2/f_1 = (r_1/r_2)(\rho_1/\rho_2)^{1/2}$, thus $f_2 = (2)^{1/2}f_1$

34. $v_1 = (2)^{1/2}v_2$, $\lambda_1 = \lambda_2/3$, thus $f_1 = 3(2)^{1/2}f_2 = 4.24f_2$

35. (a) $\lambda = 2\pi/k = 20.9$ cm, $v = \omega/k = 83.3$ cm/s;
(b) $L = 3\lambda/2 = 31.4$ cm;
(c) $L/3 = 10.5$ cm, $2L/3 = 20.9$ cm

36. $\lambda_2 = 2\lambda_1/3$, so $f_2 = 1.5f_1 = 480$ Hz

37. $P = 1/2\ \mu(\omega A)^2 v = 15$ W

38. $A^2 = 2P/\mu v\omega^2 = 0.022$ m

39. (a) $P = 1/2\ \mu(\omega A)^2 v = 6.63$ mW;
(b) Must double to $v = 120$ m/s, thus $F = \mu v^2 = 50.4$ N

40. Both, $f(x - vt)$

41. $\partial^2y/\partial t^2 = -\omega^2 A\sin(kx)\cos(\omega t)$
$\partial^2y/\partial x^2 = -k^2 A\sin(kx)\cos(\omega t)$
Thus, $\partial^2y/\partial x^2 = 1/v^2\ \partial^2y/\partial t^2$

42. $F_2/F_1 = (v_2/v_1)^2$, thus $F_2 = 3.96$ N

46. $f = 20$ Hz and $v = 120$ cm/s, so $\lambda = v/f = 6$ cm.

47. (a) $\lambda = 0.04$ m, $T = 0.05$ s; (b) $v = -\lambda/T = -0.8$ m/s.
(c) Max. $\partial y/\partial t = \omega A = (2\pi/0.05)(0.2\text{x}10^{-4}) = 2.51$ mm/s.

48. $y = A\sin(kx + \omega t + \phi) = 0.006\sin(2\pi x/0.03 + 80\pi t + \pi/2)$ m

49. (a) $v = 7.5$ m/s, then $\mu = F/v^2 = 3.2$ g/m;
(b) Max. $v_p = \omega A = 0.648$ m/s.

50. (a) $\lambda = 4$ m, $v = \omega/k = 0.5$ m/s.

51. (a) $f_2 = 2v/2L = 480$ m/s; (b) $\mu = m/L = F/v^2$, so $m = 0.91$ g.

52. (a) $f_3 = 3v/2L = 1050$ Hz, $\lambda = v/f_3 = 0.4$ m.
(b) $y = 2\text{x}10^{-3}\sin(15.7x)\cos(6600t)$ m
(c) $y = 10^{-3}\sin(15.7x \pm 6600t)$ m

53. $f \propto v$, so $f_2/f_1 = (\mu_1/\mu_2)^{1/2}$, thus $f_2 = 180/2^{1/2} = 127$ Hz.

54. (a) $f = v/2L$, $v = 2fL = 132$ m/s; (b) $F = \mu v^2 = 52.3$ N.

55. $f = v/2L$, so $f_2/f_1 = L_1/L_2$ and $L_2 = 0.450$ m, or
0.15 m from neck.

56. $v = (FL/m)^{1/2} = 80$ m/s. $f = nv/2L = 16$ Hz, 32 Hz, 48 Hz.

57. $f_3 = (3/2L)(F/\mu)^{1/2}$, leads to $\mu = 1.61$ g/m.

58. $f_2/f_1 = (F_2/F_1)^{1/2}$, thus $F_2 = 287$ N.

59. $v = (F/\mu)^{1/2} = 55.3$ m/s. Thus $f_3 = 3v/2L = 59.3$ Hz.

60. (a) $\lambda = 2\pi/k = 3$ m, thus $\Delta x = 1.5$ m;
(b) $\partial y/\partial t = -\omega A\sin(kx)\sin(\omega t) = -0.190$ m/s.

61. $\mu = F/v^2 = 5 \text{ x } 10^{-3}$ kg/m, thus $P = 0.5\mu(\omega A)^2 v = 5.68$ mW.

62. $\omega = 2\pi/T = 251$ rad/s, $v = \lambda/T = 16$ m/s, so $P = 0.5\mu(\omega A)^2 v = 8.06$ mW.

63. $v = 70.7$ m/s and $P = 0.5\mu(\omega A)^2 v$, so $A = 2.76$ mm.

64. $v = 125$ m/s and $P = 0.5\mu(\omega A)^2 v = 14.4$ mW.

65. (a) $y = 5\times10^{-3}\sin(30x)\cos(420t)$ m; (b) $5\times10^{-3}\sin(30x) = -4.63\times10^{-3}$ m, so $A = 4.63\times10^{-3}$ m.
(c) Antinodes at $kx = 2\pi x/\lambda = (2n + 1)\pi/2$, where $\lambda = \pi/15$ m, so $x = 5\lambda/4 = 0.261$ m.

66. (a) $m = \rho V = \mu L$, so $\mu = \rho A$. With $S = F/A$, $v^2 = F/\mu = S/\rho$.
(b) $v = 713$ m/s.

67. $m = \rho V = \mu L$, so $\mu = \rho A$. From $v = (F/\mu)^{1/2}$, $F = v^2 \rho A = 118$ N

68. $m = \rho V = \mu L$, so $\mu = \rho A = \rho\pi r^2$. Since $v \propto (1/\mu)^{1/2} \propto 1/r$, we have $v_2/v_1 = r_1/r_2$, thus $v_2 = 60$ m/s.

69. (a) $\phi/2\pi = \Delta x/\lambda$, so $\Delta x = 0.191$ m; (b) $\phi/2\pi = \Delta t/T$, so $\phi = 3.77$ rad

70. (a) $m = \rho V = \mu L$, so $\mu = \rho\pi r^2$. Since $f = v/2L$ and $v = (F/\mu)^{1/2}$, we have $f_2/f_1 = v_2/v_1 = (\mu_1/\mu_2)^{1/2} = r_1/r_2 = 1.40$.
(b) $F_2/F_1 = \mu_2/\mu_1 = (r_2/r_1)^2 = 0.510$.

71. (a) $y = 0.06\sin(2\pi x)\cos(10\pi t)$ m. Note $k = 2\pi$ so $\lambda = 1$ m;
(b) Nodes at $kx = 0, \pi, 2\pi\ldots$ So $x = 0.5$ m, $+1.0$ m.
(c) Antinodes at $kx = (2n + 1)\pi/2$. So $x = +0.25$ m, $+0.75$ m.
(d) $0.06\sin(2\pi/8) = 4.24 \times 10^{-2}$ m.

72. $T = 24$ ms, $\lambda = vT = 0.37$ m. $y = 3 \times 10^{-3}\sin(17x \pm 262t)$ m.

73. $m = \rho V = \mu L$, so $\mu = \rho A$. Since $f = v/2L = (F/\rho A)^{1/2}/2L$, $F \propto f^2 A$. Thus $F_2/F_1 = f_2^2 A_2/f_1^2 A_1 = 3.13$.

74. $m = \rho V = \mu L$, so $\mu = \rho A$. Since $f_3 = 3v/2L = 3(F/\rho A)^{1/2}/2L$, we have $F = (2Lf_3/3)^2 \rho A = 88.9$ N.

75. $y = 2A\sin(8.4x + \pi/6)\cos(50t)$. Nodes are at $(8.4x + \pi/6) = n\pi$, thus $x = 0.312$ m, 0.686 m.

76. $f = v/2L = (F/\mu)^{1/2}/2L$. (a) $f \propto 1/L$, so $f' = 0.8f$; (b) $f \propto 1/\mu^{1/2}$, so $f' = 1.12f$; (c) $f \propto F^{1/2}$, so $f' = 1.09f$; (d) $f' = (0.8)(1.12)(1.09)f = 0.977f$.

Problems

1. Since $\mu = \rho A$, $v = [(F/A)/\rho]^{1/2} = 160$ m/s

2. (a) $f \propto F^{1/2}$, thus $df/dF = 1/2\ F^{-1/2}$, then $df/f = dF/2F$
   (b) $df = 0.5(400)(-0.03) = -6$ Hz, so $f' = 394$ Hz
   (c) $df/f = 2/260$, thus $dF/F = 4/260$ or $+1.54\%$.

3. When $a_{max} = \omega^2 A = g$, thus $A = g/\omega^2$

4. $f = v/2L$, thus $v = 2fL = 250.9$ m/s. Use $L = v/2f$: $L_A = 7$ cm, $L_B = 13.2$ cm, $L_C = 16.1$ cm, $L_D = 21.3$ cm, all from the neck.

5. (a) $P = F_y\ (\partial y/\partial t)$, where $F_y = F\tan\theta = F(\partial y/\partial x)$ (but this is negative); (b) Same as Eq. 16.16

6. $F(y) = \mu g y$, thus $v(y) = = dy/dt = (gy)^{1/2}$
   $\Delta t = 1/g^{1/2} \int y^{-1/2}\ dy = 2(L/g)^{1/2}$

7. From Sec. 16.10: $P = dE/dt = \mu(\omega A)^2 \cos^2(kx - \omega t)\ dx/dt$. $y = A\sin(kx - \omega t)$. When $y = 0$, $\sin(kx - \omega t) = 0$, so $\cos(kx - \omega t) = +1$. When $\sin(kx - \omega t) = +1$, $\cos(kx - \omega t) = 0$.

8. $x = 0$ is a node, $x = L$ is an antinode. Thus $L = \lambda/4, 3\lambda/4, 5\lambda/4$ ...i.e. $f = v/\lambda = v/4L, 3v/4L, 5v/4L$ ..
   $f = (2n-1)/4L\ (F/\mu)^{1/2}$

9. $v = (F/\mu)^{1/2} = (F/A\rho)^{1/2}$ but $F/A = Y\ \Delta L/L$
   thus $v = (Y\Delta L/\rho L)^{1/2}$

10. Take $y = 2A\sin(kx)\cos(\omega t)$. See also Sec.16.10
    $dK/dx = 1/2\ \mu\ (\partial y/\partial t)^2 = 1/2\ \mu[2A\omega\sin(kx)\sin(\omega t)]^2$
    Time average of $\sin^2(\omega t) = 1/2$, thus
    $(dK/dx)_{av} = \mu(\omega A)^2\sin^2(kx)$
    $dU/dx = 1/2\ F(\partial y/\partial x)^2$, then $(dU/dx)_{av} = F(kA)^2\cos^2(kx)$

11. (a) Amplitude of the standing wave is 2A. Since E = K + U, $Fk^2 = \mu v^2k^2 = \mu\omega^2$, and $\sin^2(kx) + \cos^2(kx) = 1$, so $(dE/dx)_{av} = \mu(\omega A)^2$. In a length L, $E = \mu(\omega A)^2L$
(b) For one loop $L = \lambda/2$, so $E = \mu(2\pi fA)^2\lambda/2 = 2\pi^2\mu A^2f^2\lambda = 2\pi^2\mu A^2fv$.

12. Given $v = (F/\mu)^{1/2} = 245$ m/s and $\lambda = 2L$, so $f = v/\lambda = 204$ Hz. From Problem 11: $E = \mu(\omega A)^2L = (2x10^{-3})(408\pi x10^{-3})^2(0.6) = 1.972$ mJ
If the amplitude drops to 0.9A, then the energy drops to 0.81E. Half of the loss, 0.19E/2 = 0.095E, is radiated as sound in 0.1 s, thus: Radiated power = $0.095E/\Delta t = 1.87$ mW

13. From Sec. 16.10 $\eta = dE/dx = 1/2\ \mu(\omega A)^2$, thus $P = \eta v$

14. (a) $md^2s_n/dt^2 = -k(s_n - s_{n-1}) + k(s_{n+1} - s_n)$
$= k(s_{n+1} + s_{n-1} - 2s_n)$
Substitute function for s_n:
$-m\omega^2A\sin(kna - \omega t) = kA\sin[k(n+1)a - \omega t] +$
$+ kA\sin[k(n-1)a - \omega t] - 2kA\sin[kna - \omega t]$
Use $\sin A + \sin B = 2\sin[(A + B)/2]\cos[(A - B)/2]$ for first two terms, then
$-m\omega^2A\sin(kna - \omega t) = 2kA\sin(kna - \omega t)[\cos(ka) - 1] =$
$= -4kA\sin^2(ka/2)\sin(kna - \omega t)$
Thus, $\omega^2 = 4k/m\ \sin^2(ka/2)$

15. (a) $F = k(L - L_0) \approx kL$, thus $v = (F/\mu)^{1/2} = [kL/(m/L)^{1/2}]$
$= v = L(k/m)^{1/2}$
(b) $\Delta t = L/v = (m/k)^{1/2}$

16. Need $y_1 + y_2 = 0$. (a) At t = 0, $\pm 2x = 2x - 6$, so x = 1.5 m
(b) At x = 0, $\pm 3t = 3t - 6$, so t = 1 s.

17. $y = A\sin(kx + \omega t + \phi)$. $\omega = 2\pi/T = 100\pi$ rad/s, $\lambda = vT = 0.5$ m and $k = 2\pi/\lambda = 4\pi$ rad/m. At x = t = 0, $y = A\sin\phi = 3x10^{-3}$ m and $v_p = \omega A\cos\phi = -2$ m/s. Square and add to find A = 7.04 $x10^{-3}$ m. $\tan\phi = -0.15\pi$, so $\phi = 2.7$ rad ($\sin\phi > 0$, $\cos\phi < 0$).
$y(x, t) = 7.04 \times 10^{-3} \sin(4\pi x + 100\pi t + 2.7)$ m

CHAPTER 17

Exercises

1. $\lambda = v/f = 3.4$ mm

2. $\lambda = v/f = 9.7$ mm

3. $\lambda = (1500 \text{ m/s})/(4\text{x}10^6 \text{ Hz}) = 0.375$ mm

4. $D(1/340 - 1/1500) = 3.2$ s, thus $D = 1.41$ km

5. (a) $v = (B/\rho)^{1/2} = 1.43$ km/s; (b) $\lambda = v/f = 1.43$ m

6. $v = (B/\rho)^{1/2}$: (a) 314 m/s; (b) 972 m/s

7. $v = (Y/\rho)^{1/2} = 5.06$ km/s

8. $\Delta t = (100 \text{ m})(1/340 - 1/1320) = 218$ ms

9. $v = (S/\rho)^{1/2} = 3.04$ km/s

10. $v = (15.4\text{x}10^{10}/\rho)^{1/2} = 416$ km/s

11. $\partial s/\partial t = -\omega s_0 \cos(kx - \omega t)$. Using $v = \omega/k$ and $p_0 = Bks_0$, we see that $(B/v)\partial s/\partial t = p$ (See Eq. 17.14.)

12. $p = 2 \sin(5.3x + 1800t)$

13. $\lambda = v/f = 77.3$ cm. $L_1 = \lambda/4 = 19.3$ cm; $L_2 = 3\lambda/4 = 58.0$ cm

14. Plot f vs 1/2L, then slope, $v = 350$ m/s

15. Pipe: $\lambda = 4L/5 = 0.8 \text{ m} = 340/f$, thus $f = 425$ Hz.
    String: $\lambda = L = 0.6 \text{ m} = v/f$; so $v = 255$ m/s. $F = \mu v^2 = 78$ N

16. (a) $f = (40)(1200/60) = 800$ Hz; (b) No, there are many harmonics.

17. $f = nv/2L$: 8.5 Hz, 17 Hz, 25.5 Hz

18. $f_2/f_1 = (T_2/T_1)^{1/2}$, $f_2 = 983$ Hz

19. (a) $f = v/2L = 283$ Hz; (b) $L = v/2f = 51.5$ cm

20. (a) $L = v/4f = 3.4$ m; (b) $L = v/2L = 34$ cm.

21. $f_{20} = v/2L = 573.3$ Hz; $f_5 = 558.3$ Hz, thus $\Delta f = 15$ Hz

22. (a) (340/310)(1200) = 1320 Hz; (b) (340/370)(1200) = 1100 Hz

23. (a) $f' = 400(340/315) = 431.7$ Hz,
$\lambda' = v/f' = 340/431.7 = 78.8$ cm
(b) $f' = 400(340/365) = 372.6$ Hz,
$\lambda' = v/f' = 340/372.6 = 91.3$ cm

24. $\lambda = v/f_o$: (a) 85 cm; (b) 85 cm

25. (a) $f' = 200(340/300) = 227$ Hz, $\lambda' = v/f' = 340/227 = 1.50$ m
(b) $f' = 200(380/340) = 224$ Hz, $\lambda' = v'/f' = 380/224 = 1.70$ m
(c) $f' = 200(360/320) = 225$ Hz, $\lambda' = v'/f' = 360/225 = 1.60$ m

26. $v = 18.1$ m/s. $f' = 1800(340)/(340 + 18.1) = 1710$ Hz, 1900 Hz

27. (a) $f_1' = 400(325/300) = 433.3$ Hz, $f_2' = 400(355/380) =$ 373.7 Hz. Thus $\Delta f' = 59.6$ Hz;
(b) $f_1' = 400(355/300) = 473.3$ Hz, $f_{2'} = 400(325/380) =$ 342.1 Hz. Thus $\Delta f' = 131$ Hz

28. $f' = 600(320/300) = 640$ Hz, received by truck
$f'' = 640(380/360) = 676$ Hz, received by police

29. $v = 8.33$ m/s. $f' = 500(340/331.7) = 512.6$ Hz received by wall and by observer if in front of car. Beat frequency = 0. If observer is behind car: $f'' = 500(340/348.3) = 488.1$ Hz. In this case beat frequency = (512.6 - 488.1) = 24.5 Hz

30. $v = 8.33$ m/s. $f' = 500(340/348.3) = 488.1$ Hz received by wall. If observer is in front of the car: $f'' =$ 500(340/331.7) = 512.6 Hz. Beat frequencies: Zero, 24.5 Hz.

31. $640 = 600(340)/(340 - v_s)$, thus $v_s = 21.25$ m/s.
$f' = 600(340)/361.25) = 565$ Hz

32. $50 = 10\log(I_1/I_o)$, thus $I_1 = 10^5 I_0 = 10^{-7}$ W/m^2
$\beta = 10\log(2\text{x}10^4 I_1/I_o) = 10\log(2\text{x}10^9) = 93$ dB

33. (a) $\beta = 120$ dB means $I = 1$ W/m^2, so $P = IA = 4 \times 10^{-5}$ W;
(b) $\beta = 0$ dB means $I = 10^{-12}$ W/m^2, so $P = 4 \times 10^{-17}$ W

34. 100 dB = $10\log(I_2/I_o) = \log I_2 + 12$, thus $\log I_2 = -2$ and
$I_2 = 10^{-2}$ W/m^2; $P = IA = (10^{-2}$ W/m$^2)(4\pi r2) = 201$ W.

35. $8 = \log(I_1/I_o)$, so $I_1 = 10^{-4}$ W/m^2
$8.5 = \log(I_2/I_o)$, so $I_2 = 10^{-3.5}$ W/m^2
$I_1 + I_2 = 10^{-3.38}$, Since $\beta = 10\log I + 120$,
we find $I = -33.8 + 120 = 86.2$ dB

36. (a) $\beta = 10\log I + 120 = 57$ dB;
(b) $75 = 10\log I + 120$, so $\log I = -4.5$ and $I = 3.16\text{x}10^{-5}$ W/m^2

37. (a) $I = 1/2\ \rho(\omega s_o)^2 v = 1.99 \times 10^{-7}$ W/m^2;
(b) $I = p_o{}^2/2\rho v = 1.40 \times 10^{-2}$ W/m^2

38. $80 = 10\log(I_2/I_1)$, thus $I_2 = 10^8 I_1$

39. 3 dB = $10\log(I_2/I_1)$, thus $I_2 = 2I_1$
Since $I \propto s_o{}^2$, $(s_o)_2/(s_o)_1 = 1.41$

40. (a) $\beta = 10\log(10^3) = 30$ dB
(b) $I \propto p_o{}^2$, so $(p_o)^2/(p_o)^1 = (10^3)^{1/2} = 31.6$

41. (a) $P = 0.2$ W $= I(4\pi r^2)$, where $I = 1$ W/m^2, thus $r = 0.126$ m;

(b) $P = 0.2$ W $= I(4\pi r^2)$, where $I = 10^{-6}$ W/m^2, so $r = 126$ m

42. (a) $s_o/d = v_p/v_s$, where v_p is the particle velocity and v_s is the speed of sound. So $v_p = (10^{-6})(340)/(5\text{x}10^{-2}) = 6.8$ mm/s
(b) $s_o/2d = v_p/v_s$, thus $v_p = -3.4$ mm/s.

43. (a) $v = 2700/8.18 = 330$ m/s.
$s_o = p_o/\rho\omega v = 1.04 \times 10^{-5}$ m;
(b) $I = p_o{}^2/2\rho v = 0.169$ W/m^2

44. $s_o = p_o/\rho\omega v = 1.81\text{x}10^{-7}$ m

45. $\omega = 2\pi f = p_o/s_o\rho v$, thus $f = 6.35$ kHz

46. (a) $f_1 = v/4L = 60.7$ Hz; (b) $f_7 = 7v/4L = 425$ Hz.

47. $F \propto f^2$, and F must be decreased, so
$F_2 = (220/222)^2 F_1 = 589$ N.

48. (a) $v = 1800/5.3 = 340$ m/s;
(b) $p_o = \rho\omega v s_o = 5.53 \times 10^{-2}$ Pa;
(c) $\omega s_o = 0.126$ mm/s.

49. (a) $1300/929 = 1.40$, $(n + 1)/n = 1.4$ gives $n = 2.5$, so pipe is not open; but $(n + 2)/n = 1.4$ gives $n = 5$, so pipe is closed.
(b) $5v/4L = 929$, so $v = 334$ m/s.

50. At wall: $f_1 = vf_o/(v - v_s) = 658$ Hz. This is new source, at rest. Back at car: $f_2 = (v + v_o)f_1/v = 716$ Hz.

51. $960 = (v + v_o)800/(v - v_s)$ so $0.2v = 68 = 1.2v_s + v_o$ (i).
Also, $61 = v_o + v_s$ (ii).
From (i) and (ii): $v_s = 35$ m/s, $v_o = 26$ m/s.

52. $f_A = vf_o/(v - v_s)$ and $f_R = vf_o/(v + v_s)$, so
$f_A/f_R = (v + v_s)/(v - v_s) = 1.0595$, leads to $v_s = 9.82$ m/s.

53. (a) $f_A = vf_o/(v - v_s)$ and $f_R = vf_o/(v + v_s)$, so
$f_A/f_R = (v + v_s)/(v - v_s) = 475/410$, thus $v_s = 25$ m/s.
(b) Put v_s into f_A or f_R to find $f_o = 440$ Hz.

54. (a) $p_o = \rho\omega v s_o = 3.44\text{x}10^{-2}$ Pa;
(b) $s_o = p_o/\rho\omega v = 2.18\text{x}10^{-10}$ m.

55. $p_o = \rho\omega v s_o$ and $\omega = 2\pi v/\lambda$, so $p_o = 2\pi\rho s_o v^2/\lambda$ gives
$\lambda = 1.37$ m.

56. $f_1 = (v - v_o)f_o/(v - v_s) = 840$ Hz; $f_2 = (v + v_o)f_o/(v + v_s) =$
$= 768$ Hz, thus $\Delta f = -71.6$ Hz

57. $f_1 = (v + v_o)f_o/(v - v_s) = 973.3$ Hz; $f_2 = (v - v_o)f_o/(v + v_s)$
$= 663.2$ Hz, thus $\Delta f = -310$ Hz.

58. $I = P/4\pi r^2$ and $\beta = 10\log(I/I_o)$, thus $10^{10}/10^{9.4} = 10^{0.6} =$
r_2^2/r_1^2 from which $r_2 = 6.98$ m.

59. $\beta = 10\log(I/I_o)$, so $I = I_o 10^{8.5} = 3.17 \times 10^{-4}$ W/m^2 $= p_o^2/2\rho v$
so $p_o = 0.527$ Pa.

Problems

1. (a) $I = P/4\pi r^2 = 4.97\text{x}10^{-6}\ W/m^2$;
   (b) $\beta = 10\log(4.97\text{x}10^{6}) = 67$ dB;
   (c) $I = p_o^2/2\rho v$, thus $p_o = 6.6 \text{ x } 10^{-2}$ Pa;
   (d) $I = 1/2\ \rho(\omega s_o)^2 v$, thus $s_o = 9.98\text{x}10^{-8}$ m

2. (a) $s_o = p_o/\rho\omega v = 9.07\text{x}10^{-6}$ m
   (b) $I_1 = p_o^2/2\rho v = 0.114\ W/m^2$. Thus $I_5 = I_1/25$
   $\beta = 10\log I_5 + 120 = 96.6$ dB

3. (a) $dv/dT = 1/2\ (20)T^{-1/2}$, thus $dv/v = dT/2T$
   (b) Since $f \alpha v$, $\Delta f/f = \Delta T/2T$, thus $\Delta f = 10(400)/285$
   $= 14$ Hz, thus $f = 414$ Hz;
   (c) $\Delta f/f = 10/285 = 3.51\%$

4. (a) Try open: $850/607 = (n + 1)/n$ thus $n = 2.5$, so cannot be open. Closed: $850/607 = (n + 2)/n$, thus $n = 5$, i.e. closed.
   (b) $f_1 = f_5/5 = 607/5 = 121.4$ Hz.

5. $D = 4.9t_1^2$, $D = 340t_2$, given $t_1 + t_2 = 2.2$ s, thus
   $(D/4.9)^{1/2} + D/340 - 2.2 = 0$. Solve for $(D)^{1/2}$, then find
   $D = 22.3$ m

6. (a) $\lambda/4 = 18.1 + 0.6r$; $3\lambda/4 = 55.7 + 0.6r$.
   Eliminate λ to find $r = 9.17$ mm
   (b) Find $\lambda = 75$ cm, then $f = v/\lambda = 453$ Hz

7. $I \alpha 1/r^2$ and $I \alpha s_o^2$, so $s_o \alpha 1/2$.

8. $\Delta f = (3/2 - 2^{7/12})(261.63) = 0.443$ Hz

9. (a) 404 Hz or 396 Hz;
   (b) $f = 1/2L\ (F/\mu)^{1/2}$, so $F = (2Lf)^2\mu = 346$ N;
   (c) Now know $f = 404$ Hz $= v/2L$, thus $L = 42.1$ cm

10. $I = 1/2\ (\omega A)^2\rho v$ and $v = (B/\rho)^{1/2}$,
    thus $I = 1/2\ (\omega A)^2 Z$

CHAPTER 18

Exercises

1. (a) 21.1 ^oC; (b) 90.6 ^oC; (c) 37.0 ^oC

2. (a) 621 ^oF; (b) -423 ^oF; (c) 113 ^oF

3. -40 ^oF = -40 ^oC

4. (a) 33.3 ^oC; (b) 20.5 cm

5. 273x9/5 = 491 ^oR

6. Find a = 8.62 and b = -207. (a) 31.0 Ω; (b) 43.0 ^oC

7. $N/V = P/kT = 2.68 \times 10^{25}\ m^{-3}$

8. Mass of n moles is m = nM, so $\rho = nM/V$

9. PV = nRT = (m/M)RT, so $P = \rho RT/M$, where M is in kg/mol. Thus $\rho = 44.5M\ kg/m^3$. Note 1 g/mol = 10^{-3} kg/mol.
(a) 1.25 kg/m^3; (b) 1.42 kg/m^3; (c) 0.089 kg/m^3

10. (a) PV = nRT and n = m/M, thus $P \propto n \propto 1/M$. $P_2 = (M_1/M_2)P_1$ = 3.43 atm. (b) $m_2/m_1 = P_2/P_1 = 2/3.43 = 0.583$ kg.

11. (a) $V = nRT/P = 2.44 \times 10^{-2}\ m^3$ = 24.4 L; (b) 37.2 L

12. (a) n = PV/RT = 9122 moles, so m = nM = 265 kg.
(b) $P_2 = P_1(T_2/T_1)$ = 1.017 atm
(c) When the pressure drops by a factor of 1/1.017, the number of moles also drops by the same factor: n_2/n_1 = 1/1.017, thus $\Delta n = n_1 - n_2 = (1 - 0.9833)n_1 = 0.0167n_1$ = 152.5 mol. Thus, $\Delta m = \Delta n.M$ = 4.42 kg.

13. (a) $T_2/T_1 = P_2/P_1$, so T_2 = (330/300)(288K) = 316.8 K = 43.8 ^oC;
(b) $n_1 = P_1V_1/RT_1 = 1.87 \times 10^{-3}$ mol. When the pressure drops by a factor of 1/1.1, the number of moles also drops by the same factor: n_2/n_1 = 1/1.1, so
$\Delta n = n_1 - n_2 = (1 - 0.909)n_1 = 0.0909n_1$.
Thus, $\Delta m = \Delta n.M = (0.0909n_1)M$ = 4.93 mg

14. $N = PV/kT = (10\text{-}10\ Pa)(10^{-6}\ m^3)/k(300\ K) = 2.41\times10^4$ per cm^3

15. $P_2/P_1 = T_2/T_1$, so $P_2 = (303/273)(101\ kPa) = 112\ kPa$
$F = P_2A = 1120\ N$

16. (a) $T = PV/nR = 96.2\ K$; (b) $T_2 = 2T_1 = 192\ K$
(c) $P = nRT/V = 234\ kPa$

17. (a) $V = nRT/P = 16.5\ L$; (b) $V_2 = 3V_1 = 49.5\ L$

18. (a) $P_{tr} = 273.16P/(373.15) = 0.0146$ atm
(b) $T = 273.16P/P_{tr} = 503.7\ K$

19. (a) $P = (300/273.16)(40\ mm) = 43.9$ mm Hg;
(b) $T = (25/40)273.16 = 171\ K$

20. $\Delta L = \alpha L\Delta T = 4.68$ mm

21. $+10^{-5} = \alpha L\ \Delta T$, thus $\Delta T = +17.1\ ^oC$. $T = 2.9\ ^oC$ to $37.1\ ^oC$

22. $\Delta L = \alpha L\Delta T = 8.42\times10^{-3}$ cm

23. $r_s = 2[1 + 17\times10^{-6}\Delta T]$; $r_h = 1.99[1 + 24\times10^{-6}\Delta T]$
Set $r_s = r_h$ to find $\Delta T = 726.7\ ^oC$, thus $T \approx 747\ ^oC$

24. $\Delta L = \alpha L\Delta T = 5.62$ mm

25. $\Delta L = \alpha L\Delta T = (11.7\times10^{-6})(320)(55) = 20.6$ cm

26. From Example 18.4, $F/A = -\alpha Y\Delta T = 3.51\times10^7\ N/m^2$

Problems

1. (a) $P_2 = (373/273)P_1 = 1.37$ atm;
(b) $P_2V = 5R(373)$; $P_1V = nR(373)$, so $n = 5(1/1.37) =$ 3.65 mol. Thus let $(5 - 3.65) = 1.35$ mol escape.
(c) $P_3/P_2 = (273/373)$, so $P_3 = 0.732$ atm

2. $(\rho - \rho')gV = mg$, thus $\rho' = \rho - m/V$.
$\rho = PM/RT = 1.223\ kg/m^3$, where $M = 29\times10^{-3}$ kg/mol. Thus
$\rho' = 1.223 - 1/6 = 1.056\ kg/m^3 = PM/RT'$
Find $T' = 333.6\ K = 60.6\ ^oC$

3. $\Delta V = \beta V \Delta T$ so $\beta = \Delta V/V\Delta T$; $PV = nRT$, so $P\Delta V = nR\Delta T$ and $\Delta V/\Delta T = nR/P = V/T$. Find $\beta = 1/T$.

4. Ignore the expansion of the neck ($\Delta r \approx 4\text{x}10^{-5}$ cm). $\Delta V_w = \beta_w V\Delta T$; $\Delta V_g = \beta_g V\Delta T$, thus $\delta V = (\beta_w - \beta_g)V\Delta T = \pi r^2 h$ Find h = 3.88cm.

5. $Id^2\theta/dt^2 + 0.5mgL\theta = 0$, where $I = ML^2/3$. Thus, $T \alpha L^{1/2}$, and $\Delta T/T = \Delta L/2L$, where $\Delta L = \alpha L\Delta t$. Thus, $\Delta T/T = 9.35\text{x}10^{-5}$ and $\Delta T = 56.5$ s, loss

6. $\rho = m/V$, so $\rho V = m$, thus $\rho dV + Vd\rho = 0$. Then, $\Delta\rho/\rho = -\Delta V/V = -\beta\Delta T$

7. Vol. of ball + water at 5 ^oC: $V_1 = 20$ mL $+ 4\pi R^3/3 = 27.24$ mL "Overflow" volume $= \Delta V_B + \Delta V_W - \Delta V_G$, where ΔV_g is the change in volume of the glass beaker: $\Delta V = [(20)(2.1\text{x}10^{-4}) + (7.24)(35.1\text{x}10^{-6}) - (20)(27\text{x}10^{-6})](85\ ^oC) = 0.333$ mL, so $V_2 = 27.57$ mL. To find height of water in beaker note $V = \pi r^2 h$, $r_2 = r_1(1 + \alpha\ \Delta T)$, so $V_2/V_1 = (r_2/r_1)^2 h_2/h_1 \approx (1 + 2\alpha\Delta T)h_2/h_1$; $V_2/V_1 = 27.57/27.24 = (1 + 18\text{x}10^{-6}\text{x}85)h_2/h_1$, thus, $h_2 = (1.012)h_1/(1.0015)$. But $h_1 = 20/\pi(1.5)^2 = 2.829$ cm, so $h_2 = 2.859$ cm, and $\Delta h = 0.30$ mm

8. $A = LW$, so $dA = LdW + WdL$, then $dA/A = dW/W + dL/L = 2\alpha\Delta T$.

9. The change in load induced by a temperature change is $\Delta F = YA(\alpha\ \Delta T) = (2\text{x}10^{11})(4\pi\text{x}10^{-4})(11.7\text{x}10^{-6})(+10) = 29.4$ kN
(a) Load = 44.4 kN, so $F/A = 3.53 \times 10^7$ N/m^2;
(b) Load = 14.4 kN, so $F/A = 1.15 \times 10^7$ N/m^2

10. (a) $\Delta L = \alpha L\Delta T = 4.32$ mm;
(b) $F/A = -\alpha Y\Delta T = -\alpha Y(5\ ^oC) = 1.2$ MPa

11. PV/T = constant. $P_1 = P_o + \rho gh = 219$ KPa; $P_2 = P_o = 101$ kPa. $T_1 = 281$ K, $T_2 = 289$ K and $V = 4\pi r^3/3$, thus $P_1V_1/T_1 = P_2V_2/T_2$, find $r_2^3 = 1.78\text{x}10^{-8}$ m^3, so $r_2 = 2.61$ mm

12. $\Delta V_1 = \beta V\Delta T$, $B = -\Delta P/(\Delta V_2/V)$, where $\Delta V_1 + \Delta V_2 = 0$. Thus, $\beta V\Delta T - V\Delta P/B = 0$, so $\Delta P = \beta B\Delta T$.

CHAPTER 19

Exercises

1. $(4186)(0.4535)(5/9) = 1055$ J

2. $Q = mc\Delta T$, so $c = 1070$ J/kg

3. $[(0.090)(385) + (0.5)(4190)](T - 15) + (0.08)(450)(T - 180) = 0$. Thus $T = 17.7\ ^oC$

4. $m_b c_b(T_f - 210) + (m_l c_l + m_a c_a)(T_f - 20) = 0$. Find $c_l = 1680$ J/kg

5. $(0.5)(450)(80) + (0.6)(4190)(80) = 1200t$, thus $t = 183$ s

6. $Q = 0.6 x 600 x 2 = 720$ kcal $= 3.017 x 10^6$ J
   $m = Q/L = 1.23$ kg.

7. $(0.08)(2100)(10) + (0.08)(3.34x10^5) + (0.08)(4190)(100) + (0.02)(2.26x10^6) = 107$ kJ

8. $m_p c_p(T_f - 200) + (m_c c_c + m_w c_w)(T_f - T_i) = 0$. With $m_w = 0.1$ kg, $m_p = 0.2$ kg, and $m_c = 0.07$ kg, find $T_i = 9.5\ ^oC$

9. $5x10^8 = (4190)(10)(dm/dt)$, thus $dm/dt = 1.2 \ x \ 10^4$ kg/s

10. $dQ/dt = (6)(10^3)(0.8) = (dm/dt)(4190)(40)$
    Thus. $dm/dt = 28.6$ g/s.

11. (a) $1/2\ mv^2 = 463$ kJ;
    (b) $\Delta T = 2.78x10^5/(10)(450) = 61.7\ ^oC$

12. $E = (0.2)(0.5mv^2)(15) = 3\ J = (6x10^{-3})c\Delta T$, thus $\Delta T = 1.11\ ^oC$

13. (a) $Q = (0.35)mv^2$, $\Delta T = Q/(130)(0.02) = 330\ ^oC$, so $T_f = 327\ ^oC$
    (b) To reach 327 ^oC: $Q_1 = mc(327 - 30)^o = 772$ J. To melt: $Q_2 = m_2 L = (857.5 - 772)\ J = 85.5$ J. Find 3.5 g melts.

14. $mgh = mc\Delta T$, so $\Delta T = gh/c = 0.28\ ^oC$.

15. $\Delta T = 16mgh/(3.5)(4190) = 0.423\ C^o$

16. $E = 0.8(m_1 gh) = (m_1 c_1 + m_2 c_2)\Delta T$, find $\Delta T = 7.48x10^{-3}\ ^oC$

17. $W = P\Delta V = (101\ kPa)(1/920 - 1/1000) = 8.8\ J$

18. $W = P\Delta V$: (a) 600 J; (b) -300 J

19. (a) $W = P\Delta V = 2.4\ kJ$;
(b) $W = nRT\ln(P_2/P_1) = P_1V_1\ \ln(4/5) = -1.34\ kJ$

20. $W = nRT\ln(V_f/V_i)$, but $V \alpha 1/P$, so $W = nRT\ln(1/2.5)$
$= -2R(273)\ln(2.5) = -4160\ J$

21. (a) $\Delta U = Q - W = 35 - 11 = 24\ J$, thus $U_f = 229\ J$;
(b) $\Delta U = 24 = Q' - 15$, thus $Q' = 39\ J$

22. (a) $W = P\Delta V = -16.2\ J$;
(b) $\Delta U = Q - W = 416\ J$

23. (a) $W = (mg + PA)h = (19.6 + 31.4)(0.024) = 1.22\ J$;
(b) $\Delta U = 5 - 1.22 = 3.88\ J$

24. (a) For process abc, $U_c - U_a = Q - W = 4500 - W_{bc} = 500\ J$
Thus $U_c = 1100\ J$.
(b) $Q_{ca} = -3500\ J$, $\Delta U = U_a - U_c = -500\ J$, thus
$W_{ca} = Q_{ca} - \Delta U = -3000\ J$; (c) -3500 J

25. (a) $W = P_{av}\Delta V = (150\ kPa)(10^{-3}\ m^3) = 150\ J$
(b) $\Delta U = Q - W = 180 - 150 = 30\ J$; thus $U_b = 130\ J$
(c) $W = P\Delta V = (100\ kPa)(-10^{-3}\ m^3) = -100\ J$;
(d) $Q = \Delta U + W = -30 - 100 = -130\ J$

26. (a) $W = nRT\ln(V_f/V_i)$, where $n = m/M = 5/8$ mol.
$W = nRT\ln(0.3/0.12) = 1430\ J$
(b) $\Delta U = Q - W = 0$, so $Q = W = 1430\ J$

27. (a) $C_v = Mc_v = 45$ J/mol.K; $C_p = C_v + R = 53.3$ J/mol.K, so
$c_p = 2.96$ kJ/kg.K;
(b) $C_p = Mc_p = 29$ J/mol.K; $C_v = C_p - R = 20.7$ J/mol.K, so
$c_v = 0.713$ kJ/kg.K

28. $C_p = Mc_p = (32\times10^{-3}$ kg/mol$)(0.918$ kJ/kg.K$) = 29.38$ J/mol.K
$C_v = C_p - R = 21.07$ J/mol.K, thus $c_v = C_v/M = 658$ J/kg.K

29. $n = PV/RT = 3526$ moles, thus $m = 102.3$ kg.
$\Delta T = \Delta Q/mc_v = 2.04\ ^oC$

30. $PV = nRT = mRT/M$, so $m = PVM/RT = 63.88$ kg.
$Q = P\Delta t = 2.16\text{x}10^5\ J = mc_v\Delta T$, thus $\Delta T = 4.7$ K.

31. (a) $C_p = Mc_p = 29$ J/mol.K; $Q = nC_p\Delta T = 5.8$ kJ;
(b) $W = P\Delta V = nR\Delta T = 1.66$ kJ;
(c) $\Delta U = Q - W = 4.14$ kJ

32. (a) $W_{ab} = nRT\ln(30/20) = P_aV_a\ln(1.5) = 1216$ J
$W_{bc} = P\Delta V = -1000$ J; $W_{ca} = 0$
(b) $\Delta U = Q - W = 0$, thus $Q = 216$ J

33. (a) $W_1 = nRT\ln(1/2) = -2020$ J; $W_2 = 0$, so $W = -2020$ J
(b) $\Delta U_1 = 0$, $\Delta U_2 = Q_2 - 0 = +400$ J; so $\Delta U = +400$ J

34. (a) $\Delta U = Q - W = 0 - 400 = -400$ J; (b) $Q = 0$

35. (a) $W = nRT\ln(2) = W = 3.38$ kJ, $\Delta U = 0$;
(b) $Q = 0$, $\Delta U = nC\Delta T$, so $W = -\Delta U = -nC_v(T_2 - T_1)$
Show $T_2/T_1 = (V_2/V_1)^{1-\gamma} = (2)^{1-\gamma} = 0.7578$, so $T_2 = 222$ K
Then, $W = 2970$ J, $\Delta U = -2970$ J

36. (a) $W = nRT\ln(V_f/V_i) = P_iV_i\ln(20/30) = -1220$ J,
thus $W_{ext} = 1220$ J
(b) $P_1V_1^\gamma = P_2V_2\gamma$, so $P_2 = (3/2)^\gamma P_1 = 196.5$ kPa
$W = (P_1V_1 - P_2V_2)/(\gamma - 1) = (3/2)(3000 - 3931) = -1400$ J,
thus $W_{ext} = 1400$ J

37. $W = -nC_v(T_2 - T_1) = -1875$ J

38. (a) $PV_1 = nRT_1$, $P(2V_1) = 2(nRT_1)$, thus
$W = P(V_2 - V_1) = PV_1 = nRT_1 = 4870$ J
(b) $Q = nC_p\Delta T = n(C_v + R)\Delta T = 58.43\Delta T = 58.42(2T_1 - T_1)$
$= 17.12$ kJ.
$\Delta U = Q - W = 17.12\text{ kJ} - 4.87\text{ kJ} = 12.3$ kJ

39. (a) $P_1V_1^{1.4} = P_2(5V_1)^{1.4}$, so $P_2 = 10.5$ kPa;
(b) $T_2/T_1 = (P_2V_2)/P_1V_1$, so $T_2 = 158$ K

40. PV^γ = constant and $T\ \alpha\ PV$, so $T_2/T_1 = (V_2/V_1)^{1-\gamma} = 2.954$

Thus $T_2 = 866$ K $= 593$ ^oC

41. (a) $P \alpha V^{-\gamma}$, thus $T_2/T_1 = (P_2V_2)/P_1V_1 = (V_2/V_1)^{1-\gamma}$
(b) $V = nRT/P$, so $PV^\gamma = P(T/P)^\gamma = T^\gamma P^{1-\gamma}$ = constant.

42. (a) $V_1 = nRT/P_1 = 35.3$ L, then $P_1V_1^\gamma = P_2V_2^\gamma$, so $V_2 = 81.1$ L
(b) $T_2 = P_2V_2/nR = 163$ K
(c) $W = (P_1V_1 - P_2V_2)/(\gamma - 1) = 4510$ J

43. (a) See Exercise 41: $T_2 = T_1(V_1/V_2)^{\gamma-1} = 290(3/2)^{0.4} = 341$ K
(b) $P = nRT/V$: 40.2 kPa, 70.8 kPa;
(c) From Example 19.6: $W = (P_1V_1 - P_2V_2)/(\gamma - 1) = -2.1$ kJ

44. $v = (\gamma RT/M)^{1/2}$, where M is in kg/mol.
(a) 330 m/s; (b) 1020 m/s.

45. (a) $v = (\gamma P/\rho)^{1/2} = 258$ m/s;
(b) $B = \gamma P = 1.31 \times 10^5$ N/m^2

46. $dQ/dt = \kappa A\, dT/dx = (0.9)(1.68)(30/4x10^{-3}) = 11.3$ kW.

47. $(dQ/dt)/A = \kappa dT/dx = (1)(0.03) = 30$ mW/m^2

48. $P = \sigma T^4(4\pi R^2) = 3.9x10^{26}$ W

49. (a) $dQ/dt = \kappa A\, dT/dx = (400)(\pi r^2)(60/0.04) = 75.4$ W;
(b) $T(x) = T(0) + x(dT/dx) = 60 - (0.1)(60/0.4) = 45$ ^oC

50. (a) $dQ/dt = \kappa_1 A(80 - T)/L = \kappa A(T - 10)/L$; so
$400(80 - T) = 240(T - 10)$, thus $T = 53.8$ ^oC
(b) $dQ/dt = 6.6$ W

51. From p. 384, $R = 0.14/\kappa = 0.156$, for 1 inch. Thus for
$R = 6$ need $6/(0.156) = 38.5$ in

Problems

1. (a) $Q = mc\Delta T = (7.8 \text{ kg})(450)(40) = 140$ kJ;
(b) $\Delta V = 3\alpha V\Delta T = 3(11x10^{-6})(0.1)^3(40) = 1.32x10^{-6}$ m^3
$W = P\Delta V = 0.132$ J;
(c) $\Delta U = Q - W \approx 140$ kJ

2. Heat available if water drops to 0 ^oC: (4190)(8) = 33.52 kJ
Heat needed to get ice to 0 ^oC: (2)(2100)(10) = 42 kJ
The final state is a mixture of water and ice at 0 ^oC.
[When the water reaches 0 ^oC, the rise in temperature of the ice is $\Delta T = Q/mc$ = 33.5 kJ/(2 kg)(2100 J.kg.K) = 8 ^oC, thus the ice reaches -2 ^oC. To raise 2 kg of ice to 0 ^oC, we need 8.4 kJ which is obtained by the freezing of $m = Q/L$ = 8.4 kJ/333 kJ/kg = 0.025 kg of water.]

3. $dQ/dt = \kappa\ (2\pi rL)(dT/dr)$; so $(dQ/dt) \int dr/r = (2\pi\kappa L) \int dT$
then $dQ/dt = 2\pi\kappa L(T_b - T_a)/\ln(b/a)$

4. $dQ/dt = -\kappa(4\pi r^2)(dT/dr)$; thus $(dQ/dt)\int dr/r^2 =$
$-(4\pi\kappa)\int dT$.
$(dQ/dt)(1/b - 1/a) = 4\pi\kappa(T_b - T_a)$.

5. $(dQ/dt)/A = \kappa_g\ \Delta T_g/t_g = \kappa_a\ \Delta T_a/t_a$;
$T_1 - T_2 = 2\Delta T_g + \Delta T_a = [2t_g/\kappa_g + t_a/\kappa_a](dQ/dt)/A$, so
$dQ/dt = A(T_1 - T_2)/[2t_g/\kappa_g + t_a/\kappa_a]$ = 565 W

6. $\int dQ = \int nCdT = (nk/T_D^3) \int T^3\ dT$ = 2.74 J

7. $PV = nRT$, $PdV + VdP = 0$, so $dP/dV = -P/V$
Thus, $B = -V(dP/dV) = P$

8. (a) $W_{AB} = 2P_oV_o$; $W_{BC} = 0$; $W_{CA} = -P_{av}(2V_o) = -1.5P_oV_o$
(b) W_{net} = Area = $0.5P_oV_o$
(c) $\Delta U = Q - W = 0$, so $Q = W = 0.5P_oV_o$
(d) $0.5P_oV_o = 0.5nRT = 0.75R(293)$ = 1830 J

9. $dQ/dt = \kappa AT/y$ and $dm/dt = (dQ/dt)/L = \rho dV/dt$
$= \rho A\ dy/dt$.
Thus $dy/dt = (dQ/dt)/\rho LA = \kappa T/L\rho y$

10. $W = \int P\ dV = RT \int dV/(V - b) - a \int dV/V^2$
$= RT\ln[(V_f - b)/(V_i - b)] + a(1/V_f - 1/V_i)$

CHAPTER 20

Exercises

1. $v_{rms} = (3RT/M)^{1/2}$, where the unit of M is kg/mol.
   (a) 1350 m/s; (b) 604 m/s; (c) 181 m/s

2. $v_{rms} = (3kT/m)^{1/2}$: (a) 12.0 km/s; (b) 780 m/s

3. (a) $K_{av} = 3kT/2 = 4.14 \times 10^{-16}$ J;
   (b) 7.04×10^5 m/s

4. $T = Mv^2/3R$: (a) 1.41×10^5 K; (b) 1.61×10^5 K; (c) 10,060 K

5. (a) $v_{rms} = (3RT/M)^{1/2}$, where the unit of M is kg/mol.
   $v_{rms} = 2740$ m/s; (b) 967 m/s

6. Let $M_2 = 235 + 6 \times 19 = 349$ u; and $M_1 = 352$ u, then
   $v_2/v_1 = (M_1/M_2)^{1/2} = 1.004$.

7. $v = (3kT/m)^{1/2} = 2.73$ km/s

8. $T \propto v^2$: (a) 1200 K; (b) 4800 K

9. (a) $3kT/2 = 2.07 \times 10^{-14}$ J; (b) 3.52×10^6 m/s

10. $T = PV/nR = 180.4$ K; $K_{av} = 1.5kT = 3.74 \times 10^{-21}$ J

11. $(3kT/m)^{1/2} = 273$ m/s

12. (a) $K_{av} = 1.5kT = 6.28 \times 10^{-21}$ J; (b) $P = nRT/V = 3.78 \times 10^4$ Pa

13. (a) $v_{rms} = (3RT/M)^{1/2} = 517$ m/s. (M is in kg/mol.)
    (b) $P = nRT/V = 1.48$ MPa

14. $V = NkT/P = 1.87 \times 10^{-16}$ m^3

15. Volume for 1 mole = $RT/P = 2.25 \times 10^{-2}$ m^3
    Volume for 1 molecule, $d^3 = 2.25 \times 10^{-2}$ m$^3/N_A = 3.73 \times 10^{-26}$ m^3
    Thus, $d = 3.34 \times 10^{-9}$ m.

16. $K_{av} = 1.5kT = 5.655 \times 10^{-21}$ J,
    $mg\Delta y = 5.655 \times 10^{-24}$ J, thus $\Delta y = 10.9$ m.

17. $v_2/v_1 = (T_2/T_1)^{1/2} = (373/273)^{1/2} = 1.17$

18. (a) $n_1 = n_2$, $m = nM$, so $m_2/m_1 = M_2/M_1 = 28/32 = 7/8$.
(a) $m_1 = m_2$, $n = m/M$, so $n_2/n_1 = M_1/M_2 = 32/28 = 8/7$.

19. $E = 3kT/2$, so $T = 1.93 \times 10^5$ K

20. (a) 7.45 cal/K = 31.19 J/K. $C_V = 1.5R$, thus n = 2.5 moles
(b) $U = 1.5nRT = 9.04$ kJ

21. (a) $35 = 3nR/2$, so $n = 2.81$ mol;
(b) $U = 3nRT/2 = (35)(273) = 9.56$ kJ;
(c) $C_p = 5R/2 = 20.8$ J/mol.K

22. (a) $C_p - C_V = R$, so $n = 21.6/R = 2.6$ moles.
(b) $nC_p = 2.6(3.5R) = 75.7$ J/K

23. (a) $Q_V = C_V\Delta T = \Delta U = 3R\Delta T/2 = 1250$ J;
(b) $Q_p = C_p\Delta T = 5R\Delta T/2 = 2080$ J. $\Delta U = 3RT/2 = 1250$ J.

24. $C = Mc$, where $c = 0.6$ J/g.K. Thus
$M = C/c = 3R/0.6 = 41.6$ g/mol.

25. (a) $Q_V = 3R\Delta T/2$, so $\Delta T = 316$ K;
(b) $Q_p = 5R\Delta T/2$, so $\Delta T = 310$ K

26. (a) $1.5RT = 3.74$ kJ; (b) $2.5RT = 6.24$ kJ;
(c) $3.5RT = 8.73$ kJ

27. $C_V = Mc_V = 3R/2$, and $c_V = 0.6195$ J/g.K, thus
$M = 20.1$ g/mol, neon

28. Sound $\lambda = v/f = 1.65\text{x}10^{-2}$ m
Mean free path: $MFP = 1/(1.414 n_V \pi d^2)$. From Example 20.4:
For 1 mole, $V = RT/P = 2.246\text{x}10^{-2}$ m^3, thus
$n_V = N_A/V = 2.68\text{x}10^{25}$ $/m^3$. Finally $MFP = 9.33\text{x}10^{-8}$ m
Therefore the ratio $\lambda/MPF = 1.77\text{x}10^5$.

29. Note $n_V = 2.7\text{x}10^{25}$ molecules/m^3. From Eq. 20.18,
$d = (1.41 n_V \pi\lambda)^{-1/2} = 0.304$ nm

30. $MFP = 10^{-3}$ m $= 1/(1.414 n_V \pi d^2)$; $n_V = 2.50\text{x}10^{21}$ molecules per m^3.
Thus volume of 1 mole $V = 240.9$ m^3. Then, $P = RT/V = 10.4$ Pa

31. $n_V = N/V = P/kT$, so Eq. 20.18 becomes $\lambda = kT/(1.41\pi d^2P)$

32. (a) $n_V = N/V = P/kT = 2.41\text{x}10^{12}\ m^{-3} = 2.41\text{x}10^{6}\ cm^{-3}$
(b) $\lambda = 1/(1.414 n_V \pi d^2) = 1.04\text{x}10^{6}$ m

33. From Eq. 20.18, $\lambda = 2.25 \times 10^{19}$ m

Problems

1. (a) 3.74 m/s; (b) 3.95 m/s; (c) 4 m/s

2. $f = Av^2e^{-\alpha v^2}$, where $\alpha = m/2kT$.
$df/dv = 0$ leads to $v^2 = 1/\alpha$, or $v = (2kT/m)^{1/2}$

3. Let $\alpha = m/2kT$, then $A = 4\pi N(\alpha/\pi)^{3/2}$.
$A \int v^2 \exp(-\alpha v^2)\, dv = A\ (\pi/16\alpha^3)^{1/2} = N$

4. Let $\alpha = m/2kT$, then $A = 4\pi N(\alpha/\pi)^{3/2}$.
(a) $(A/N) \int v^3 \exp(-\alpha v^2)\, dv = (A/N)(1/2\alpha^2) = (8kT/\pi m)^{1/2}$
(b) $(A/N) \int v^4 \exp(-\alpha v^2)\, dv = 3/2\alpha = 3kT/m$.

5. In Eq. 20.16, find $A = 3.0\text{x}10^{-3}$ and $\alpha = (m/2kT) = 5.61\text{x}10^{-6}$.
Then use $\Delta n = Av^2\exp(-\alpha v^2)\Delta v$: (a) 283; (b) 1780; (c) 110

6. $F = PA = (101\text{ kPa})(0.16\ m^2) = 1.616\text{x}10^4$ N
In 1 sec, $\Delta p = F\Delta t = 1.616\text{x}10^4$ m.kg/s
If there are N strikes in 1 sec; $\Delta p = N(2mv_x)$; m = 32 u.
$3v_x^2 = (v_{rms})^2 = 2kT/m$; thus $v_x = (2kT/3m)^{1/2}$
Thus, $N = \Delta p/2mv_x = 7.0\text{x}10^{26}$ strikes per second

7. $P = RT/(V - b) - a/V^2$.
$dP/dV = 0$ leads to $2a(V - b)^2 = RTV^3$ (i)
$d^2P/dV^2 = 0$ leads to $3a(V - b)^3 = RTV^4$ (ii)
(ii)/(i) yields $V_c = 3b$. Substitute into (i) to find
$T_c = 8a/27bR$, then finally $P_c = a/27b^2$.

8. (a) $T_1 = PV/R = 36.5$ K; (b) $T_2 = 2T_1 = 72.9$ K;
$\Delta U = nC_V\Delta T = (1.5R)(36.4) = 455$ J, $U_2 = 555$ J
(c) $T_3 = 0.5T_2 = 36.5$ K; $\Delta U = nC_V\Delta T = -455$ J; so $U_3 = 100$ J

CHAPTER 21

Exercises

1. (a) $Q_H = 4W = 800$ J; (b) $Q_C = Q_H - W = 600$ J

2. $W = Q_H - Q_C = 250$ J, $P = 250/0.4 = 625$ W

3. (a) $P = W/T = 500$ W; (b) $COP = Q_C/W = 100/25 = 4$

4. (a) $W = Q_C/COP = 80/3.5 = 22.9$ J; (b) $Q_H = 102.9$ J

5. $Q_H = 4W = 40$ kWh

6. $dQ_H/dt = (dW/dt)/\epsilon = 333.3$ MW, and $dQ_C/dt = 233.3$ MW
   $0.8(233.3) = (dm/dt)c\Delta T$, thus $dm/dt = 5560$ kg/s

7. (a) $dQ_H/dt = 30/0.22 = 136$ kW;
   (b) $dQ_C/dt = dQ_H/dt - dW/dt = 106$ kW;
   (c) $(1.36\text{x}10^8$ J/s$)(3600$ s/h$)/(1.3\text{x}10^8$ J/gal$) = 3.77$ gal/h

8. (a) $dW/dt = 1$ kW; $dQ_C/dt = 1.758$ kW; since $COP = Q_H/W$, we find $COP = 2.76$; (b) 2.76 kW

9. $\epsilon_C = 1 - 283/363 = 22\%$

10. $\epsilon_C = 1 - 300/400 = 1/4$, so $Q_H = 4W = Q_C + W$, thus $W = 110$ J

11. (a) $\epsilon_C = 1 - 278/295 = 5.76\%$;
    (b) $Q_H = W/\epsilon = Q_C + W$, thus $dQ_C/dt = 16.4$ MW

12. $\epsilon_C = 1 - 350/600 = 0.417$;
    $dQ_H/dt = W/\epsilon_C = 500/0.417 = 1200$ W, thus $dQ_C/dt = 700$ W

13. (a) \$12.00;
    (b) $\epsilon_C = 1 - 273/293 = 0.0683$. $dW/dt = \epsilon_C\, dQ_H/dt$
    $= 0.341$ kW, thus $W = 8.18$ kW.h, which costs \$0.82

14. $Q_H/Q_C = 298/268$, thus $Q_H = 111$ J

15. (a) $Q_H/Q_C = 293/273$, thus $Q_H = 1073$ J
    (b) $W = Q_H - Q_C = 73$ J

16. (a) $Q_H = Q_C/(1 - \epsilon_C) = 200/0.65 = 308$ J
    (b) $T_H = T_C/(1 - \epsilon_C) = 300/0.65 = 462$ K

17. $\epsilon_C = 1 - 293/472 = 0.381$; thus $dQ_H/dt = (dW/dt)/\epsilon_C = 945$ W, and $dQ_C/dt = 945 - 360 = 585$ W. For a 0.2 s cycle, $Q_H = 189$ J, $Q_C = 117$ J

18. (a) Carnot COP $= Q_H/W = T_H/(T_H - T_C) = 298/25 = 11.92$ so actual COP $= 7.152 = (Q_C + W)/W$, thus $W = 16.3$ J.
(b) $Q_H = Q_C + W = 116.3$ J

19. (a) $\epsilon_C = 1 - 313/553 = 0.434$; thus $W = \epsilon_C Q_H = 434$ J;
(b) COP $= Q_H/W = 1/\epsilon_C = 2.30$

20. $\epsilon_1 = [(T_H - T_C) + 5]/(T_H + 5)$, so $\Delta\epsilon_1 = 5/(T_H + 5)$
$\epsilon_2 = [(T_H - T_C) - 5]/T_H$, so $\Delta\epsilon_2 = -5/T_H$
Thus a decrease in T_c has a greater effect on ϵ.

21. (a) $\Delta S = mL/T = 6.06$ kJ/K;
(b) $\Delta S = mL/T = 1.22$ kJ/K;
(c) $\Delta S = mc\ln(373/273) = 1.31$ kJ/K

22. Since the heat "available", $mc(100) = 419$ kJ, is greater than $mL = 334$ kJ, the final state is water at T_f:
$mL + mc(T_f - 0) + mc(T_f - 100) = 0$, thus $T_f = 10.1\ ^oC = 283.1$ K.
$\Delta S_i = mL/273 = (334\text{x}10^3)/273 = 1223$ J/K
$\Delta S_w = mc\ln(T_f/273) + mc\ln(T_f/373) = -1006$ J/K
$\Delta S = 217$ J/K

23. (a) $\Delta S = mL/T = -61.2$ J/K; (b) 0

24. (a) $\Delta S_B = mc\ln(293/473) = (0.1)(450)\ln(293/473) = -21.6$ J/K
(b) $\Delta S_L = mc\Delta T/293 = 27.6$ J/K; (c) $\Delta S_u = +6$ J/K

25. (a) $m_1c_1(T_f - T_1) + m_2c_2(T_f - T_2) = 0$, find $T_f = 20.8\ ^oC$;
(b) $\Delta S_1 = m_1c_1\ln(T_f/T_1) = -3.10$ J/K;
(c) $\Delta S_2 = m_2c_2\ln(T_f/T_2) = +3.43$ J/K;
(d) $\Delta S_u = 3.43 - 3.10 = 0.33$ J/K

26. $\Delta S = mc_i\ln(273/268) + mL/273 + mc_w\ln(278/273)$
$= 38.8 + 1223 + 76 = 1340$ J/K

27. $\Delta S = Q/T$: (a) −3 J/K; (b) 4.8 J/K; (c) 0; (d) 1.8 J/K

28. $\Delta E = mgh + 1/2\ mv^2$, then $\Delta S = \Delta E/T = 18.38\ kJ/293 = 62.7\ J/K$

29. (a) $dS = dQ/T = nC_p\ dT/T$, thus $\Delta S = 5nR/2\ \ln(T_2/T_1)$;
(b) $PV = nRT$, so $dT/T = dP/P$.
$dS = dQ/T = nC_v\ dT/T = (3nR/2)\ dP/P$, thus
$\Delta S = (3nR/2)\ln(P_2/P_1)$

30. (a) $\Delta S = nC_v\ln(T_2/T_1) = (3nR/2)\ln(T_2/T_1)$
(b) $\Delta U = Q - W = 0$, so $Q = W = nRT\ln(V_2/V_1)$ and
$\Delta S = nR\ln(V_2/V_1)$

31. (a) $\Delta U = 0$, so $Q = W = nRT\ln(V_2/V_1)$;
$\Delta S = \Delta Q/T = 2R\ln(2) = 11.5\ J/K$;
(b) 11.5 J/K

32. (a) $\epsilon_C = 1 - 378/523 = 0.277$. Actual $\epsilon = 0.139 = W/Q_H$, thus
$dQ_H/dt = 1439$ kW, and $dQ_C/dt = 1239$ kW.
In 1 h, $Q_C = 4.46 \times 10^9$ J.
(b) In 1 h, $Q_H = 5.18 \times 10^9$ J, $\Delta S_H = Q_H/T_H = -9.91 \times 10^6$ J/K
(c) $\Delta S_C = Q_C/T_C = 1.18 \times 10^7$ J/K

33. (a) $\epsilon_C = 1 - 300/550 = 0.454$, thus $W = 454$ J, $Q_C = 546$ J
$\Delta S_H = Q_H/T_H = 1.82$ J/K, $\Delta S_C = Q_C/T_C = -1.82$ J/K, others are zero; (b) 0

Problems

1. (a) $PV = nRT$, so $P\Delta V = nR\Delta T$. Note $C_p = 5R/2$, and $C_v = 3R/2$.
$Q_1 = nC_p\Delta T = C_pP\Delta V/R = 2.5(3P_o)(2V_o - V_o) = Q_1 = 7.5P_oV_o$,
$Q_2 = nC_v\Delta T = (nC_v/R)(2P_oV_o - 6P_oV_o) = -6P_oV_o$,
Similarly, $Q_3 = nC_p\Delta T = -2.5P_oV_o$, and $Q_4 = nC_v\Delta T = 3P_oV_o$;
(b) $W = \text{Area} = 2P_oV_o$;
(c) $\epsilon = W/Q_{in} = W/(Q_1 + Q_4) = 0.19$

2. $Q_1 = nRT_1\ln(V_b/V_a) = 500R\ln2$; $Q_2 = nC_v\Delta T = -300R$
$Q_3 = nRT_3\ln(V_d/V_c) = -300R\ln2$; $Q_4 = nC_v\Delta T = 300R$
For 1 cycle $W = \Sigma Q = 200R\ln2$, and
$\epsilon = W/Q_{in} = W/(Q_1 + Q_4) = 0.214$

3. (a) $V_a = V_c = nRT_c/P_c = 0.04157\ m^3$.
$P_a/T_a = P_c/T_c$, thus $P_a = 160$ kPa.
Since $P_c = P_b$, $P_aV_a{}^\gamma = P_cV_b{}^\gamma$, which gives
$V_b = 1.6^{5/7}\ V_a = 1.4V_a$.
$W_{bc} = P\Delta V = P_c(V_c - V_b) = P_c(V_a - 1.4V_a) = -1663$ J; $W_{ca} = 0$
$W_{ab} = \int PdV = \int K\ dV/V^\gamma = K(V_b{}^{1-\gamma} - V_a{}^{1-\gamma})/(1 - \gamma)$, where K $= P_aV_a{}^\gamma$. Find $W_{ab} = P_aV_a{}^\gamma(1/1.4^{0.4} - 1)/(1 - \gamma) = 2094$ J
$W_{net} = 2094 - 1663 = 431$ J;
(b) $Q_{ab} = 0$; $Q_{bc} = nC_p\Delta T = C_pP\Delta V/R = -5820$ J (in)
$\epsilon = W/Q_{in} = 431/5820 = 0.074$

4. (a) $Q_{bc} = nC_p\Delta T = nC_pP\Delta V/nR = 3.5P(V_b - V_a) = 1750$ J (out)
From PV = nRT find $T_c = 250/R$ and $T_a = 2T_c = 500/R$, thus
$Q_{ca} = nC_v\Delta T = 5R(T_a - T_c) = 1250$ J (in)
$Q_{ab} = W_{ab} = nRT\ln(V_b/V_a) = PV\ln 2 = 693$ J; (in)
(b) $W = \Sigma Q = 693 + 1250 - 1750 = 193$ J
(c) $\epsilon = W/Q_{in} = 193/(Q_{ab} + Q_{ca}) = 193/1943 = 9.9\%$

5. $|Q_{bc}| = nC_p(T_c - T_b)$; $|Q_{da}| = nC_v(T_d - T_a)$
$\epsilon = 1 - |Q_{da}|/|Q_{bc}| = 1 - (T_d - T_a)/\gamma(T_c - T_b)$

6. Note $TV^{\gamma-1}$ = constant for paths ab and cd and $V_a = V_d$.
At each point PV = RT.
$T_d - T_a = [T_cV_c{}^{\gamma-1} - T_bV_b{}^{\gamma-1}]/V_a{}^{\gamma-1} = [P_cV^\gamma - P_bV^\gamma]/RV_a{}^{\gamma-1}$
$T_c - T_b = P_c(V_c - V_b)/R$. Substitute into
$\epsilon = 1 - (T_d - T_a)/\gamma(T_c - T_b)$ to find given expression.
For the given values, $\epsilon = 0.558$.

7. $P_aV_a = RT_a$, so $V_a = R/10^3\ m^3$; and $P_cV_c = RT_c$, so
$V_c = 3R/10^3\ m^3$
From $T_aV_a{}^{\gamma-1} = T_dV_d{}^{\gamma-1}$ we find $V_d = (5/3)^{3/2}V_a = 2.15R/10^3$
From $T_bV_b{}^{\gamma-1} = T_cV_c{}^{\gamma-1}$ we find $V_b = (3/5)^{3/2}V_c = 1.39R/10^3$
$W_{ab} = nRT_a\ln(V_b/V_a) = 500R\ \ln(1.39) = 1369$ J
$W_{cd} = nRT_c\ln(V_d/V_c) = 300R\ \ln(2.15/3) = -831$ J
Thus, $W_{net} = 538$ J
(b) $\epsilon = 1 - 300/500 = 0.4$

8. Since $\Delta U = Q - W = 0$ for a complete cycle,
$W = Q = \int dQ = \int T\, dS$

9. At the hot reservior $\Delta S_H = -Q_H/T_H < 0$

10. (a) $W = Q_H - Q_C = 300$ J, so $Q_C = 900$ J
$\epsilon = 1 - 900/1200 = 1/4$
(b) $\Delta S_H = -1200/800 = -1.5$ J/K; $\Delta S_C = +\ 900/300 = +3$ J/K
$\Delta S_u = +1.5$ J/K
(c) $\epsilon_C = 1 - 3/8 = 5/8$, so $W_C = \epsilon_C Q_H = 750$ J
(d) $T_C \Delta S_u = (300)(1.5) = 450$ J. The equals $W_C - W$.

11. (a) $\Delta S_w = C\ln(373/273) = 0.312C$, where $C = Mc = 74.9$ J/mol.K.
$\Delta S_R = -C\Delta T/T_H = -100C/373 = -0.268C$
$\Delta S_u = 0.044C = 3.3$ J/K;
(b) $\Delta S_w = 0.312C$; $\Delta S_R = -50(1/323 + 1/373) = -0.289C$;
$\Delta S_u = 0.023C = 1.73$ J/K;
(c) $\Delta S_w = 0.312C$; $\Delta S_R = -25(1/298 + 1/323 + 1/348 + 1/373) = 0.300C$, thus $\Delta S_u = 0.012C = 0.90$ J/K;
(d) Let $\Delta T \to 0$

12. (a) $\epsilon_1\epsilon_2 = (1 - T_1/T_H)(1 - T_C/T_1)$;
(b) $\epsilon_1\epsilon_2 = \epsilon_C - (T_1/T_H + T_C/T_1)$.

13. $Q_C = mc_w\Delta T_w + mL + mc_i\Delta T_i = 1.97\times10^4$ J,
$W = Q_C/COP = Q_C/4 = \quad 4.92 \times 10^4$ J

14. For an adiabatic process $(T_f/T_i) = (V_i/V_f)^{\gamma-1}$, thus
$\ln(T_f/T_i) = (\gamma - 1)\ln(V_i/V_f)$. Also, $(\gamma - 1)C_v = C_p - C_v = R$.
Find $\Delta S = nC_v\ln(T_f/T_i) + nR\ln(V_f/V_i) = 0$.

CHAPTER 22

Exercises

1. (a) $\underline{F} = 90(6/16 - 10/4)\underline{i} = -191\underline{i}$ N
   (b) $\underline{F} = (9\text{x}10^9)[10/4 - 15/36)(10^{-8})\underline{i} = 188\underline{i}$ N;

2. (a) $67.5\underline{i} - 80\underline{j}$ μN;
   (b) $F_x = -F_{32}\cos\theta - F_{34}$; $F_y = F_{32}\sin\theta$. $\underline{F} = -84.8\underline{i} + 13\underline{j}$ μN

3. (a) $F_x = 0$, $F_y = F_{32}\sin 60^o$, so $F = 208\underline{j}$ N;
   (b) $F_x = (F_{23} - F_{21})\text{cod}60^o$; $F_y = -(F_{21} + F_{23})\sin 60^o$
   $\underline{F} = 80\underline{i} - 277\underline{j}$ N

4. (a) $F_x = kQ^2[2/16 + 0.8(4/25)]\text{x}10^4 = 0.364$ mN
   $F_y = kQ^2[6/9 - 0.6(4/25)]\text{x}10^4 = 0.822$ mN
   (b) $F_x = kQ^2[6/16 + 0.8(3/25)]\text{x}10^4 = 0.678$ mN
   $F_y = kQ^2[-6/9 + 0.6(3/25)]\text{x}10^4 = -0.857$ mN

5. (a) $27/x^2 = 3/(1 - x)^2$, leads to $x = 0.75$ m;
   (b) $27/(1 + d)^2 = 3/d^2$, leads to $d = 0.5$ m or $x = 1.5$ m

6. $kQ^2 = GmM$, so $Q = (GmM/k)^{1/2} = 5.72\text{x}10^{13}$ C

7. $r^2 = kQ^2/F$ leads to $r = 1.52 \text{ x } 10^{-14}$ m

8. (a) $kqQ/r^2 = 4.61\text{x}10^3$ N; (b) $a = F/m = 6.88\text{x}10^{29}$ m/s^2

9. (a) $ke^2/r^2 = 4.21 \text{ x } 10^{-8}$ N; (b) $ke^2/r^2 = 2.90 \text{ x } 10^{-9}$ N

10. (a) $kqQ(-0.222\underline{i} - 0.250\underline{j})$ or $F = 0.334kqQ$ at 48.4^o below -x axis;
    (b) Need $F = 0.334kqQ$ opposite to (a). Thus $r = 2.73$ m at 48.4^o below the -x axis, or, (-1.82 m, -2.04 m).

11. By symmetry $q_1 = q_2$ and they must be negative.
    Consider force on the 2 μC on the left.
    $F = k(2\ \mu C)[|q_1|/1 + |q_2|/9 - (1\ \mu C)/4 - (4\ \mu C)/16]\text{x}10^4$
    Find $q_1 = q_2 = -27/80$ μC

12. ΣF_x: $T\sin\theta = kQ^2/d^2$, where $d = 2L\sin\theta$.
    ΣF_y: $T\cos\theta - mg = 0$
    Find $Q^2 = mgd^2 \tan\theta/k$, thus $Q = 0.395$ μC

13. $F_{elec}/F_{grav} = kQ_1Q_2/GM_1M_2 = 2.85 \times 10^{-18}$

14. (a) $kqQ/x^2 = k(9qQ)/(4 - x)^2$, so x = 1 m.
    If net force on 9Q is zero: $9Q^2/16 = 9Qq/9$, so $q = -9Q/16$.
    (b) $1/d^2 = 9/(4 + d)^2$, so d = 2 m, or x = -2 m.
    If net force on Q is zero: $qQ/4 = 9Q^2/16$, thus $q = -9Q/4$.

15. $k(q)(2q)/r^2 = 0.2$, thus $q = \pm 133$ nC, $2q = \pm 267$ nC.

16. (a) $F_1 = F_2 = kqQ/d^2$, so $F = 2kqQ/d^2 = 0.36\underline{i}$ mN;
    (b) $F_1 = kqQ/d^2$, $F_2 = kqQ/5d^2$ at $\alpha = \cos^{-1}(2/(5)^{1/2})$ to x axis.
    $F_x = 2kqQ/5(5)^{1/2}d^2$; $F_y = [1 - 1/5(5)^{1/2}]kqQ/d^2$
    $\underline{F} = 0.0322\underline{i} + 0.164\underline{j}$ mN;
    (c) $F_y = 0$, $F_x = (2kqQ/d^2)\cos 45^o = 0.127\underline{i}$ mN;
    (d) $F_x = 2kqQ/5(5)^{1/2}d^2$; $F_y = [-1 + 1/5(5)^{1/2}]kqQ/d^2$
    $\underline{F} = 0.0322\underline{i} - 0.164\underline{j}$ mN

17. Let $F_o = k(e^2/9)/3r^2 = 5.926$ N
    [u]: $F_x = 4F_o - 2F_o\cos 60^o = 3F_o$; $F_y = 2F_o\sin 60^o = (3)^{1/2}F_o$
    Thus $F_u = (12)^{1/2}F_o = 20.5$ N
    [d]: $F_x = 0$; $F_y = 4F_o \sin 60^o = 20.5$ N

18. $F = k(10^{-11})/25 = 3.6\text{x}10^{-3}$ N
    $F_x = 0.8F = 2.88\text{x}10^{-3}$ N; $F_y = -0.6F = -2.16\text{x}10^{-3}$ N

19. $kq^2/r^2 = 5.76 \times 10^5$ N

Problems

1. $100kq_1q_3 = 15$, $100kq_2q_3 = 9$, $100kq_1q_2 = 5.4$
   From $q_1/q_2 = 15/9$, find $q_2^2 = (5.4/15)10^{-11}$, and then
   $q_2 = 1.90\ \mu C$. Then find $q_1 = -3.16\ \mu C$, $q_3 = -5.27\ \mu C$

2. (a) $F_y = 0$; $F_x = (2kqQ/r^2)\cos\theta = 2kqQx/(a^2 + x^2)^{3/2}$;
   (b) Set $dF/dx = 0$ to find $x = \pm a/(2)^{1/2}$;
   (c) $2kqQ/x^2$

3. (a) $F_x = 0$; $F_y = -2F\sin\theta = -2kQqa/(a^2 + x^2)^{3/2}$;
   (b) x = 0

4. (a) $F = kqQ[1/(y - a)^2 - 1/(y + a)^2] = 4kqQay/(y^2 - a^2)^2$
(b) $F \approx 4kqQa/y^3$

5. $F = kq(Q - q)/r^2$. Set $dF/dq = 0$ to find $q = Q/2$

6. (a) $-q_1q_2 = -15\text{x}10^{-12}$, $(q_1 - q_2)^2 = 4\text{x}10^{-12}$
Find $q_2^2 + 2\text{x}10^{-6}q_2 - 15\text{x}10^{-12} = 0$, thus $q_2 = 3\ \mu C, -5\mu C$
The possible charges are ($3\ \mu C$, $-5\mu C$) or ($-3\ \mu C$, $5\ \mu C$)
(b) $-q_1q_2 = -15\text{x}10^{-12}$, $(q_1 - q_2)^2 = 60\text{x}10^{-12}$
Find $q_2^2 + 7.746\text{x}10^6q_2 - 15\text{x}10^{-12} = 0$, thus
$q_2 = 1.60\ \mu C, -9.35\ \mu C$. The possible charges are
($1.60\ \mu C$, $9.35\ \mu C$) or ($-1.60\ \mu C$, $-9.35\ \mu C$).

7. (a) $kQ^2/r^2 = 10$, thus $Q = 3.33\ \mu C$. $N = Q/e = 2.08 \times 10^{13}$;
(b) In 10 g, number of atoms $n = mN_A/M$, whereas number of electrons is Zn, where $Z = 29$. Thus, fraction
$f = N/Zn = 7.57 \times 10^{-12}$

8. (a) $(kq^2/L^2)(\underline{i} + \underline{j} + \underline{k}) + (kq^2/2L^2)(2\underline{i} + 2\underline{j} + 2\underline{k})/2^{1/2} + (kq^2/3L^2)(\underline{i} + \underline{j} + \underline{k})/3^{1/2} = (1.90kq^2/L^2)(\underline{i} + \underline{j} + \underline{k})$
(b) $-(0.485kq^2/L^2)(\underline{i} + \underline{j} + \underline{k})$

9. (a) $ke^2/r^2 = mv^2/r$, so $v = (ke^2/mr)^{1/2}$;
(b) $L = mvr = nh/2\pi$, leads to $r = (nh/2\pi)^2/mke^2$
(c) 5.3×10^{-11} m, 2.12×10^{-10} m, 4.77×10^{-10} m

10. (a) $q_1 + q_2 = 8\mu$, $q_1q_2 = 15\mu^2$, lead to $q = 3\ \mu C$, $5\ \mu C$
(b) $q_1 - q_2 = 8\mu$, $q_1q_2 = 15\mu^2$, lead to $q_2 = 1.57\ \mu C$,
$q_1 = 9.57\ \mu C$ and $q_2 = -9.57$, $q_1 = -1.57\ \mu C$

CHAPTER 23

Exercises

1. Use $mg = eE$: (a) 5.58×10^{-11} N/C; (b) 1.02×10^{-7} N/C

2. (a) $F = eE = 1.92\text{x}10^{-17}$ N; (b) $a = F/m = 1.15\text{x}10^{10}$ m/s^2

3. (a) $\underline{E} = F/q_1 = 2500\underline{i}$ N/C; (b) $F = q_2E = -1.6 \times 10^{-5}\ \underline{i}$ N

4. (a) $4/d^2 = 9/(1 + d)^2$, thus $d = 2$ m, or $x = -2$ m.
   (b) $4/x^2 = 1/(1 - x)^2$, thus $x = 2/3$ m

5. (a) $E_y = -12kQ/(2)^{1/2}L^2 = -7.64 \times 10^{10}\ Q/L^2$;
   (b) $E_x = -8kQ/5(5)^{1/2}L^2$; $E_y = -12kQ[(1 + 1/5(5)^{1/2}]/L^2$
   $\underline{E} = (-6.44\underline{i} - 118\underline{j})\text{x}10^9\ Q/L^2$

6. $k(Q_1 + Q_2) = 10.8$, $k(Q_1/9 - Q_2/1) = -0.8$
   Find $Q_1 = 1$ nC; $Q_2 = 0.2$ nC

7. $2eE = mg$, $E = 3.06 \times 10^6$ N/C

8. (a) $\underline{E}_1 = (kq_1/r^2)\underline{i} = 4.22\text{x}10^3\underline{i}$ N/C
   (b) $\underline{E}_2 = (kq_2/r^2)\underline{i} = 9.84\text{x}10^3\underline{i}$ N/C
   (c) $\underline{F}_2 = q_2\underline{E}_1 = -2.95\text{x}10^{-5}\underline{i}$ N
   (d) $\underline{F}_1 = q_1\underline{E}_2 = +2.95\text{x}10^{-5}\underline{i}$ N

9. (a) $E_x = -E\cos\theta$, $E_y = E\sin\theta$, where $E = 9000$ N/C
   $\underline{E} = -8050\underline{i} + 4020\underline{j}$ N/C;
   (b) $E = 3460$ N/C, $\underline{E} = 1920\underline{i} - 2880\underline{j}$ N/C

10. $E_1 = 2.12\text{x}10^3$ N/C, $E_2 = 27\text{x}10^3$ N/C.
    $E_x = -E_1/(17)^{1/2} + 2E_2/(5)^{1/2} = 23.6\text{x}10^3$ N/C
    $E_y = -4E_1/(17)^{1/2} + E_2/(5)^{1/2} = 10.0\text{x}10^3$ N/C

11. (a) $k(2\ \mu C)/(5\ \text{cm})^2 = 7.2 \times 10^6$ N/C
    $E_x = -14.4\text{x}10^6 + 7.2\text{x}10^6\cos 60^o = -10.8\text{x}10^6$ N/C
    $E_y = 7.2\text{x}10^6 \sin 60^o = 6.24\text{x}10^6$ N/C
    (b) $\underline{F} = q\underline{E} = 32.4\underline{i} - 18.7\underline{j}$ N;
    (c) Not at all

12. (a) $Q_1 = Q_2$; (b) $Q_1 = -4Q_2$; (c) $Q_2 = -9Q_1$

13. $\underline{E} = kQ\underline{r}/r^3$, $E_x = \underline{E}.i = kQx/r^3$, etc.

14. $E = kq/r^2$: (a) $2.25\text{x}10^{21}$ N/C; (b) $5.13\text{x}10^{11}$ N/C

15. $(x < 0)$ $E = kq[-1/x^2 + 1/(6 - x)^2]$
$(0 < x < 6)$ $E = kq[1/x^2 + 1/(6 - x)^2]$
$(x > 6)$ $E = kq[1/x^2 - 1/(x - 6)^2]$

16. (a) $(x < 0)$ $E = kq[-2/x^2 + 1/(6 - x)^2]$
$(0 < x < 0)$ $E = kq[2/x^2 + 1/(6 - x)^2]$
$(x > 6)$ $E = kq[2/x^2 - 1/(x - 6)^2]$
(b) $2/x^2 = 1/(x - 6)^2$, so $x = 20.5$ m

17. $\underline{E}_1 = 500\underline{i}$ N/C; $\underline{E}_2 = 720(0.6\underline{i} + 0.8\underline{j})$ N/C
$\underline{E} = 932\underline{i} + 576\underline{j}$ N/C
$\underline{F} = q\underline{E} = (4.66\underline{i} + 2.88\underline{j})\text{x}10^{-3}$ N

18. (a) $\underline{E} = -7.2\text{x}10^3\underline{i} - 16.2\text{x}10^3\underline{j}$ N/C; (b) Nothing; (c) Nothing

19. (a) $(2kq/r^2)[1 - (1 + a^2/r^2)^{-3/2}]$;
(b) $2kqa^2(a^2 - 3r^2)/r^2(r^2 - a^2)^2$
(c) For $r >> a$, $E \alpha 1/r^4$ in both cases

26. Inner: -16μC, outer: $+8$ μC

27. (a) $\Delta t = v/a = (0.1c)m/eE = 1.71$ ns;
(b) $\Delta x = 1/2\ at^2 = 2.56$ cm;
(c) $K = 1/2\ mv^2 = 4.1 \text{ x } 10^{-16}$ J

28. $\Delta K = (eE)\Delta x$, so $E = \Delta K/e\Delta x = 4450$ N/C

29. (a) $\Delta x = v^2/2a = mv^2/2eE = 13.9$ cm;
(b) $\Delta t = \Delta x/v_{av} = 0.348$ μs

30. $t = L/v = 2\text{x}10^{-8}$ s, $\Delta y = 1/2\ at^2 = 8\text{x}10^{-3}$,
From $a = eE/m$, find $E = 228$ N/C

31. (a) $v = (2K/m)^{1/2} = 9.37\text{x}10^5$ m/s, $a = eE/m = 5.27\text{x}10^{17}$ m/s^2
Thus $t = v/a = 1.78 \text{ x } 10^{-12}$ s;
(b) $\Delta x = 1/2\ at^2 = 8.33 \text{ x } 10^{-7}$ m

32. (a) $ke^2/(2r)^2 = mv^2/r$, so $v = 1.13\text{x}10^6$ m/s

(b) $T = 2\pi r/v = 2.79\text{x}10\text{-}16$ s

33. $0 = (v_o\sin45^o)^2 - 2(eE/m)\Delta y$, leads to $v_o = 3.75\text{x}10^6$ m/s.

34. (a) $t = L/(v_o\cos\theta) = 5.77\text{x}10^{-8}$ s, $a = eE/m = -9.58\text{x}10^{12}$ m/s^2
$y = (v_o\sin\theta)t + 1/2\ at^2 = 7.11$ mm
(b) $v_y = v_o\sin30^o + at = -1.53\text{x}10^5$ m/s, $v_x = 6.93\text{x}10^5$ m/s
$\tan\theta = v_y/v_x$ leads to $\theta = -12.5^o$

35. I and III: $\sigma/2\epsilon_0$; II: $5\sigma/2\epsilon_0$; IV: $-\sigma/2\epsilon_0$

36. $E = \sigma/\epsilon_0 = 1000$, so $\sigma = 8.85\text{x}10^{-9}$ C/m^2

37. (a) $F = qE = q(\sigma/2\epsilon_o) = 2.26$ N;
(b) $kq/r^2 = \sigma/2\epsilon_o$, leads to r = 12.6 cm from q

38. From Example 23.6: $E = k\lambda[1/a - 1/(a + L)]$ where a = 0.15 m
Find $E = 4.8\text{x}10^4$ N/C.

39. From Example 23.8: $E = 2\pi k\sigma[1 - y/(a^2 + y^2)^{1/2}] =$
2.02×10^4 N/C

40. $\underline{E} = 2k\lambda(\underline{i}/x + \underline{j}/y)$

41. From Example 23.6: $\underline{E} = kQ(-\underline{i} + \underline{j})/d(d + L)$
$= (-3\underline{i} + 3\underline{j})\text{x}10^6$ N/C

42. (a) $\Delta U = -pE\cos60^o - (-pE) = 1.33\text{x}10^{-25}$ J
(b) $\tau = pE\sin60^o = 2.30\text{x}10^{-25}$ N.m

43. (a) $qd = 8 \times 10^{-11}$ C.m;
(b) $\Delta U = 0 - (-pE) = 8 \times 10^{-6}$ J

44. (a) $F_y = eE_y \approx 2ekp/r^3 = 1.43\text{x}10^{-10}$ N
(b) $F_x = eE_x \approx ekp/r^3 = 7.14\text{x}10^{-11}$ N

45. (a) $E = kqQ/r^2 = 3.2 \times 10^5$ N/c; (b) $a = eE/m$
$= 5.62 \times 10^{16}$ m/s^2.

46. $E = mg/e = 5.58 \times 10^{-11}$ N/C, downward.

47. (a) $E_x = k(2.2\text{ nC} + 3.5\text{ nC})/4 = 12.8$ N/C;
(b) $E_2 = k(3.5\text{ nC})/20$ and $\tan\theta = 0.5$. Thus, $E_x = E_2\cos\theta =$

$= 1.41$ N/C and $E_y = E_1 - E_2\sin\theta = 4.25$ N/C.

48. (a) $E_x = -k(5.2\text{ nC})(1/4 + 1/16) = -14.6$ N/C.
(b) $\underline{F} = 4.39 \times 10^{-8}\underline{i}$ N.

49. From Q's to center $d = 0.5L/\cos 30^o$. Note $E_x = 0$.
$E_y = 2kQ\cos 60^o/d^2 + kQ/d^2 = 2kQ/d^2 = 6kQ/L^2$.

50. (a) $0 = v_o - (eE/m)t$, so $t = 8.54$ ns.
(b) $0 = v_o^2 - 2a\Delta x$, so $\Delta x = 1.28$ cm

51. $T\sin\theta = qE$ and $T\cos\theta = mg$ give $\tan\theta = qE/mg$, so $q = 80.1$ nC

52. $kQ/2.5^2 + kq/1.5^2 = 0$, thus $q = -7.2$ nC.

53. $E_x = -E_1 + E_2\cos\theta$ and $E_y = -E_2\sin\theta$ where $\tan\theta = 0.5$.
$\underline{E} = 1.30\underline{i} - 4.02\underline{j}$ N/C.

54. (a) $\Delta x = 0.5(eE/m)t^2$, so $E = 9{,}880$ N/C;
(b) $v = at$, so $t = 3.17\ \mu s$.

Problems

1. Since one part of path is completely out of the (uniform) field $\int \underline{E}.d\underline{\ell}$ cannot be zero.

2. (a) $E = 2\pi Rk\lambda x/(x^2 + R^2)^{3/2}$;
(b) Set $dE/dx = 0$ to find $x = \pm R/(2)^{1/2}$;
(c) $2\pi R\lambda k/x^2$

3. $F_x = pdE/dx$, so $\underline{F} = -(Cp/x^2)\underline{i}$

4. (a) $dE_y = k(\lambda Rd\alpha)\cos\alpha/R^2$, where α is to the vertical.
Integration from $\alpha = -\theta/2$ to $\theta/2$ leads to $E = 2k\lambda\sin(\theta/2)/R$.
(b) For $\theta = \pi$, $E = 2k\lambda/R$.

6. $E(r) = kQ(r)/r^2$, where $Q(r) = 4\pi\rho r^3/3$. Thus $E(r)\ \alpha\ r$.

7. $dE_y = (k\lambda dx)\cos\theta/r^2$, where $\cos\theta = y/r$ and $r^2 = y^2 + x^2$.
$E = k\lambda y \int dx/(x^2 + y^2)^{3/2} = k\lambda y(1/y^2)[x/(x^2 + y^2)^{1/2}]$
$= 2kQ/y(L^2 + 4y^2)^{1/2}$. Find: (a) kQ/y^2; (b) $2kQ/yL$

8. $D = (\ell/2 + L)(eE/m)(\ell/v_o^2) = (\ell/2 + L)\tan\theta$

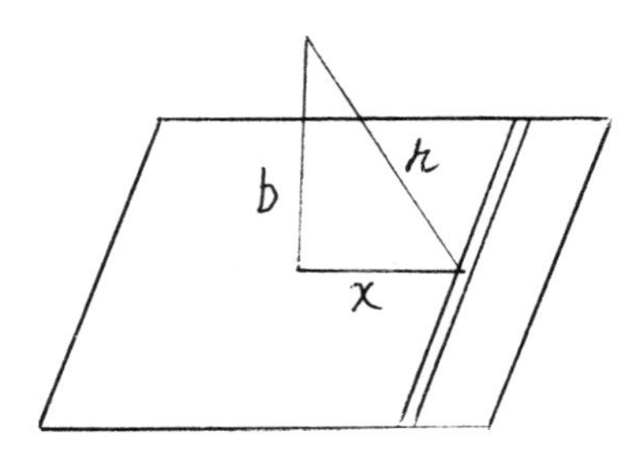

9. Replace $\lambda = \sigma dx$, thus

$dE_y = dE(b/r) = 2k\sigma b\ dx/r^2$

$E_y = 2k\sigma b \int dx/(b^2 + x^2)$

$= 2k\sigma[\tan^{-1}(x/b)]_{-\infty}^{\infty} = \sigma/2\epsilon_0$

10. $m(4\pi^2 R/T^2) = q(2k\lambda/R)$, thus $T = \pi R(2m/kq\lambda)^{1/2}$

11. $\tau = I\alpha$: $-pE\sin\theta = I\ d^2\theta/dt^2$. For small θ, $\sin\theta \approx \theta$, $d^2\theta/dt^2 + (pE/I)\theta = 0$, thus $\omega^2 = pE/I$ and $f = 1/2\pi\ (pE/I)^{1/2}$

12. (a) $dE = k\lambda dx/x^2$, $E = k\lambda/R$

(b) $dE_x = dE\sin\theta = k\lambda x\ dx/(x^2 + R^2)^{3/2}$

$E_x = k\lambda[-1/(x^2 + R^2)^{1/2}] = k\lambda/R$

$dE_y = dE\cos\theta = k\lambda R\ dx/(x^2 + R^2)^{3/2}$

$E_y = k\lambda R[x/R^2(x^2 + R^2)^{1/2}] = k\lambda/R$

Thus $E = (2)^{1/2}k\lambda/R$ at 45°.

13. (a) $2kQx/(x^2 + a^2)^{3/2}$; (b) $2kQ/x^2$;

(c) From $dE/dx = 0$, $x = \pm a/(2)^{1/2}$

14. (a) $E \approx kQx/R^3$

(b) $F = -qE = -(kqQ/R^3)x$. Since $F = -Cx$, motion is SHM.

(c) $d^2x/dt^2 + (kqQ/mR^3)x = 0$, thus $\omega^2 = kqQ/mR^3$.

15. $4\pi r^3(\rho - \rho_A)g/3 = 6\pi\eta r v_T$, thus

$r^2 = 9\eta v_T/2g(\rho - \rho_A)$. Substitute for r into

$m_{eff}g = qE$ to find the given expression.

16. $a = eE/m = 1.76\text{x}10^{14}\ m/s^2$ (down).

$t = L/v_o\cos 30^o = 4.62\text{x}10^{-2}/v_o$

$\Delta y = \pm(5\text{x}10^{-3}) = (v_o\sin 30^o)t - (0.5)(1.76\text{x}10^{14})t^2$

From this obtain $v_o = 2.59\text{x}10^6$ m/s, $3.22\text{x}10^6$ m/s

17. Use $R = v_o^2\sin 2\theta/a$, where $a = eE/m$. Find $\theta = 25.7^o$

18. Integrate $dE_x = -k(\lambda R\ d\theta)\sin\theta/R^2$ from 0 to $\pi/2$ to find

$E_x = -2k\lambda/R\ [-\cos\theta] = -2k\lambda/R$.

19. $dE = k(\lambda\ dx)/x^2$, $E = 2\lambda \int dx/x^2 = 2k\lambda/a$ in the +x direction.

20. $E_x = 0$ by symmetry. $dE_y = k\lambda(R\ d\theta)\sin\theta/R^2 = (k\lambda_o/2R)(1 - \cos 2\theta)$. Integrating 0 to π gives $E_y = \pi k\lambda_o/2R$.

CHAPTER 24

Exercises

1. ϕ = EAcos60^o = 10.2 N.m^2/C

2. ϕ = EAcos53^o = 0.867 N.m^2/C

3. Projected area is πR^2, so $\phi = \pi R^2 E$

4. $\phi = \underline{E}.A$ = (70$\underline{i}$ + 90$\underline{k}$ N/C).(144x10$^{-4}\underline{k}$ m^2) = 1.30 N.m^2/C

5. $\phi = (q_1 + q_2)/\epsilon_o$ = -2.26 x 10^5 N.m^2/C (inward)

6. $Q = \epsilon_o\phi = \epsilon_o(6)(3\text{x}10^4)$ = 1.59 μC

7. (a) Q/ϵ_o = 6.78 x 10^6 N.m^2/C;
   (b) 1.13 x 10^6 N.m^2/C;
   (c) No for (a), yes for (b)

8. Zero for "near" faces; $(1/8)(1/3)Q/\epsilon_o = Q/24\epsilon_o$ for "far" faces

9. (a) $Q = 4\pi R^2\sigma$ = 8.04x10^{-12} C, $E = kQ/R^2$ = 11.3 N/C;
   (b) $E = kQ/r^2$ = 7.23 N/C

10. (a) Inside: E = 1.44x10^5/r^2; Outside: E = 0.72x10^5/r^2
    (b) Inner: -16 μC; Outer: +8 μC

11. $E = kQ/R^2 = k(4\pi R^2\sigma)/R^2 = 4\pi k\sigma = \sigma/\epsilon_o$

12. (a) $E = kQ/R^2$, so $Q = ER^2/k$ = -3.56x10^{-11} C
    (b) No. A spherically symmetric distribution

13. (a) Zero; (b) σ/ϵ_0

14. (a) $2\sigma/\epsilon_o$; (b) σ/ϵ_0

15. (a) σ/ϵ_0; (b) Zero

16. (a) $(a + bL)L^2 - aL^2 = bL^3$;
    (b) $bL^3 = Q/\epsilon_o$, so $Q = bL^3\epsilon_o$

17. $(2\pi rL)E = (2\pi RL)\sigma/\epsilon_o$, thus $E = R\sigma/\epsilon_o r$
    For net E = 0 , need $a\sigma_1 + b\sigma_2 = 0$, i.e. $\sigma_2 = -a\sigma_1/b$

18. $(2\pi rL)E = (2\pi RL)\sigma/\epsilon_o$, thus $E = R\sigma/\epsilon_o r$
(a) $E = a\sigma/r\epsilon_o$; (b) $E = (a - b)\sigma/r\epsilon_o$

19. $E = 2k\lambda/r$, thus need $\lambda_2 = -\lambda_1$

20. (a) $E = 2k\lambda/r$; (b) zero

21. (a) kQ/r^2; (b) zero

22. (a) $kQ/r^2 = k(4\pi a^2\sigma)/r^2 = a^2\sigma/r^2\epsilon_o$;
(b) $(a^2 - b^2)\sigma/r^2\epsilon_o$

23. $E = kQ/r^2 = R^2\sigma/r^2\epsilon_o$. Thus need, $\sigma_a a^2 + \sigma_b b^2 = 0$,
or, $\sigma_a/\sigma_b = -b^2/a^2$

24. (a) $\sigma_1 = -Q/4\pi R_1{}^2$, $\sigma_2 = +Q/4\pi R_2{}^2$
(b) kQ/r^2; (c) kQ/r^2
(d) Yes, σ_2 is still uniform

25. $E = 2k\lambda/r = 245$ N/C

26. (a) $E = 2k\lambda/r = 1350$ N/C; (b) $E_a - E_b = -900$ N/C.

27. $E = kQ/r^2$, thus $Q = 4\pi R^2\sigma = -9.68$ nC and $\sigma = -7.70\text{x}10^{-8}$ C/m^2.

28. (a) $kQ_1/r^2 = 1.29\text{x}10^4$ N/C; (b) $k(Q_1 + Q_2)/r^2 = 9900$ N/C.

29. $\lambda H/\epsilon_o$

30. (a) $Q/6\epsilon_o - 500(0.4\text{ m})^2 = -38.6$ N.m^2/C
(b) $Q/6\epsilon_o + 500(0.4\text{ m})^2 = 122$ N.m^2/C.

31. Zero for xy, xz and yz planes, $Q/24\epsilon_o$ for each of the other three.

32. (a) $2kQ/r^2$; (b) $k(2Q - 3Q)/r^2 = -kQ/r^2$.

33. (a) $E = kQ/r^2$ and $Q = \rho(4\pi r3/3)$, so $E = \rho r/3\epsilon_o$ and
$\rho = 1.06\ \mu$C/m^3.
(b) $Q = 4\pi\rho R^3/3 = 4.45$ nC, so $E = 1000$ N/C.

34. $EL^2/4\epsilon_o$.

35. Within charged sphere $E = kq/r^2$ where $q = \rho(4\pi r^3/3)$ so $E = \rho r/3\epsilon_0$. For whole sphere $q = e = \rho(4\pi R^3/3)$ so $E = ker/R^3$ directed inward since ρ is negative. Add the field due to positive point charge at the center to get the expression.

Problems

1. (a) $4\pi r^2E = 4\pi\rho r^3/3\epsilon_0$, so $E = \rho r/3\epsilon_0$;
   (b) $4\pi r^2E = 4\pi\rho R^3/3\epsilon_0$, so $E = \rho R^3/3\epsilon_0 r^2$

2. $Q(r) = \int \rho\ 4\pi r^2\ dr = \pi Ar^4$, total $Q = \pi AR^4$
   (a) $4\pi r^2E = \pi Ar^4/\epsilon_0$, thus $E = Ar^2/4\epsilon_0 = kQr^2/R^4$
   (b) $4\pi r^2E = \pi AR^4/\epsilon_0$, thus $E = AR^4/4\epsilon_0 r^2 = kQ/r^2$

3. (a) $-4\pi\sigma R_1{}^2$;
   (b) $-4\pi\sigma(R_2{}^2 - R_1{}^2)$;
   (c) $4\pi r^2E = -(4\pi R_2{}^2\sigma)/\epsilon_0$, thus $E = -\sigma R_2{}^2/\epsilon_0 r^2$

4. The "far field" is $\sigma/2\epsilon_0$, thus $F/A = \sigma E = \sigma^2/2\epsilon_0$ N/m^2

5. (a) $(2\pi rL)E = (\pi r^2L\rho)/\epsilon_0$, thus $E = \rho r/2\epsilon_0$;
   (b) $(2\pi rL)E = (\pi R^2L\rho)/\epsilon_0$, thus $E = \rho R^2/2\epsilon_0 r$

6. (a) $4\pi r^2E = \rho 4\pi(R^3 - a^3)/3\epsilon_0$, so $E = \rho(R^3 - a^3)/3\epsilon_0 r^2$
   (b) $4\pi r^2E = \rho 4\pi(r^3 - a^3)/3\epsilon_0$, so $E = \rho(r^3 - a^3)/3\epsilon_0 r^2$

7. Point charge: $E_1 = ke/r^2$
   Uniform charge: $4\pi r^2E = \rho 4\pi r^3/3\epsilon_0$, thus $E_2 = \rho r/3\epsilon_0$
   But $\rho = -3e/4\pi R^3$, thus $E_2 = -ker/R^3$. Total field is
   $E = ke(1/r^2 - r/R^3)$

8. $4\pi r^2g = -4\pi Gm$, thus $g = -Gm/r^2$. Thus, $F = Mg = -GmM/r^2$.

9. (a) kQ/r^2; (b) kQ/r^2

10. $dQ = \rho 4\pi r^2\ dr = 4\pi Ar\ dr$, thus $Q(r) = 2\pi A(r^2 - a^2)$
    $4\pi r^2E = 2\pi A(r^2 - a^2)/\epsilon_0$, thus $E = A(1 - a^2/r^2)/2\epsilon_0$

11. (a) $(2\pi rL)E = [\rho\pi L(r^2 - a^2)]/\epsilon_0$, so $E = \rho(r - a^2/r)/2\epsilon_0$;
    (b) $(2\pi rL)E = [\rho\pi L(R^2 - a^2)]/\epsilon_0$, so $E = \rho(R^2 - a^2)/2\epsilon_0 r$

12. $dQ = \rho 4\pi r^2\, dr = \rho_0(1 - r/R)4\pi r^2\, dr$, thus
$Q(r) = 4\pi\rho_0(r^3/3 - r^4/4R)$
$4\pi r^2 E = Q(r)/\epsilon_0$, thus $E = \rho_0(r/3 - r^2/4R)/\epsilon_0$

13. Inside: $2AE = 2Ax\rho/\epsilon_0$, thus $E = \rho x/\epsilon_0$
Outside: $E = \rho t/2\epsilon_0$

14. $4\pi r'^2 E = \rho 4\pi r'^3/3\epsilon_0$, thus $\underline{E}_1 = (\rho\underline{r}'/3\epsilon_0)$.
Similarly, $\underline{E}_2 = -(\rho\underline{r}/3\epsilon_0)$
$\underline{E}_T = \rho(\underline{r}' - \underline{r})/3\epsilon_0 = \rho\underline{d}/3\epsilon_0$.

CHAPTER 25

Exercises

1. (a) $q\Delta V = 1.87 \times 10^{28}$ eV; (b) 5×10^7 s $\approx$ 1.58 y

2. (a) (80)(3600) = 2.88×10^5 C; (b) qV = 3.46×10^6 J

3. $W_{ext} = q(V_f - V_i)$, thus $4\text{x}10^{-7} = (-5 \text{ nC})(-20 - V_i)$. Find V_i = 60 V

4. (a) $V_B - V_A = -\underline{E}.\Delta\underline{s} = -(-180k)(10^{-1}\underline{k}) = +18$ V
   (b) $\Delta V = Ed$, so $d = \Delta V/E = 27/180 = 0.15$ m

5. $V_B - V_A = -\int E_x\, dx - \int E_y\, dy = +6$ V

6. $V(x) - V(0) = -\int E_x\, dx$.
   (a) $-A \ln(x/x_o)$; (b) $(e^{-Bx} - 1)A/B$

7. $1/2\ mv^2 = eV$, $V = 2.847\text{x}10^{-12}v^2$
   (a) 3.10×10^{-7} V; (b) 3.57×10^{-4} V; (c) 2.56×10^3 V

8. $1/2\ mv^2 = eV$, $V = 5.22\text{x}10^{-9}v^2$
   (a) 5.68×10^{-4} V; (b) 0.655 V; (c) 4.70 MV

9. $v = (2eV/m)^{1/2}$: (a) 2.05×10^6 m/s; (b) 4.80×10^4 m/s

10. V = Ed = 3 kV

11. $W_{ext} = q\Delta V = (-2\ \mu C)(-10\ V) = 2 \times 10^{-5}$ J

12. (a) $V_B - V_A = -\underline{E}.d\underline{s} = -(600\ V/m)(-0.04\ m) = +24$ V
    (b) $U_B - U_A = q(V_B - V_A) = (-3\ \mu C)(24\ V) = -7.2\text{x}10^{-5}$ J

13. V = Ed = (F/q)d = 150 V

14. $V = mv^2/2q$: (a) 9.33×10^6 V; (b) 1.19×10^7 V

15. V = Ed: (a) 216 V; (b) 52 kV

16. (a) $E = \Delta V/d = 4$ kV/m;
    (b) $W = \Delta K = -q\Delta V = +1.92\text{x}10^{-17}$ J;
    (c) $\Delta V = +120$ V; (d) $\Delta U = -1.92\text{x}10^{-17}$ J

17. $W = \Delta K + \Delta U = 1/2\ mv^2 + q(-6000) = 1.6 + 0.09 = 1.69$ J

18. (a) $U = ke^2/r = 2.3x10-13$ J;

(b) $\Delta K = -\Delta U$, $mv^2 = -ke^2(1/4 - 1)/10^{-15}$, so $v = 1.02x10^7$ m/s

19. $\Delta K = -\Delta U = k(48e)(44e)/r = 6.95x10^{-11}$ J

20. (a) $V = k(10^{-4})(4/4(2)^{1/2} + 2/4 - 3/4) = 4.11x10^5$ V

(b) $U = qV = -0.823$ J

(c) Six terms lead to $U = k10^{-10}[-1/2 - 7/2(2)^{1/2}] = -2.68$ J

21. $V = (9x10^3)(0.6 + 2.2 - 3.6 + 4.8)/(0.0707) = 5.09x10^5$ V

$W = \Delta U = q\Delta V = (-5\ \mu C)(5.09x10^5 V) = -2.55$ J

22. (a) $\Delta V = -2kQ/3 = -3x10^4$ V;

(b) $\Delta K = -q\Delta V$, so $1/2\ mv^2 = 2kqQ/3$, thus $v = 20$ m/s.

23. (a) $\Delta V = 2kQ(1/3 - 1/5) = 12$ kV;

(b) From $\Delta K = -q\Delta V = +0.06$ J, find $v = 2$ km/s

24. $kQ/r^2 = 200$; $kQ/r = 600$: Find $r = 3$ m, $Q = 2x10^{-7}$ C

25. (a) For $-Q$ at $x = 1$ m: $4/x = 1/(1 - x)$, thus $x = 0.8$ m.

Or, $4/x = 1/(x - 1)$, thus $x = 4/3$ m.

(b) $4/x = 9/(1 - x)$ so $x = 4/13 = 0.308$ m

Or, $4/d = 9/(1 + d)$, so $d = 0.8$ m or $x = -0.8$ m

26. $V_A = 7.637x10^5$ V; $V_B = 5.927x10^5$

$W = q\Delta V = (-4\ \mu C)(-1.71x10^5\ V) = -0.684$ J

27. $V = kQ/r$: (a) 3.00 MV;

(b) $r = 2.236x10^{-2}$ m, so $V = 2.01$ MV; (c) 1.80 MV

28. (a) $V = (9x10^3)(6/0.05 - 4/0.07) = 5.66x10^5$ V

(b) $W = qV = +1.13$ J

29. $U = (ke^2/9d)(4 - 2 - 2) = 0$

30. (a) $k(3.2\ \mu C)/0.0583 = 4.94x10^5$ V

(b) $k(-4\ \mu C)/0.0583 = -6.17x10^5$ V

(c) $U = kq_1q_2/r = -1.98$ J

31. (a) $V(0) = (9x10^3)(6/0.03 - 2/0.025 + q_3/0.025) = 0$

Find, $q_3 = -3\ \mu C$

(b) $V(0) = (9x10^3)(6/0.03 - 2/0.025 + q_3/0.025) = -4x10^5$
Find, $q_3 = -4.11\ \mu C$

32. (a) $V_o = -1.65$ MV, $V_A = -0.45$ MV, thus $V_A - V_o = 1.20$ MV
(b) $W_{ext} = qV_o = (-2\ \mu C)(-1.65\ MV) = +3.3$ J

33. (a) $U = k(46e)^2/r = 6.59 \times 10^{-11}$ J;
(b) $\Delta K = -\Delta U$, so $K = 6.59 \times 10^{-11}$ J;
(c) $10^6/(0.3K) = 5.06 \times 10^{16}$ per second

34. (a) $V(x) = -\int E_x\, dx$ leads to $V(x) = \sigma(x_o - x)/2\epsilon_o$
(b) $\Delta x = 2\epsilon_o V/\sigma = 5.06$ cm

35. $\Delta V = -\underline{E}.\Delta\underline{s} = -(319\underline{i} - 241\underline{j}).\Delta\underline{s}$.
(a) $\Delta\underline{s} = 0.03\underline{i}$ m, so $\Delta V = -9.57$ V;
(b) $\Delta\underline{s} = -0.03\underline{j}$ m, so $\Delta V = -7.23$ V

36. (a) From $\Delta K = -(-e)\Delta V$, find $\Delta V = -157$ V
(b) $E_x = -\Delta V/\Delta x = 5.23x10^4$ V/m

37. (a) $2kq[1/x - 1/(x^2 + a^2)^{1/2}]$; (b) $2kq[1/y - y/(y^2 - a^2)]$

38. (a) $r = 18/V = 36$ m, 18 m, 12 m, 9 m, 7.2 m, 6 m, 5.14 m

39. $r = 18/V$. (a) $r_1 = 1$ m, $r_2 = 18/19$ m, so $\Delta r = -5.26$ cm;
(b) $r_2 = 18/17$ m, so $\Delta r = +5.88$ cm

40. $K = k(2e)(79e)/r = 6.72x10^{-13}$ J, so $r = kqQ/K = 5.42x10^{-14}$ m

41. (a) $k(2e)(90e)/r = 5.6 \times 10^{-12}$ J; (b) 5.6×10^{-12} J

42. (a) $2kQ/(x^2 + a^2)^{1/2}$;
(b) $E_x = -dV/dx = 2kQx/(x^2 + a^2)^{3/2}$

43. (a) $2kQy/(y^2 - a^2)$;
(b) $E_y = -dV/dy = 2kQ(y^2 + a^2)/(y^2 - a^2)^2$

44. (a) For $x > a$: $V = 2kQa/(x^2 - a^2)$;
(b) $E_x = -dV/dx = 4kQax/(x^2 - a^2)^2$

45. $E_r = -dV/dr = kQr/R^3$

46. $V(r) = V(r_o) - 2k\lambda \ln(r/r_o)$; $E_r = -dV/dr = +2k\lambda/r$.

47. $E_y = -dV/dy = 2\pi k\sigma[1 - y/(a^2 + y^2)^{1/2}]$

48. From Eq. 25.14:
$\underline{E} = (3y^2z - 6x^2y)\underline{i} + (6xyz - 2x^3 - 5z^3)\underline{j} + (3xy^2 - 15yz^2)\underline{k}$

49. (a) 1.44×10^6 V; (b) 27.2 V; (c) No change

50. (a) $V = kQ/(a^2 + y^2)^{1/2}$; (b) $E_y = kQy/(a^2 + y^2)^{3/2}$

51. V = ER: (a) 30 V; (b) 30 kV; (c) 3 MV

52. (a) $V = kQ/R = 4\pi k\sigma R$, so $\sigma = \epsilon_0 V/R = 4.43\text{x}10^{-6}$ C/m^2
(b) $Q = VR/k = Ne$, so $N = 1.39\text{x}10^{11}$ electrons
(c) $E = kQ/R^2 = 5\text{x}10^5$ V/m

53. Within inner sphere: $V_1 = kQ(1/a - 2/b)$, $E = 0$.
$a < r < b$: $V = V_1 + kQ/r$, $E = kQ/r^2$
$r > b$: $V = -kQ/r$, $E = -kQ/r^2$

54. $\Delta V = -\underline{E}.\Delta\underline{s} = -(420\underline{i})(2\cos 30^o\underline{i} + 2\sin 30^o\underline{j}) = -727.5$ V
$W_e = +q\Delta V = 1.82$ mJ

55. (a) $V = 2kQ/r = 20$ V; (b) $U = qV = -40$ nJ.

56. $\underline{d} = -3\underline{i} + \underline{j} + 7\underline{k}$ m, so d = 7.68 m. $U = kQ_1Q_2/d = 11.7$ mJ.

57. (a) $kQ_1/r_1 + kQ_2/r_2 = 4.40$ V; (b) $U = qV = -22$ nJ.

58. $\Delta V = -\underline{E}.(\underline{r}_A - \underline{r}_B) = -(-2\underline{i} + 3\underline{j} - 5\underline{k}).(-4\underline{i} + 3\underline{j} - 4\underline{k}) = -37$ V

59. $V = 2kQ/(2^{1/2}R) = 1.414kQ/R = 255$ V

60. $kQ/r = 3.8$ kV, so Q = 105.6 nC and $\sigma = Q/4\pi R^2 = 0.84\ \mu$C/m^2.

61. $Q_1/R_1 = Q_2/R_2$, so $Q_2 = 4\pi R_2{}^2\sigma_2 = 10.3$ nC.

62. $E = kQ_1/1^2 - kQ^2/1^2 = -27$ N/C, and $V = k(Q_1 + Q_2)/1 = 63$ V
Find $Q_1 = 2$ nC and $Q_2 = 5$ nC.

63. $4kQ^2/L + 2kQ^2/(1.414L) = 5.414kQ^2/L = 1.39\ \mu$J.

64. $E_x = -\partial V/\partial x = -6x + 15 = 0$, so x = 2.5 m.

65. Each initial drop $kQ/R_1 = 1000$. Single large drop has double the volume: $R_2{}^3 = 2R_1{}^3$, i.e. $R_2 = 1.26R_1$ and final charge is

2Q. Thus, $V = 2kQ/R_2 = 2kQ/1.26R_1 = 1590$ V.

66. $Q_1/R_1 = Q_2/R_2$ and $Q_1 + Q_2 = 30$ nC. Find $Q_1 = 9$ nC, $Q_2 = 21$ nC

67. $kQ/R = 320$ and $kQ/(R + 0.15) = 220$, thus $R = 0.330$ m,
$Q = 11.7$ nC.

68. All parts are at R from center. Total charge is $\pi R\lambda$, so
$V = kQ/R = k\pi\lambda = 62.2$ V.

69. (a) $U = qV = q(kQ/R) = 2\pi kq\lambda = 0.170\ \mu J$;
(b) $\Delta K = -\Delta U$, gives $0.5mv^2 = 0.170\ \mu J$, so $v = 18.4$ cm/s.

70. From Example 25.5 $V = 2\pi k\sigma[(a^2 + y^2)^{1/2} - y] = 13.98$ V.
$W = qV = 69.9$ nJ.

71. $V = kQ/R$ and $E = kQ/R^2$, so $V = ER = 1.2$ MV.

72. (a) $\Delta K = -q\Delta V = +eE\ell = 1.23 \times 10^{-17}$ J;
(b) $v = 5.19 \times 10^6$ m/s.

73. $\Delta K = -e\Delta V$, so $\Delta V = 2.67 \times 10^4$ V.

74. $V = (F/e)d = 410$ V.

Problems

1. $E = U = kqQ/r = 5.60\text{x}10^{-12}$ J; Also $m_1v_1 = m_2v_2$.
$E = 1/2\ m_1v_1^2 + 1/2\ m_2v_2^2 = 1/2\ m_1v_1^2(1 + m_1/m_2)$
Thus $K_1 = Em_2/(m_1 + m_2)$. Similarly, $K_2 = m_1E/(m_1 + m_2)$
$K_\alpha = 5.51 \times 10^{-12}$ J, $K_{Th} = 9.4 \times 10^{-14}$ J

2. Use Example 25.5: $V = 2\pi k\sigma[(b^2 + y^2)^{1/2} - (a^2 + y^2)^{1/2}]$

3. (a) $U_1 = -6ke^2/d = -30.6$ eV;
(b) $U_2 = U_1 + 12ke^2/d(2)^{1/2} = 12.7$ eV

4. $P = (1.6\text{x}10^{-19}\text{ C})(20\text{ kV})(4\text{x}10^{16})(0.3) = 38.4$ W
$\Delta E = (0.5\text{ kg})(134\text{ J/kg.K})(10\ ^oC) = 670$ J; $\Delta t = \Delta E/P = 17.4$ s

5. (a) $k(Q_1/R_1 - Q_2/R_2)$; (b) $k(Q_1 - Q_2)/R_2$;
(c) $kQ_1(1/R_1 - 1/R_2)$; (d) $Q_1 = 0$

6. From Example 25.7: $U = kQ^2/2r = k(\sigma A)^2/2r$.

$F = -dU/dr = k(\sigma A)^2/2r^2$. Using $A = 4\pi r^2$ and $k = 1/4\pi\epsilon_0$:
$F/A = k\sigma^2 A/2r^2 = \sigma^2/2\epsilon_0$ N/m^2

7. (a) $V_b - V_a = -\int E_r\, dr = -\int (2k\lambda/r)\, dr = -2k\lambda \ln(b/a)$
(b) From ΔV find $k\lambda = 59.48$, then $E = 2k\lambda/a = 3.97 \times 10^6$ V/m

8. $V = \int k\lambda dx/(a + x) = kQ/L \ln[(L + a)/a]$

9. $dV = k(\lambda dx)/(y^2 + x^2)^{1/2}$. See integral tables.
$V = k\lambda \ln[L/y + (1 + L^2/y^2)^{1/2}]$

10. $V(r) - V(R) = -\int (kQr/R^3)\, dr$, where $V(R) = kQ/R$
$V(r) = (kQ/R^3)(3R^2/2 - r^2/2)$

11. Note $q(r) = 4\pi\rho r^3/3 = Qr^3/R^3$, and $\rho = 3Q/4\pi R^3$
$V(r) = kq(r)/r = kQr^2/R^3$
$dU = V(r)dq = (3kQ^2/R^6)\, r^4\, dr$. Find $U = 3kQ^2/5R$

12. (a) $V = kq(1/r_+ - 1/r_-) = kq(r_- - r_+)/r_+r_-$
Use $(r_- - r_+) \approx 2a\cos\theta$ and $r_= r_- \approx r^2$. Thus $V = kp\cos\theta/r^2$
(b) $E_r = 2kp \cos\theta/r^3$; $E_\theta = kp \sin\theta/r^3$

13. $\underline{p} = p\underline{i}$, $\underline{r} = x\underline{i} + y\underline{j}$, so $r^2 = x^2 + y^2$. Thus
$V = kpx/r^3 = kpx(x^2 + y^2)^{-3/2}$
$E_x = -\partial V/\partial x = kp(2x^2 - y^2)/r^5$;
$E_y = -\partial V/\partial y = 3kpxy/r^5$

14. $\underline{r}/r = (x/r)\underline{i} + (y/r)\underline{j}$, so $\underline{p}.\underline{r}/r = x/r$
$\underline{E} = (k/r^3)[3(x/r)p[(x/r)\underline{i} + (y/r)\underline{j}] - p\underline{i}] =$
$= (kp/r^5)(3x^2 - r^2)\underline{i} + (3kpxy/r^5)\underline{j}$
$= [kp(2x^2 - y^2)/r^5]\underline{i} + (3kpxy/r^5)\underline{j}$

15. (a) 3.38×10^{-2} eV; (b) -3.38×10^{-2} eV;
(b) -6.76×10^{-2} eV; (d) 6.76×10^{-2} eV

16. Consider three charges.
$U_1 = q_1V_1 = kq_1(q_2/r_{12} + q_3/r_{13})$
$U_2 = q_2V_2 = kq_2(q_1/r_{12} + q_3/r_{23})$
Thus kq_1q_2/r_{12} appears twice. To avoid double counting,
$U = \Sigma\ 1/2\ q_iV_i$.

CHAPTER 26

Exercises

1. (a) $C = \epsilon_0 A/d = 50$ pF; (b) $Q = CV = 600$ pC

2. (a) $C/L = 2\pi\epsilon_0/\ln(b/a) = 24.1$ pF; (b) $Q = CV = 1.45$ nC

3. (a) $A = Cd/\epsilon_0 = 54.2\ \text{cm}^2$;
   (b) $V = Q/C = 167$ V;
   (c) $E = V/d = 8.33 \times 10^5$ V/m

4. (a) $V = Ed = 24$ V; (b) $C = Q/V = 2.5$ nF
   (c) $A = Cd/\epsilon_0 = 0.226\ \text{m}^2$

5. (a) $\sigma = \epsilon_0 E = -8.85 \times 10^{-10}\ \text{C/m}^2$;
   (b) $C = R_1R_2/k(R_2 - R_1) = 91.7$ mF;
   (c) 712 μF for the earth

6. Given $\lambda = 4$ nC/m.
   $\lambda = Q/L = CV/L = 2\pi\epsilon_0 V/\ln(b/a)$
   From $\ln(b/a) = 2\pi\epsilon_0 V/\lambda = 0.375$, find $b = 1.45$ mm

7. "Effective area" is 4A, so $C = 4\epsilon_0 A/d$

8. (a) $d = \epsilon_0 A/C = 2.21$ cm, thus $V = Ed = 66.3$ kV
   (b) $Q = CV = 1.59\ \mu$C

9. $C = (1.6\times10^{-7}\ \text{C})/20\ \text{V} = 8$ nF

10. $C = R_1R_2/k(R_2 - R_1) = 66.7$ pF, $Q = CV = 0.8$ nC

11. $Q_1' + Q_2' = 80\ \mu$C; $Q_1'/C_1 = Q_2'/C_2$
    $Q_1' = 32\ \mu$C, $Q_2' = 48\ \mu$C, $V_1 = V_2 = 8$ V

12. $C = 6\epsilon_0 A/d$, where $A = (\pi R^2/2)(1 - \theta/180)$
    (a) 33.4 pF; (b) 25.1 pF; (c) 8.35 pF

13. (a) $C = R_1R_2/k(R_2 - R_1) = 4.58$ pF; (b) $N = CV/e = 1.43 \times 10^8$

14. (a) $C_{eq} = 0.0714\ \mu$F, $Q_1 = Q_2 = 0.857\ \mu$C,
    $V_1 = 8.57$ V, $V_2 = 3.43$ V
    (b) $V_1 = V_2 = 12$ V, $Q_1 = 1.2\ \mu$C, $Q_2 = 3\ \mu$C

15. From $12.4 = 10 + (1/4 + 1/C_1)-1$, find $C_1 = 6\ \mu F$

16. From $1/2.77 = 1/4 + 1/(2 + C_2)$, find $C_2 = 7\ \mu F$

17. (a) Two in parallel with two in series; (b) Four in series

18. $C = 15/41\ \mu F = 0.366\ \mu F$

19. $Q_1' + Q_2' = 48\ \mu C$; $Q_1'/C_1 = Q_2'/C_2$
$Q_1' = 16\ \mu C$, $Q_2' = 32\ \mu C$; 8 V for both

20. $Q_1' + Q_2' = 240\ \mu C$; $Q_1'/C_1 = Q_2'/C_2$
$Q_1' = 60\ \mu F$, $Q_2' = 180\ \mu F$, $V = 30$ V for both

21. (a) $Q_1' + Q_2' = 86\ \mu C$; $Q_1'/C_1 = Q_2'/C_2$
$Q_1' = 32.25\ \mu C$, $Q_2' = 53.75\ \mu C$, $V = 10.75$ V for both;
(b) $Q_1' + Q_2' = 14$; $Q_1'/C_1 = Q_2'/C_2$
$Q_1' = 5.25\ \mu C$, $Q_2' = 8.75\ \mu C$, $V = 1.75$ V for both

22. 17 values (μF): 4/7, 2/3, 4/5, 6/7, 1, 10/7, 4/3, 12/7, 2, 7/3, 14/5, 3, 14/3, 4, 5, 6, 7

23. $1/2\ CV^2 = 1.6\text{x}10^{-11}$ J, so $C = 0.22$ pF

24. (a) $1/2\ (100\ \mu F)V^2 = 20$ mJ; (b) $1/2\ (25\ \mu F)V^2 = 5$ mJ

25. (a) $C = \epsilon_0 A/d = 14.2$ pF; (b) $1/2\ CV^2 = 4.08$ nJ;
(c) $E = V/d = 9.6$ kV/m; (d) $1/2\ \epsilon_0 E^2 = 408\ \mu J/m^3$

26. (a) $C = Q/Ed = 125$ pF; (b) $1/2\ CV^2 = 3.6\ \mu J$

27. $1/2\ \epsilon_0 (V/d)^2 = 0.192\ J/m^3$

28. (a) $U_1 = 1/2\ C_1 V^2 = 0.6$ mJ, $U_2 = 1$ mJ
(b) $3V_1 = 5V_2$, $V_1 + V_2 = 20$, thus $V_1 = 12.5$ V, $V_2 = 7.5$ V,
$U_1 = 0.234$ mJ, $U_2 = 0.141$ mJ

29. Initial: $C_{eq} = 10/7\ \mu F$, so $Q = 200/7\ \mu C$, $U = Q^2/2C$:
$U_1 = 204\ \mu J$, $U_2 = 81.6\ \mu J$,
Final: $Q_1' + Q_2' = 400/7\ \mu C$; $Q_1'/C_1 = Q_2'/C_2$
Find $Q_1' = 16.33\ \mu C$; $Q_2' = 40.82\ \mu C$, thus
$U'_1 = 66.7\ \mu J$, $U'_2 = 167\ \mu J$

30. Initial: $U_1 = 1/2\ C_1V^2 = 1.6$ mJ; $U_2 = 4$ mJ
Final: $Q_1' + Q_2' = 120\ \mu C$; $Q_1'/2 = Q_2'/5$
$U_1' = 0.294$ mJ, $U_2' = 0.734$ mJ

31. $C_{eq} = 40/11\ \mu F$, $Q = 800/11\ \mu C$.
(a) $U_5 = Q^2/2C_5 = 529\ \mu J$;
(b) In the parallel combination: $Q_1 + Q_2 = 800/11$, where Q_1 applies to 2 μF and 4 μF, and Q_2 to the 12 μF. The two in series are equivalent to 4/3 μF, so $3Q_1/4 = Q_2/12$.
Find $Q_1 = 80/11\ \mu C$, $Q_2 = 720/11\ \mu C$ and $U_4 = Q_1^2/2C_4 = 6.61\ \mu J$

32. (a) $U = 1/2\ CV^2 = 1.56$ nJ
(b) $d = \epsilon_0 A/C = 7.08$ mm, thus Volume = Ad = $2.83\text{x}10^{-4}\ m^3$
u = U/Volume = $55.1\ \mu J/m^3$

33. (a) $\epsilon_0 A/(d - \ell)$; (b) No change

34. $C_f = 0.5C_i$. (a) None; (b) Halved; (c) Halved

35. $C_f = 0.5C_i$. (a) Doubles; (b) No change; (c) Doubles

36. (a) $U = 1/2\ CV^2$, and $V_4 = V_5$, so $U_4 = (4/5)U_5 = 160$ mJ
$1/2\ (5\ \mu F)V^2 = 0.2$, so $V = 282.8$ V
(b) $V_3 = 2V/3 = 188.5$ V, thus $U_3 = 53.3$ mJ

37. $u = 1/2\ \epsilon_0 E^2 = 8.96 \times 10^5\ J/m^3$

38. $U = u\text{x}Volume = (\epsilon_0 E^2/2)(10^3\ m^3) = 63.7\ \mu J$

39. $u = 1/2\ \epsilon_0 (V/d)^2$, thus $V = 6.38$ V

40. $d = \epsilon_0 A/C = 4.72\text{x}10^{-3}$ m; $u = 1/2\ \epsilon_0 (V/d)^2 = 4.58\text{x}10^{-4}\ J/m^3$

41. In parallel: $C_0(\kappa_1 + \kappa_2)/2$

42. In series: $C = (2\kappa_1\kappa_2)C_0/(\kappa_1 + \kappa_2)$

43. See Example 26.10: (a) $V_0[1 - (\kappa - 1)\ell/\kappa d] = 1.72$ V;
(b) $C_0/[1 - (\kappa - 1)\ell/\kappa d] = 4.66$ pF.

44. $Q_0 = C_0V_0 = 1.2\ \mu C$; $Q = \kappa C_0V_0 = 4.8\ \mu C$, so $\Delta Q = 3.6\ \mu C$

45. (a) $C = \kappa C_0 = 6\epsilon_0 A/d$, so $A = 0.94\ cm^2$; (b) $V = Ed = 15$ kV

46. (a) 1.5; (b) 1.33; (c) 2

47. (a) $C = R/k$, so $R = Ck = 3.78$ cm; (b) $Q = CV = 4.2$ nC;
(c) $\sigma = Q/4\pi R^2 = 234$ nC/m^2

48. $A = Cd/\epsilon_0 = \pi r^2$, so $r = 2.94$ cm.

49. $C = \sigma A/V = \epsilon_0 A/d$, thus $d = \epsilon_0 V/\sigma = 7.08$ mm.

50. (a) $C = 2\pi\epsilon_0 L/\ln(b/a)$ gives $b/a = 1.56$, so $a = 0.449$ cm;
(b) $CV/L = 3.0$ nC/m.

51. $C = R_1R_2/k(R_2 - R_1) = 32.6$ pF, $Q = CV = 1.63$ nC; $\sigma = Q/A$.
(a) 20.3 nC/m^2; (b) 10.7 nC/m^2.

52. (a) Two in parallel with two in series.
(b) One in parallel with three in series.

53. $U = 0.5CV^2$, so $(V + 2)^2 = 1.21V^2$, thus $V = 20$ V.
Original $Q = CV = 100$ μC.

54. (a) $C_{eq} = 2.01$ μF, $U = 0.5CV^2 = 402$ μJ.
(b) $C_{eq} = 8.6$ μF, $V^2 = 2U/C$ leads to $V = 9.67$ V.

55. (a) $C = \kappa\epsilon_0 A/d = 1.65$ nF; (b) From Table 26.1, 1600 V.

56. $1/C_1 + 1/C_2 = 2.1$ and $C_1 + C_2 = 10$ lead to
$C_1^2 - 10C_1 + 21 = 0$, so $C_1 = 3$ μF and $C_2 = 7$ μF.

57. $Q_1' + Q_2' = 520$ μC and $Q_1'/C_1 = Q_2'/C_2 = 16$. $Q_1' = 320$ μC
$Q_2' = 200$ μC, so $C_2 = Q_2'/16 = 12.5$ μF.

58. (a) $C = R_1R_2/k(R_2 - R_1) = 0.272$ nF; (b) $A = Cd/\epsilon_0 = 615$ cm^2.

59. $R_2^3 = 2R_1^3$, so $R_2 = 1.26R_1$. $C = R/k$, $C_2 = 1.26C_1 = 0.063$ pF

60. $U = 0.5CV^2$, $P = \Delta U/\Delta t = 7.2$ kW

61. $E = V/R = 3 \times 10^4$ V/m, so $u = 0.5\epsilon_0 E^2 = 3.98$ mJ/m^3.

62. From Table 26.1: $d = V/E = 5\times10^{-4}$ m. Then $C = \kappa\epsilon_0 A/d$ gives
$A = 305$ cm^2.

63. $C_0 = \epsilon_0 A/d = 3.2$ pF and $C = \kappa\epsilon_0 A/2d = 8$ pF, thus $\kappa = 5$.

Problems

1. $C = \kappa C_0 = \kappa\epsilon_0 A/d$. If V is the potential difference across the dielectric slab, then,
$U = 1/2\ CV^2 = 1/2\ \kappa C_0(Ed)^2 = 1/2\ \kappa\epsilon_0 E^2\ (Ad)$.
Thus, $u = U/Ad = 1/2\ \kappa\epsilon_0 E^2$

2. $U_1 = 1/2\ CV^2 = \epsilon_0 AV^2/2(d - \ell)$; $U_2 = \epsilon_0 AV^2/2d$, thus
$\Delta U = \epsilon_0 AV^2\ell/2d(d - \ell)$

3. $U_1 = 1/2\ CV^2 = \epsilon_0 AV^2/2(d - \ell)$;
$U_2 = (\epsilon_0 A/2d)[dV/(d - \ell)]^2 = \epsilon_0 AdV^2/2(d - \ell)^2$, thus
$\Delta U = \epsilon_0 A\ell V^2/2(d - \ell)^2$

4. (a) $C_0 = 35.4$ pF, $C_{eq} = 17.7$ pF, so 6 V and 212 pC for each.
(b) $C_1 = C_0$, $C_2 = 5C_0$, so $C_{eq} = 5C_0/6$, and $Q = 354$ pC for each. $V_1 = 10$ V, $V_2 = 2$ V.

5. $1/C_{eq} = 2/C + 1/(C_{eq} + C)$, so $2C_{eq}^2 + 2CC_{eq} - C^2 = 0$.
Find $C_{eq} = 0.366C$

6. Let $C = 1\ \mu F$.
$Q_2/4C + Q_1/2C = V$ (i)
$2Q_1/2C + Q_3/3C = V$ (ii)
$2Q_2/4C - Q_3/3C = V$ (iii)
$Q_1 - Q_2 - Q_3 = 0$ (iv)
Substitute for Q_3 from (iv) and V from (i) into (ii) and (iii) to find $7Q_2 = 10Q_1$ and then $Q_1 = 7CV/6$ and $Q_2 = 10CV/6$.
Thus $C_{eq} = (Q_1 + Q_2)/V = 17C/6 = 2.83\ \mu F$.

7. $U = Q^2/2C$ where $C = \epsilon_0 A/x$, thus $U = (Q^2/2\epsilon_0 A)x$
$F_x = -dU/dx = -\ Q^2/2\epsilon_0 A$ (attractive)

8. $U = 1/2\ C_D V_D^2 = 1/2\ (\kappa C_0)(V_0/\kappa)^2 = U_0/\kappa = CV^2/2\kappa$.

9. $U = 1/2\ C_D V_D^2 = 1/2\ (\kappa C_0)V_0^2 = \kappa U_0 = \kappa CV^2/2$.

10. Let $(b - a) = d$, then $\ln(b/a) = \ln(1 + d/a)$
For $d << a$: $\ln(1 + d/a) \approx d/a$.
$C = 2\pi\epsilon_0 L/\ln(b/a) \approx \epsilon_0 2\pi La/(b - a) = \epsilon_0 A/d$

11. (a) $C = R_1R_2/k(R_2 - R_1)$. If $R_2 >> (R_2 - R_1) = d$, then $R_1R_2 \approx R^2$, and $C = 4\pi\epsilon_0 R^2/d = \epsilon_0 A/d$
(b) When $R_2 >> R_1$, $R_2/(R_2 - R_1) \approx 1$, thus $C = R_1/k$.

12. (a) $u = \epsilon_0 E^2/2 = \epsilon_0(2k\lambda/r)^2/2 = \lambda^2/8\epsilon_0\pi^2r^2$
(b) $dU = u(2\pi rL)dr$, $U = k\lambda^2L \int dr/r = k\lambda^2L \ln(b/a)$
(c) $C = L/2k\ln(b/a)$, and $Q = \lambda L$, so $U = Q^2/2C = k\lambda^2L \ln(b/a)$

13. See Eq. 26.15: $\sigma_b = \sigma_f(1 - 1/\kappa) = \sigma_f(\kappa - 1)/\kappa$

CHAPTER 27

Exercises

1. (a) In 1 s, $N = \Delta Q/e = 1.19 \times 10^{16}$.
   (b) $J = I/A = 2.42 \times 10^3$ A/m^2

2. (a) $J = I/A = 0.318$ A/m^2;
   (b) $J = nev$, so $n = J/ev = 3.97\text{x}10^{11}$ m^{-3}

3. (a) $J = I/A = nev$, so $v = I/Aev = 1.07 \times 10^{-5}$ m/s;
   (b) $E = \rho J = 1.49 \times 10^{-3}$ V/m

4. (a) $J = I/A = 6.37\text{x}10^6$ A/m^2
   (b) $E = \rho J = 0.108$ V/m
   (c) $J = nev$, so $v = J/ne = 4.68\text{x}10^{-4}$ m/s
   (d) $\Delta t = d/v = 6.41\text{x}10^7$ s = 2.03 y.

5. (a) $J = I/A = 2.83 \times 10^6$ A/m^2;
   (b) $E = \rho J = 4.81 \times 10^{-2}$ V/m

6. (a) $J = I/A = 7.21\text{x}10^6$ A/m^2
   (b) $v = J/ne = 5.30\text{x}10^{-4}$ m/s

7. $I = Q/T = ev/2\pi r = 1.06$ mA

8. $Q = \int I\,dt = [2t3/3 - 3t^2/2 + 5t] = 61.5$ C

9. (a) $J = I/A = 6.79 \times 10^6$ A/m^2;
   (b) $v = J/ne = 4.24 \times 10^{-4}$ m/s;
   (c) $E = \rho J = 0.19$ V/m

10. $\rho = VA/IL = 1.14\text{x}10^{-6}$ Ω.m

11. $I = VA/\rho L = 6.85 \times 10^{-5}$ A

12. $L_1A_1 = L_2A_2$, but given $A_2 = A_1/2$, and $L_2 = 2L_1$
    $R_2 = \rho L_2/A_2 = 4R_1$

13. $R = \rho L/A = \rho \ell/\pi(b^2 - a^2)$

14. $R = R_o[1 + \alpha\,\Delta t] = 1.27$ Ω.

15. $\Delta T = (R/R_o - 1)/\alpha = 128$ ^{o}C, so $T = 148$ ^{o}C

16. $\alpha = (R/R_o - 1)/\Delta T = 2.08\text{x}10^{-3}\ ^oC^{-1}$

17. $L\ \alpha\ 1/\rho$, so $L_{Cu}/L_{Al} = 2.8/1.7 = 1.65$

18. $\Delta R = R_{o1}\alpha_1\Delta T + R_{o2}\alpha_2\Delta T = 0$, so $R_{o1}/R_{o2} = -\alpha_2/\alpha_1 = 5/4$
$R_{o1}/(R_{o1} + R_{o2}) = 5/9 = 55.5\%$.
Percentage of carbon = 4/9 = 44.4%.

19. $d\ \alpha\ (\rho)^{1/2}$, so $d_{Cu}/d_{Al} = (1.7/2.8)^{1/2} = 0.78$

20. G = I/V = 33 mS

21. $\rho = RA/L = 2.83 \text{ x } 10^{-8}\ \Omega.m$, Al

22. $R = R_o(1 + \alpha\Delta T) = 0.591\ \Omega$

23. $R_2/R_1 = (L_2/A_2)(A_1/L_1) = 3.6$, thus $R_2 = 5.04\ \Omega$

24. $R_1 = 3\ \Omega$, $R_2 = 3.53\ \Omega$. Then, $\alpha = (R_2/R_1 - 1)/\Delta T$
$= 2.21\text{x}10^{-3}\ ^oC^{-1}$

25. $\Delta R/R = \alpha\Delta T$, so $\Delta T = \pm 0.1/\alpha$.
(a) 45.6 ^{o}C; (b) -5.6 ^{o}C

26. (a) $R = \rho L/A = 0.413\ \Omega$
(b) $P_w/(P_w + P_s) = 0.413/4.413 = 9.36\%$

27. $V = (10^{-2}\ \Omega)(1200\ A) = 12\ V$

28. $V = IR = \rho IL/A$. $V_{14} = 1.23$ V, $V_{18} = 1.03$ V

29. (a) 80 (C/s)(3600 s) = $2.88 \text{ x } 10^5$ C;
(b) $U = QV = 3.46\text{x}10^6$ J, $\Delta t = U/P = 1.38\text{x}10^5$ s = 38.4 h

30. It costs 6 cents for $3.6\text{x}10^6$ J. For one toast $U = 2.52\text{x}10^4$ J
Cost for one toast: $(2.52\text{x}10^4/3.6\text{x}10^6)6 = 0.042$ cents

31. $P = I^2\rho L/A$: $P_{14} = 5.23$ W, $P_{18} = 13.2$ W

32. $U = P\Delta t = (120\ W)(3600\ s) = 4.32\text{x}10^5$ J
At 25% efficiency, 1 L provides $7.5\text{x}10^6$ J, thus
Number of liters = $(4.32\text{x}10^5\ J)/(7.5\text{x}10^6\ J/L) = 57.6$ mL.

33. $I^2 = P/R$, so I = 0.0612 A. Then $N = Q/e = It/e = 2.3 \text{ x } 10^{19}$

34. $P = I^2R = I^2\rho L/A = 14.8$ W

35. $P = I^2R = 720$ kW

36. (a) $P_{mech} = Fv = mgv = 490$ W $= 0.66$ hp
(b) $P_{elec} = IV = 2400$ W. Efficiency $= 490/2400 = 20.4\%$

37. (a) $I = P/V = 10$ A, $P = I^2R = 500$ W;
(b) $I = 0.5$ A, $P = 1.25$ W

38. $\Delta E = mc\Delta T = (1.5 \text{ kg})(4186 \text{ J/kg})(70\ ^oC) = 4.395\text{x}10^5$ J
$P = \Delta E/\Delta t = (4.395\text{x}10^5 \text{ J})/(480 \text{ s}) = 916$ W
$P = IV$, so $I = 7.63$ A

39. $R = \rho L/A = 6.37\ \Omega$, and $m = 62.8$ g.
$\Delta E = mc\Delta T = (62.8 \text{ g})(4.186 \text{ J/g})(30\ ^oC) = 7890$ J
$P = \Delta E/\Delta t = 7890 \text{ J}/240 \text{ s} = 32.9$ W
$P = V^2/R$, so $V = 14.5$ V

40. $R = V^2/P$: $R_1 = (120)^2/70 = 206\ \Omega$; $R_2 = (120)^2/100 = 144\ \Omega$.
Note that $(120 \text{ V})^2/350\ \Omega = 41$ W (not exactly 40 W)

41. $R = V^2/P$: $R_1 = (120)^2/50 = 288\ \Omega$; $R_2 = (120)^2/100 = 144\ \Omega$

42. $R = V/I = U/qI = \text{N.m.s/C}^2 = \text{kg.m}^3/\text{A}^2.\text{s}^3$

43. $R = \rho L/A = 0.6\ \Omega$

44. $6.52[1 + 3.9\text{x}10^{-3}(T - 20)] = 6.45[1 + 4.5\text{x}10^{-3}(T - 20)]$ gives
$T = 39.5\ ^oC$.

45. Since $L = L_o(1 + \alpha_E\Delta T)$ and $A = A_o(1 + \alpha_E\Delta T)^2$,
$R = \rho L/A = (\rho_o L_o/A_o)[1 + \alpha_R\Delta T]/[1 + \alpha_E\Delta T]$ where
$\alpha_R = 4 \times 10^{-3}\ ^oC^{-1}$ and $\alpha_E = 2 \times 10^{-5}\ ^oC^{-1}$.
Error due to $1/[1 + \alpha_E\Delta T]$: (a) If $\Delta T = 100\ ^oC$, $\alpha_E\Delta T = 2\text{x}10^{-3}$
so error is 0.20%; (b) 1.96%.

46. $J = E/\rho = I/A$, so $\rho = EA/I = 2.83 \times 10^{-8}\ \Omega$.m.

47. $m = DLA$, where D is density; $R = \rho L/A$, thus $A^2 = m\rho/RD$
which gives $A = 7.856\text{x}10^{-7} \text{ m}^2 = \pi r^2$, so $r = 0.500$ mm. Also,
$L = RA/\rho = 3.00$ m.

48. $R/R_o = (1 + \alpha\Delta T)$, so $1/0.96 = [1 + \alpha(T - 20)]$, then $T = 124\ ^oC$

49. (a) $n = 3\rho N_A/M = 1.81 \times 10^{29}$ electrons/m^3.
(b) $J = I/A = nev_d$, thus $v_d = 0.226$ mm/s.

50. Volume $= AL = \rho L^2/R$ and mass $m = DAL = D\rho L^2/R = 151$ kg.

51. $V = I(\rho L/A) = J\rho L$, thus $J = V/\rho L = 1.12 \times 10^7\ A/m^2$.

52. $J = I/A = nev_d$, thus $A = \pi r^2 = I/nev_d$, so $r = 0.756$ mm.

53. $\Delta R/R_o = \alpha(T - 20) = 0.015$, so $T = 23.8\ ^oC$.

54. (a) $I = dq/dt = 10t - 4$ A; (b) $J = I/A = 3 \times 10^4\ A/m^2$.

Problems

1. (a) $L = AR/\rho = 41.9$ m;
(b) $R = R_o(1 + \alpha\Delta T) = 17.15\ \Omega$
Then $I = V/R = 7.00$ A

2. (a) 5 V/0.6 A = 8.33 Ω; (b) Zero; (c) No

3. (a) $J = E/\rho$ leads to $dV = \rho I dr/A$.
$V = \int dV = \int \rho I\ dr/2\pi r L = \rho I/2\pi L\ \ln(b/a) = IR$
Thus, $R = \rho/2\pi L\ \ln(b/a)$;
(b) 37.4 μΩ

4. From $J = E/\rho$ find $dV = \rho I dr/A$
$V = \int dV = \int (\rho I/4\pi r^2)\ dr = \rho I/4\pi\ (1/a - 1/b) = IR$
Thus, $R = \rho(1/a - 1/b)/4\pi = \rho(b - a)/4\pi ab$.

5. $dQ = \sigma(2\pi r\ dr)$, $dI = dQ/T = \omega\ dQ/2\pi$
$I = \int dI = \int \sigma\omega r\ dr = \sigma\omega a^2/2$
[Or, more simply: $Q = \sigma\pi a^2$, $I = Q/T = \omega Q/2\pi = \sigma\omega a^2/2$]

6. (a) $J = E/\rho = V/\rho L = 4.44 \times 10^6\ A/m^2$
Since $J = nev$, find $v = 4.8 \times 10^{-4}$ m/s
(b) $\tau = m/ne^2\rho = 4.1 \times 10^{-14}$ s
(c) $\lambda = v_{rms}\tau = 4.1$ nm

7. Number of atoms $N = Q/e = (60\ C)/e = 37.5 \times 10^{19}$

Mass of one Ag atom = 108 u = 1.79x10-25 kg.

Mass deposit = $(37.5\text{x}10^{19})(1.79\text{x}10^{-25}$ kg) = $1.34\text{x}10^{-4}$ kg = 67.5 mg

8. (a) Q = (20)(200 A.h)(3600 s) = $1.44\text{x}10^{7}$ C

Electrical energy U = QV = $1.73\text{x}10^{8}$ J

W = Fs = (180 N)s, so s = 960 km.

Thus Δt = 960 km/(60 km/h) = 16 h.

(b) W = (180 + mgsin10^{o})s, so s = 144 km and Δt = 2.4 h.

9. (a) $R = (\rho_1 + \rho_2)L/A$, $I = V/R = 10A/(\rho_1 + \rho_2)L$ = 1.883 A

$P_1 = I^2R_1 = (1.883)^2(0.2165) = 0.77$ W,

$P_2 = I^2R_2 = (1.883)^2(5.093) = 18.1$ W

(b) $E_1 = IR_1/L = 0.0102$ V/m, $E_2 = IR_2/L = 0.240$ V/m

CHAPTER 28

Exercises

1. $V_{ba} = \xi - Ir$: $9.5 = \xi - (9.5/4)r$, $10 = \xi - (10/6)r$
   $r = 0.706\ \Omega$, $\xi = 11.2$ V

2. $\xi = 12.4$, thus $11.2 = 12.4 - 80r$, thus $r = 0.015\ \Omega$

3. $8.4 = \xi - 6r$; $7.2 = \xi - 8r$. Thus $\xi = 12$ V, $r = 0.6\ \Omega$

4. $8 = \xi/R$; $6 = \xi/(R + 2)$. Thus, $R = 6\ \Omega$, $\xi = 48$ V

5. (a) $P = [V/(R + r)]^2R$, so $50 = 256x4/(4 + r)^2$, thus
   $r = 0.525\ \Omega$
   (b) $100 = 256R/(R + 0.525)^2$, so $R^2 - 1.51R + 0.276 = 0$
   $R = 0.215\ \Omega$, $1.295\ \Omega$

6. (a) $I = (14.2 - 12.4)/0.05 = 36$ A, $P = I^2r = 64.8$ W
   (b) (36 A)(12.4 V) = 446 W

7. $R = V^2/P = (11.2\text{ V})^2/(50\text{ W}) = 2.51\ \Omega$, $I = (P/R)^{1/2} = 4.46$ A
   $V_{ba} = 11.2 = 12 - (4.46)r$, $r = 0.18\ \Omega$

8. (a) $P = 100R/(R + 1)^2$: $1.25P = 100(1.5R)/(1.5R + 1)^2$,
   thus $R = 0.236\ \Omega$.
   (b) $0.75P = 100(1.5R)/(1.5R + 1)^2$, thus $R = 4.83\ \Omega$

9. (a) $7.82\ \Omega$;
   (b) $I_o = 1.28$ A and $I_1 = 4.4I_2$, so $I_2 = 1.28/5.4 = 0.237$ A
   and $I_1 = 1.043$ A. Find $V_1 = 1.043$ V; $V_2 = 0.474$ V.
   Finally, $V_4 = V_1 - V_2 = 0.569$ V

10. $4R/3 = 16\ \Omega$, thus $R = 12\ \Omega$

11. 2 Ω, 3 Ω, 4 Ω, 5 Ω, 6 Ω, 7 Ω, 9 Ω, 6/5 Ω, 4/3 Ω, 12/7 Ω,
    12/13 Ω, 26/7 Ω, 13/3 Ω, 26/5 Ω, 20/9 Ω, 14/9 Ω

12. Two in series, then placed in parallel.
    Two in parallel, then placed in series.

13. 1 Ω, 2 Ω, 3 Ω, 4 Ω; 1 Ω, 2 Ω, 4 Ω, 5 Ω
    1 Ω, 2 Ω, 2 Ω, 5 Ω; 1 Ω, 2 Ω, 4 Ω, 4 Ω.

14. $V_{ba} = \xi_1 - I_1r_1 = \xi_2 - I_2r_2$, also $I_1 = -I_2$.
Find $I_1 = (\xi_1 - \xi_2)/(r_1 + r_2) = 0.25$ A, thus $V_{ba} = 1.52$ V
$P = (0.25)^2(0.2) = 12.5$ mW

15. $I = P/V$: 0.5 A, 0.083 A, 8.33 A; 12.5 A

16. (a) $V = (PR)^{1/2} = 7.07$ V for each, so 21.2 V for three in series; (b) 7.07 V

17. (a) $V = (PR)^{1/2} = 8.94$ V for each, so 8.94 V;
(b) $I = 8.94\ V/4\Omega = 2.236$ A through single R, and 1.118 A through each R in parallel. Thus $V = 1.118 \times 4 + 8.94 = 13.4$ V

18. (a) $R_{eq} = 2 + 3R_2/(3 + R_2)$, but $(\xi_1 - \xi_2) = (2\ A)R_{eq}$, thus $R_{eq} = 3\ \Omega$. Solve to find $R_2 = 1.5\ \Omega$
(b) $P_1 = 1.33$ W; $P_2 = 2.66$ W
(c) $V_1 = 10$ V, $V_2 = 8$ V
(b) $P_1 = I\xi_1 = 24$ W, $P_2 = -I\xi_2 = -12$ W

19. (a) 0.5 A; (b) $P_1 = 0.5$ W, $P_2 = 1$ W; (c) 4.5 W, -3 W

20. (a) $I = 1$ A, so $V_P = \xi_1 + Ir_1 = 6$ V
(b) $V_1 = \xi_1 + Ir_1 = 6$ V, $V_2 = \xi_2 - Ir_2 = 7.5$ V
(c) $P = I^2R = 1.5$ W

21. $I_7 = 0$; $I_3 = 3.33$ A; $I_4 = 2.5$ A

22. $I_2 = 2$ A; $V_{2+4} = 12$ V, thus $R = 6\ \Omega$.
Circuit is equivalent to three 3 Ω resistors in parallel with 1 Ω in series, thus for whole circuit $R_{eq} = 2\ \Omega$.
The current in each 3 Ω branch is 4 A, thus $I_{tot} = 12$ A.
Finally, $\xi = I_{tot}R_{eq} = 24$ V

23. $\xi - (I/2)r - IR = 0$, so $I = \xi/(R + r/2)$
$P = I^2R = \xi^2R/(R + r/2)^2$. Set $dP/dR = 0$ to find $R = r/2$

24. $R_{eq} = R_1 + R_2R_L/(R_2 + R_L)$; $I = V/R_{eq}$, and $I_L = IR_2/(R_2 + R_L)$
$P_L = I_l^2R_L$. Set $dP_L/dR_L = 0$ to find $R_L = R_1R_2/(R_1 + R_2)$.

25. Left: $-\xi_1 + I_1R_1 - 5I_3 = 0$; Right: $5I_3 + \xi_2 + 5R_2 = 0$;

$I_2 = I_1 + I_3$.

(a) $I_1 = 3$ A, $V_1 = 6$ V, $I_2 = -1$ A, $V_2 = 5$ V,

$I_3 = -4$ A; $V_3 = 20$ V;

(b) $V_A - V_B = -I_2R_2 - \xi_2 = -20$ V

26. Left: $\xi_1 - I_1R_1 + I_2R_2 - \xi_2 = 0$;

Right: $\xi_2 - I_2R_2 - I_3R_3 + \xi_3 = 0$; $I_3 = I_1 + I_2$

$I_1 = 3$ A, $I_2 = 1$ A, $I_3 = 4$ A. $V_1 = 6$ V, $V_2 = 2$ V, $V_3 = 12$ V

27. Left: $\xi_1 - I_1R_1 + I_2R_2 - \xi_2 = 0$;

Right: $\xi_2 - I_2R_2 - I_3R_3 + \xi_3 = 0$; $I_3 = I_1 + I_2$

$I_1 = 2$ A, $V_1 = 8$ V, $I_2 = 1$ A, $V_2 = 3$ V, $I_3 = 3$ A, $V_3 = 9$ V

28. Left: $-5I_1 + 12 - 5I_3 - 7 = 0$;

Right: $7 + 5I_3 - 5I_2 + 13 = 0$; $I_1 = I_2 + I_3$

$I_1 = 2$ A, $I_2 = 3$ A, $I_3 = -1$ A.

$V_A - V_B = -2I_1 + 12 = +8$ V

29. $I_2R = 4$ V; $\xi - 5I_1 = 8$, $\xi - 2I_1 = 14$, $I_3 = I_1 + I_2$

Find $I_1 = 2$ A, $I_2 = 4$ A, then $\xi = 18$ V, $R = 1\ \Omega$.

30. (a) $R_{eq} = 4.5\ \Omega$, $I = 48/4.5 = 10.67$ A;

$V = (I/2)3 = 16$ V.

(b) $R_{eq} = 4.275\ \Omega$, $I = 11.23$ A,

$V = 3(5/8)(11.23) = 21.1$ V

31. $V^2(1/R_1 + 1/R_2) = 4V^2/(R_1 + R_2)$, thus $(R_1 + R_2)^2 = 4R_1R_2$

or, $R_2^2 - 6R_2 + 9 = 0$, so $R_2 = 3\ \Omega$

32. (a) By symmetry, $I = 0$ through the central resistor.

For each resistor, we have $V/2$.

(b) $1/2R + 1/2R = 1/R_{eq}$, thus $R_{eq} = R$.

33. (a) $R_{eq} = 10/3\ \Omega$, so $I = 3\xi/10$, $I_3 = \xi/5$.

$P_3 = I_3^2R_3$ leads to $\xi = (75)^{1/2} = 8.66$ V;

(b) $P_1 = 13.5$ W, $P_2 = 3$ W

34. (a) The loop rule immediately yields $\xi_1 = 3$ V, $\xi_2 = 13$ V

(b) -4 V

35. Infer that $T_{1/2} = 1$ ms $= 0.693RC$, thus $R = 1.44 \times 10^5\ \Omega$

36. $\tau = 8RC/3$.

37. From Eq. 28.11 find $0.1 = \exp(-2/RC)$, thus $C = 86.8\ \mu F$

38. (a) $V_C = 0.632\xi = 126$ V; (b) $V_R = 0.368\xi = 73.6$ V
(c) $U = 1/2\ CV^2 = (C/2)\xi^2(1 - e^{-0.5})^2 = 0.155$ J
(d) $P = V^2/R = \xi^2(e^{-1})/R = 73.6$ mW

39. (a) $Q = Q_o e^{-1} = 0.37$ mC, $I = I_o e^{-1} = 0.37$ mA;
(b) $U = Q^2/2C = 1.71$ mJ;
(c) $RC = 1$ s, so $P = (I_o e^{-0.5})^2 R = 9.2$ mW;
(d) $dU/dt = -(\xi^2/R)\exp(-2t/RC) = -\xi^2 e^{-1}/R = 9.2$ mW.

40. (a) $dQ/dt = (Q_o/RC)\exp(-t/RC) = \xi/R$ (at $t = 0$)
(b) $Q_o = (\xi/R)t$, so $t = Q_o R/\xi = RC$.

41. (a) $0.1 = \exp(-5/\tau)$, so $\tau = 2.17$ s
$0.5 = \exp(-t/2.17)$, thus $t = \tau \ln 2 = 1.50$ s;
(b) $I/I_o = \exp(-10/2.17) = 0.01$ or 1 %

42. Given $0.05 = \exp(-0.02/\tau)$, so $\tau = 6.67$ ms. Also, $RC = 0.5$ ms. $\tau = \kappa RC$, thus $\kappa = 13.3$.

43. (a) $I_3 = 0$ and $I_1 = I_5 = 5/3$ A;
(b) $V_C = -4.33$ V, so $Q = CV = 21.7\ \mu C$

44. (a) $I = 1/20{,}000 = 50\ \mu A$; (b) $(2\times10^4\ \Omega/V)(50\ V) = 1\ M\Omega$.

45. $(10^{-3})(R_1 + 50) = 1$, thus $R_1 = 950\ \Omega$,
$(10^{-3})(R_2 + 10^3) = 10$, thus $R_2 = 9\ k\Omega$,
$(10^{-3})(R_3 + 10^4) = 50$, thus, $R_3 = 40\ k\Omega$

46. (a) $1/20{,}000 = 5\times10^{-5}$ A.
(b) $5\times10^{-5}(R_s + 40) = 250$, thus $R_s \approx 5\ M\Omega$ (Ignore $40\ \Omega$)
(c) Across galvanometer: $V = Ir_G = (50\ \mu A)(40\ \Omega) = 2$ mV.
$(5 - 5\times10^{-5})R_{SH} = 2\times10^{-3}$, thus $R_{SH} = 4\times10^{-4}\ \Omega$

47. (a) $(5\times10^{-5})(R_s + 20) = 10$, thus $R_s = 200\ k\Omega$;
(b) $(499\ mA)R_{SH} = 10^{-3}$, thus $R_{SH} = 2\ m\Omega$

48. (a) 0; (b) $R_2V/(R_1 + R_2)$; (c) $R_2V/(2R_1 + R_2)$;
(d) $R_2V/(3R_1 + R_2)$

49. (a) I_R = 100 V/10.1 Ω = 9.9 A, V_R = 99 V; (b) 9.9 A, 100 V

50. (a) R_{eq} = 10 Ω, so I_{tot} = 10 A. Since I_A = 10 A, we have V_A = 1.0 V. Thus V_R = 99 V and I_R = 9.9 A
(b) I_A = 10 A, V_V = 99 V

51. If I is the current through 4 Ω, the loop rule gives:
12 - 3.6(I + 0.8) - 4I = 0, thus I = 1.2 A. Also 4I = 0.8R, thus R = 6 Ω.

52. (a) $I = \xi/(R + r)$, $R = \xi/I - r$ = 4.5 Ω. Thus,
$P_R = I^2R$ = 7.03 W.
(b) $P_r = I^2r$ = 0.469 W.

53. $V_{ab} = \xi - Ir$, so $11.4 = \xi - \xi r/(R + r) = \xi R/(R + r)$,
thus ξ = 12.4 V.

54. First $1.4 = \xi/R$; Second: $R_{eq} = 2R/(R + 2)$, so
$1.82 = \xi(R + 2)/2R$, i.e. $1.82 = 1.4(R + 2)/2$, so R = 0.6 Ω.

55. (a) $-\xi + 3I = -2I + 5 = 1$, so I = 2 A and ξ = 5 V;
(b) $-3I + \xi = -5 + 2I = 1$, so I = 3 A, and ξ = 10 V.

56. $R_1 + R_2 = 8$ and $R_1R_2/(R_1 + R_2)$ 1.5. Find R_1 = 2 Ω, R_2 = 6 Ω.

57. (a) Use $P = V_2/R$ for 2 Ω to find V = 4.47 V;
(b) Use $P = (V/7)^2R$ for 5 Ω to find V = 9.9 V.

58. $U = U_o\exp(-2t/RC)$ gives t = 3.77 s.

59. Note $(1/2)^3 = 1/8$. (a) $Q/Q_o = 1/8$, so Q = 62.5 μJ;
(b) $I/I_o = 1/8$, so I = 41.0 μA;
(c) $Q^2/2C$ = 78.1 μJ; (d) I^2R = 0.103 mW.

60. $0.3 = \exp(-7.2/RC)$, leads to RC = 5.98 s.

61. $0.25 = \exp(-62\times10^{-3}/RC)$, so RC = 44.7 ms.

62. Initially: $I_o = \xi/2R$, so $P_A = P_B = \xi^2/4R$.
Finally: $I_o = 2\xi/3R$, $P_A = 4\xi^2/9R$; $P_B = P_C = \xi^2/9R$.
(a) For R_A: Ratio is 16/9; (b) For R_B: Ratio is 4/9.

63. Left: $\xi_1 + I_1R_1 - I_2R_2 = 0$; Right: $-\xi_2 + I_2R_2 + I_3R_3 = 0$

Junction: $I_1 + I_2 = I_3$. Find $I_1 = 84.7$ mA, $I_2 = 65.8$ mA, $I_3 = 150.5$ mA.

64. $I = \xi/12$ and $3I = \xi(12 + R)/12R$ give $R = 6\ \Omega$.

65. $I = \xi/12$ and $1.5I = \xi(12 + R)/12R$ give $R = 24\ \Omega$.

66. $1.44 = \xi - 0.8r$ and $1.7 = \xi + 0.5r$ give $\xi = 1.6$ V and $r = 0.2\ \Omega$.

67. $R_1C_1 = 12 = R_1R_2(C_1 + C_2)/(R_1 + R_2)$. Find $R_2 = 6 \times 10^5\ \Omega$.

68. (a) $U = U_0[1 - \exp(-t/RC)]^2$ gives $\exp(-t/RC) = 0.4$, so $t = 5.94$ s.
(b) $P = P_0\exp(-2t/RC)$ gives $t = 3.31$ s.

69. (a) $I = 24/800 = 0.03$ A. Across 500 Ω, $V = 15$ V. Since there is no current in 200 Ω, p.d. across C is 15 V. So $Q = 900\ \mu C$
(b) $Q = Q_0\exp(-t/RC)$, where $RC = 42$ ms. Find $t = 58.2$ ms.

70. $4I_1 - 6 = 2$ V, thus $I_1 = 2$ A, and $I_2 = I_1 + 2 = 4$ A.
$14 - I_2R = 2$ V, thus $R = 3\ \Omega$.

71. (a) $R_G = 50\ \Omega$. $(5\text{ A} - 1.2\text{ mA})R_{SH} = 60$ mV, so $R_{SH} = 12$ mΩ.
(b) $(R_s + 50)(1.2\text{ mA}) = 5$ V, thus $R_s = 4.12$ kΩ.

72. $(4.9\text{ k}\Omega + R_G)(1\text{ mA}) = 5$ V, thus $R_G = 100\ \Omega$.
$(R_s + 100)(1\text{ mA}) = 1$ V, thus $R_s = 900\ \Omega$.

73. $(5\text{ A} - I_G)(0.04) = I_GR_G$; $(1.2\text{ k}\Omega + R_G)I_G = 5$ V. Eliminate I_GR_G to find $R_G = 50\ \Omega$, then $I_G = 4$ mA

74. Ideal $I_0 = \xi/R$, but actually $I_1 = (\xi/R)(0.99) = = \xi/(R + r)$. Find $r = 10.1$ mΩ.

75. (a) $I = 5/12$ A, so $I^2R_2 = 0.694$ W; (b) $I\xi_3 = 4.17$ W (By ξ_3);
(c) $- 5(5/12) - 5 = -7.08$ V.

76. $P_1 = \xi^2(R_1 + R_2)/R_1R_2$ and $P_2 = \xi^2(R_1 + R_2)$ where $P_1 = 4.5P_2$. Find $R_2 = 1\ \Omega$ or $4\ \Omega$.

77. $R_{eq} = 2.608\ \Omega$ so $I_0 = 4.60$ A. Then $4I_1 = 7.5I_2$ gives $I_1 = 3$ A and $I_2 = 1.6\text{ A} = I_3$. $P_1 = 36$ W, $P_2 = 5.63$ W, $P_3 = 13.6$ W.

78. $R_{eq} = 4.71\ \Omega$ so $I_o = 3.29\ A = I_1$. $V_2 = V_3 = 7.60\ V$, so $I_2 = 2.17\ A$ and $I_3 = 1.12\ A$

Problems

1. $V_G = (20\ \Omega)(2\ mA) = 0.04\ V.$
   $0.04 = (R_1 + R_2 + R_3)(0.098)$
   $0.04 + (2x10^{-3})R_3 = (R_1 + R_2)(0.998)$
   $0.04 + (2x10^{-3})(R_2 + R_3) = R_1(4.998)$
   Find $R_1 = 8.16\ m\Omega$; $R_2 = 32.8\ m\Omega$; $R_3 = 367\ m\Omega$

2. $2R + RR_e/(R + R_e) = R_e$; $R_e^2 - 2RR_e - 2R^2 = 0$,
   Find $R_e = [(3)^{1/2} + 1]R$

3. By symmetry $V_2 = V_4 = V_5$ and $V_3 = V_6 = V_8$. We obtain the two-dimensional pattern below which leads to 5R/6.

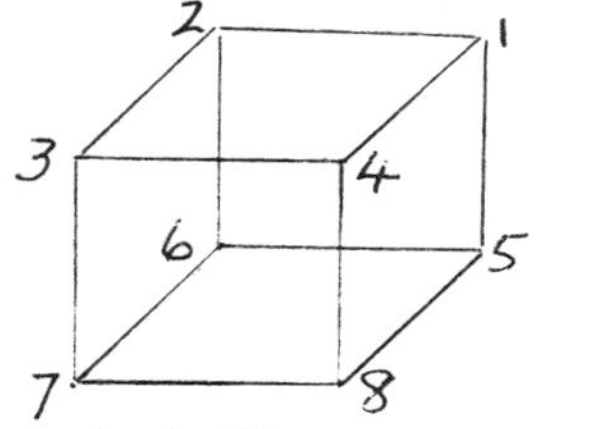

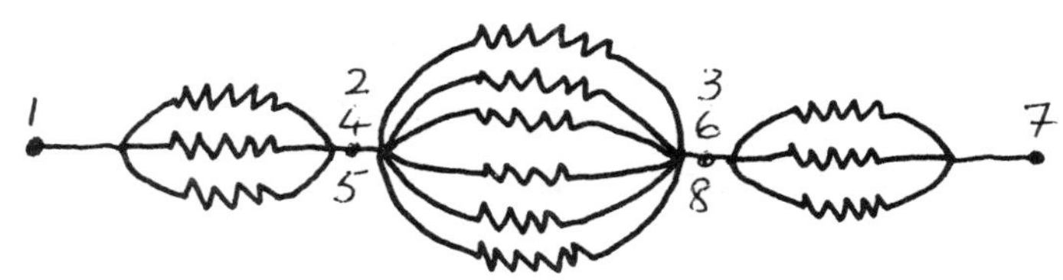

4. (a) $V_2 = V_4$ and $V_6 = V_8$ give 7R/12; (b) $V_3 = V_6$ and $V_4 = V_5$. Pattern shows $V_3 = V_6 = V_4 = V_5$, so ignore two R's. Find 3R/4

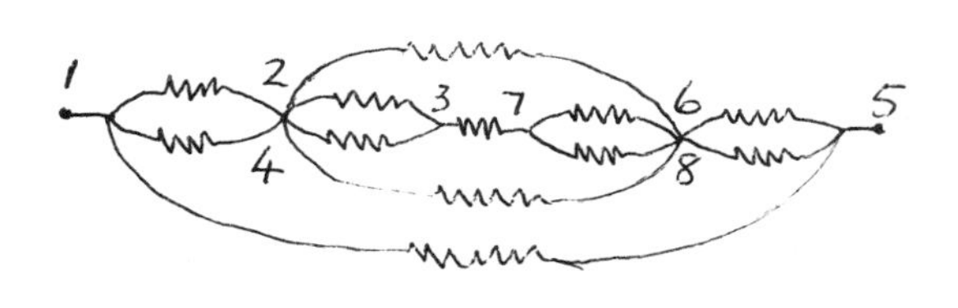

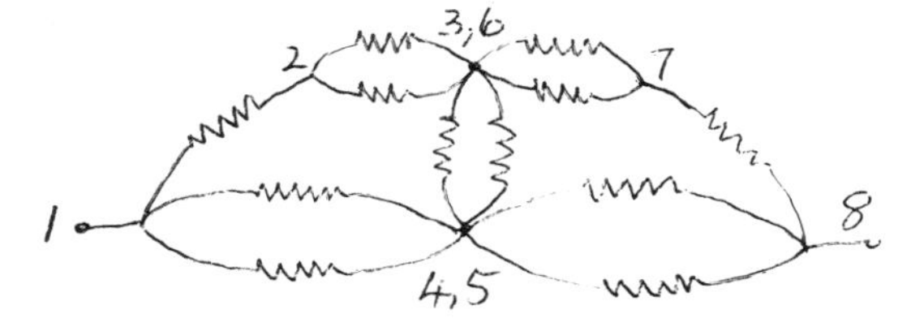

5. $R = \rho d/A$, $C = \kappa\epsilon_0 A/d$, thus $RC = \kappa\epsilon_0\rho$

6. $\exp(-t/RC) = (V_o - V_e)/V_o$; $\exp[-(t + T)/RC] = (V_o - V_f)/V_o$,
   $\exp(T/RC) = (V_o - V_e)/(V_o - V_f)$, so
   $T = RC \ln[(V_o - V_e)/(V_o - V_f)]$

7. $20 - 4I_1 - (I_1 + I_G) = 0$; $20 - 3I_2 - 6(I_2 - I_G) = 0$; and $-4I_1 + 20I_G + 3I_2 = 0$.
   Find $I_1 = 3.918\ A$, $I_2 = 2.495\ A$, $I_G = 0.409\ A$

8. $RC = 320\ ms$, $V_o = 50\ \mu C/40\ \mu F = 1.25\ V$
   (a) $I = (1.25/8000)\exp(-10/320) = 1.51x10^{-4}\ A$;

(b) $Q = (50\ \mu C)\exp(-1/32) = 48.5\ \mu C$; (c) $P = I^2R = 0.182$ mW
(d) $U\ \alpha\ V^2$, so $0.1 = \exp(-2t/0.32)$, $t = 0.368$ s.

9. (a) $I = 4$ A in each branch. $V_a = 12$ V, $V_b = 4$ V, so $\Delta V = 8$ V
(b) Same as (a), 8 V; C charges up;
(c) $R_{eq} = 8/3\ \Omega$, $\tau = 26.7\ \mu s$

10. $I_4 = 3.5$ A (left), $I_2 = 5$ A (up); $I_3 = 0.67$ A (up)

11. $I_1 = 5$ A (down), $I_4 = 2$ A (left), $I_2 = 1.5$ A (up)

12. (a) $I_1 = \xi/R_1$, $I_2 = \xi/R_2$; (b) $I_1 = \xi/R_1$, $I_2 = 0$
(c) $U = 1/2\ C\xi^2$; (d) $(R_1 + R_2)C$

13. (a) $I_1 = \xi/R_1$, $I_2 = 0$, $I_C = I_1$; (b) $\xi/(R_1 + R_2)$, $I_C = 0$.
(c) $\xi - I_1R_1 - Q/C = 0$, $Q/C = I_2R_2$, $I_1 = I_C + I_2$
$\xi - (I_C + Q/R_2C)R_1 - Q/C = 0$, where $I_C = dQ/dt$
$dI_C/I_C = -dt/\tau$, $\ln I_C = -t/\tau +$ constant; then
$I_C = (\xi/R_1)\exp(-t/\tau)$

14. (a) $U\ \alpha\ V^2$, so need $V = (1/2)^{1/2}\xi = \xi(1 - e^{-t/8})$
Thus, $\exp(-t/8) = 0.293$, thus $t = 9.82$ s
(b) $dQ/dt = I = I_o\exp(-2/8) = 0.779$ mA
(c) $P = I^2R = (0.779\ \text{mA})^2R = 60.7$ mW
(d) $\int P\,dt = I_o^2R \int \exp(-2t/\tau)\,dt = 0.367$ W

15. $\xi_1/r_1 - I_1 = IR/r_1$; $\xi_2/r_2 - I_2 = IR/r_2$; $I = I_1 + I_2$
$(\xi_1/r_1 + \xi_2/r_2)(1/r_1 + 1/r_2)^{-1} - I(1/r_1 + 1/r_2)^{-1} = IR$
Express the above as: $\xi_{eq} - Ir_{eq} = IR$

16. Let $D = (R_2 + R_4 + R_5)(R_1 + R_3 + R_5) - R_5^2$
$\alpha_1 = [R_3(R_2 + R_4 + R_5) + R_4R_5]/D = 8/17$;
$\alpha_2 = [R_4(R_1 + R_3 + R_5) + R_3R_5]/D = 39/68$. Find $R_{eq} = 2.66\ \Omega$.

17. Series: $P_s = (2\xi)^2R/(2r + R)^2$;
Parallel: $\xi - Ir/2 - IR = 0$, so $I = \xi/(R + r/2)$
$P_p = I^2R = \xi^2R/(R + r/2)^2$
Let $f = P_s/P_p = (2R + r)^2/(2r + R)^2$
(a) Let $R = r + \delta$, find $f > 1$, thus series;
(b) Let $r = R + \delta$, find $f < 1$, thus parallel

CHAPTER 29

Exercises

1. (a) $r = mv/eB = 6.26$ m; (b) $T = 2\pi r/v = 1.31\ \mu s$

2. $v = (2K/m)^{1/2} = 1.874\text{x}10^7$ m/s; (a) $r = mv/eB = 2.13$ cm; (b) $a = v^2/r = 1.65\text{x}10^{16}\ \text{m/s}^2$; (c) $T = 2\pi m/eB = 7.16$ ns

3. (a) $p = qrB = 1.6 \times 10^{-20}$ kg.m/s; (b) $K = p^2/2m = 4.8\text{x}10^5$ eV

4. $\underline{v}_1 = 10^6\underline{i}$, $\underline{F}_1 = +0.05\underline{k}$ N
   $\underline{v}_2 = (-\underline{i} + \underline{j})10^6/(2)^{1/2}$; $\underline{F}_2 = -0.035\underline{k}$ N
   $\underline{v}_3 = (\underline{i} - \underline{k})10^6/(2)^{1/2}$; $\underline{F}_3 = 0.035(\underline{i} + \underline{k})$ N

5. Since the forces are along $\pm z$, we infer $B_z = 0$.
   Since the forces have the same magnitude and opposite directions the field must be directed midway between $\underline{v}_1$ and $\underline{v}_2$. Thus $\underline{B}$ is directed at 30° to the +y axis.

6. (a) $F = qvB\sin 45^o$ and RHR lead to 0.0106j N;
   (b) Know $B_y = 0$. Then $\sin\theta = F/qvB = 4/15$, so $\theta = 15.5^o$
   Direction of B is $(45 - 15.5) = 29.5^o$ to the +z axis in the xz plane. (Note that the charge is negative.)

7. $\underline{F} = q\underline{v} \times \underline{B} = -0.16\underline{i} - 0.32\underline{j} - 0.64\underline{k}$ N

8. $\underline{F} = q\,\underline{v} \times \underline{B}$: $(3\underline{i} + \underline{j} + 2\underline{k}) = (-2)(-\underline{i} + 3\underline{j}) \times (B_y\underline{j} + B_z\underline{k})$;
   So, $3\underline{i} + \underline{j} + 2\underline{k} = 2B_y\underline{k} - 2B_z\underline{j} - 6B_z\underline{i}$, which gives
   $\underline{B} = \underline{j} - 0.5\underline{k}$ T

9. $(-2\underline{i} + 6\underline{j})\text{x}10^{-13} = (-e)(v_x\underline{i} + v_y\underline{j})\underline{\text{x}}(-1.2\underline{k})$
   Find $\underline{v} = (-3.13\underline{i} - 1.04\underline{j})\text{x}10^6$ m/s

10. (a) $B_z = 0.25$ T, $B_y = 0$, B_x is unknown;
    (b) $B_x = 0$, thus $\underline{B} = 0.25\underline{k}$ T

11. With $\underline{v}$ in the xy plane, $\underline{F}$ is along z, so $B_z = 0$.
    If $\underline{v}$ is along the z axis, $\underline{F}$ is along x, so $B_x = 0$.
    $-1.28\text{x}10^{-13}\ \underline{k} = (1.6\text{x}10^{-13})(2\underline{i} + 3\underline{j}) \times (B_y\underline{j})$. $\underline{B} = -0.4\underline{j}$ T

12. $F = ILB = (10^3)(1)(5\text{x}10^{-5}) = 0.05$ N, upward

13. $\underline{F}_1 = IdB_1\underline{j}$, $\underline{F}_2 = -IdB_1\underline{i}$, $\underline{F}_3 = IdB_1(\underline{i} - \underline{j})$

14. $\underline{F}_1 = 0$, $\underline{F}_2 = IdB_2\underline{k}$ N; $\underline{F}_3 = -IdB_2\underline{k}$ N

15. $\underline{F}_1 = IdB_3\underline{k}$; $\underline{F}_2 = 0$; $\underline{F}_3 = -IdB_3\underline{k}$

16. With I along -z, the magnetic force is along +x.
Equate force components along incline: $mg\sin\theta = ILB\cos\theta$
Thus, $I = mg\tan\theta/LB = 5.9$ A.

17. $F/L = IB = 0.064$ N/m, in the vertical plane 30° below the horizontal, directed south.

18. $F = ILB\sin 150^o = 0.72$ N, in the +z direction.

19. (a) $F = ILB$, so $B = F/IL = 1.85 \times 10^{-2}\underline{j}$ T;
(b) $F = ILB\sin 30^o$, so $B = 3.7\text{x}10^{-2}$ T at 30° to the +z axis in the yz plane. Thus, $\underline{B} = (1.85 \times 10^{-2}\ \underline{j} + 3.21 \times 10^{-2}\ \underline{k})$T

20. (a) $\underline{F}_1 = \underline{F}_3 = 0$; $\underline{F}_2 = -\underline{F}_4 = NIcB\underline{k} = 3\underline{k}$ N
$\underline{\tau} = \underline{\mu} \times \underline{B} = NIacB\underline{j} = 0.06\underline{j}$ N.m
(b) $\underline{F}_3 = -\underline{F}_1 = NIaB\underline{j} = 1.2\underline{j}$ N;
$\underline{F}_2 = -\underline{F}_4 = NIcB\underline{i} = 3\underline{i}$ N; $\underline{\tau} = 0$

21. (a) With $\underline{B} = 0.5\underline{i}$ T; $\underline{F}_1 = -\underline{F}_3 = NIaB\sin 30^o\underline{k} = 8\underline{k}$ N,
$\underline{F}_2 = -\underline{F}_4 = -NIcB\underline{j} = -40\underline{j}$ N;
(b) $\underline{\mu} = NIac(\cos 60^o\underline{i} - \sin 60^o\underline{j}) = 8\underline{i} - 13.8\underline{j}$ A.m^2;
(c) $\underline{\tau} = \mu B\sin 60^o\underline{k} = 6.93\underline{k}$ N.m

22. (a) $\tau = NIAB = 0.16$ N.m; (b) $P = \tau\omega = \tau(40\pi$ rad/s$) = 20.1$ W

23. $\tau = IAB\sin 30^o = 1.88 \times 10^{-4}$ N.m

24. (a) $\underline{\mu} = 0.5Id^2\ \underline{k}$; (b) $\underline{\tau} = -0.5Id^2B\ \underline{j}$

25. (a) $\underline{\tau} = IA(0.6\underline{i} - 0.8\underline{j})\text{x}(0.2\underline{i} - 0.4\underline{k}) =$
$(4.5\underline{i} + 3.4\underline{j} - 2.25\underline{k}) \times 10^{-3}$ N.m;
(b) $U = -IA\underline{n}.\underline{B} = -1.7 \times 10^{-3}$ J

26. $\kappa = NIAB/\phi = 2.13\text{x}10^{-8}$ N.m/deg $= 1.22 \times 10^{-6}$ N.m/rad

27. $\phi = NIAB/\kappa = 3.13^o$

28. (a) $v = erB/m = 1.53\times10^7$ m/s; (b) $T = 2\pi r/v = 8.2\times10-8$ s
(c) $K = 1/2\ mv^2 = 1.95\times10^{-13}$ J

29. $\mu = eL/2m = 1.85 \times 10^{-23}$ A.m^2

30. (a) $r_d/r_p = p_2/p_1 = 1$; (b) $r_d/r_p = m_d/m_p = 2$
(c) $r_d/r_p = (m_d/m_p)^{1/2} = 1.41$

31. (a) $r_e/r_p = m_e/m_p = 5.45 \times 10^{-4}$;
(b) $r_e/r_p = (m_e/m_p)^{1/2} = 2.33 \times 10^{-2}$

32. $p = mv = qrB$ and $K = (qrB)^2/2m$.
(a) $B_\alpha/B_p = (m_\alpha/m_p)(q_p/q_\alpha) = 2$; (b) $B_\alpha/B_p = q_p/q_\alpha = 1/2$
(c) $B_\alpha/B_p = (q_p/q_\alpha)(m_\alpha/m_p)^{1/2} = 1$

33. $d = T(v\cos30^o) = (2\pi m/qB)(v\cos30^o) = 3.1$ mm

34. (a) $r = mv/qB = 1.56\times10^4$ m; (b) East

35. (a) $f = qB/2\pi m = 11.4$ MHz;
(b) $K = (qrB)^2/2m = 4.41\times10^{-15}$ J $= 27.6$ keV;
(c) $p = mv = qrB = 3.84 \times 10^{-21}$ kg.m/s

36. $1/mv^2 = qV$, so $v = (4eV/m)^{1/2} = 1.16\times10^6$ m/s;
$r = mv/qB = 4.04$ cm

37. $v = (2eV/m)^{1/2}$, thus $r = mv/qB = (2mV/e)^{1/2}/B$.
$r_{20} = 5.09$ cm, $r_{22} = 5.34$ cm, so $\Delta r = 2.5$ mm and $d = 5$ mm

38. From Eq. 29.14: $\Delta r = \Delta mE/qB_1B_2 = 3.89$ cm and $d = 7.8$ cm.

39. (a) From $\underline{E} = -\underline{v} \times \underline{B}$, find $\underline{B} = -10^{-4}\underline{k}$ T;
(b) $r = mv/qB = 11.4$ cm

40. $v = (2eV/m)^{1/2} = 1.38\times10^6$ m/s; $\underline{B} = (E/v)\underline{k} = 7.22\times10^{-3}\ \underline{k}$ T.

41. (a) $B = (2mK)^{1/2}/er = 0.914$ T;
(b) 200 crossings, so $10^7/200 = 50$ kV;
(c) $f = eB/2\pi m = 13.9$ MHz

42. (a) $\omega = eB/m = 8.62\times10^7$ rad/s
(b) $K = (qrB)^2/2m = 3.49\times10^{-12}$ J $= 21.8$ MeV

43. (a) $r = (2mK)^{1/2}/qB = 31.3$ cm;
(b) $T = 2\pi m/qB = 4.1\text{x}10^{-8}$ s.
Number of revolutions = 0.5(12 MeV)/60 keV = 100 rev.
Thus total time $\Delta t = 4.1\ \mu s$

44. $1/2\ mv^2 = qV$, $mv^2/r = qvB$ lead to
$q/m = 2V/B^2r^2 = 1.8\text{x}10^{11}$ C/kg.

45. $1/2\ mv^2 = qV$, $mv^2/r = qvB$, lead to
$r = (2mV/q)^{1/2}/B = 6.12$ cm

46. (a) $v_d = V_H/Bw = 1.88$ mm/s; (b) $n = BI/Vqt = 3.13\text{x}10^{27}\ m^{-3}$

47. $B = V_Hnqt/I = 4.08$ T

48. $n = BI/V_Hqt = 7.14 \times 10^{28}\ m^{-3}$

49. (a) 9.6×10^{-18} N, East; (b) 9.6×10^{-18} N, Vertically up.

50. Since $\underline{F}$ is along x axis, $\underline{B}$ must be along $\pm$z axis.
$F\underline{i} = -e(-v\underline{j}) \times B\underline{k}$, so $\underline{B} = +0.02\underline{k}$ T.

51. $F = evB = 5.18 \times 10^{-18}$ N at 45° N of E.

52. $\underline{F} = 2e\underline{v} \times \underline{B} = (-35.2\underline{i} - 41.6\ \underline{j} + 9.60\underline{k}) \times 10^{-14}$ N.

53. Maximum $F = I\ell B$, so $\underline{B} = 333\ \mu T$ at 30° E of N.

54. $\underline{F}_1 = -1.2\underline{k}$ N; $\underline{F}_2 = 1.2\underline{j}$ N; $\underline{F}_3 = 1.2\underline{k}$ N; $\underline{F}_4 = 0$.

55. $\underline{F}_1 = 1.2\underline{i}$ N; $\underline{F}_2 = 1.2\underline{j}$ N; $\underline{F}_3 = (-1.2\ \underline{i} -1.2\underline{j})$ N; $\underline{F}_4 = 1.2\underline{j}$ N

56. $\underline{F} = I\underline{\ell} \times \underline{B}$, so $(4\underline{i} - 6\underline{j}) \times 10^{-5} = 2(-25\underline{k}) \times (B_x\underline{i} + B_y\underline{j})$.
Find $\underline{B} = 1.2\underline{i} + 0.8\underline{j}\ \mu T$.

57. (a) $F = I\ell B\sin 70^o = 2.71$ mN East; (b) $F = I\ell B = 2.88$ mN
at 20° above horizontal line due north.

58. $\underline{\mu} = NI(\pi r^2)(-\underline{k})$; $\underline{\tau} = \underline{\mu} \times \underline{B} = -1.18\underline{j}$ N.m

59. $0.5mv^2 = eV$ and $mv^2/r = evB$ give $B = (2mV/e)^{1/2}/r =$
9.07×10^{-4} T.

60. (a) $T = 2\pi m/eB = 0.328\ \mu s$; (b) $K = (erB)^2/2m = 15.5$ keV.

61. $\underline{E} = -\underline{v} \times \underline{B} = -(v\underline{i}) \times (-0.05\underline{k})$, where $v = (2eV/m)^{1/2} = 1.45\times10^7$ m/s. Find $\underline{E} = -7.25 \times 10^5 \underline{j}$ V/m.

62. $\underline{E} = -\underline{v} \times \underline{B}$, so $-10^4\underline{j} = +(5\times10^6\underline{k}) \times \underline{B}$. Find $\underline{B} = -2\underline{i}$ mT.

63. (a) $\underline{F}_1 = -\underline{F}_3 = 0.64\underline{j}$ N; $\underline{F}_2 = -\underline{F}_4 = 0.64\underline{k}$ N.
(b) $\underline{\mu} = IA(0.707\underline{i} + 0.707\underline{j})$, $\underline{\tau} = \underline{\mu} \times \underline{B} = -128\underline{k}$ N.m.
OR: $\tau = (F_1 + F_3)(d/2) = 0.128$ N.m; CW looking along -z axis.

64. Since $\underline{F}_1$ is along -y we can say $B_y = 0$ and $B_z > 0$. Since $\underline{F}_2$ is along -x axis, we can say $B_x = 0$. Then $F_2 = ev_2B\sin30^o = -4.8\times10^{-14}$ N gives B = 0.3 T, so $\underline{B} = 0.3\underline{k}$ T.

65. (a) $F = evB\sin\theta$ tells us $\underline{B}$ could be at $\theta = 25^o$ or 155° to the +x axis. (It lies on one of two possible cones.)
(b) Since $\underline{F}$ is along the +z axis, $\underline{B}$ is in the xy plane. Furthermore, the RHR tells us $\theta = \pm155^o$ to +x axis is the proper choice. Thus, $\underline{B} = -0.589\underline{i} \pm 0.275\underline{j}$ T.

66. (a) $r = mv/eB$ and $v = (2eV/m)^{1/2}$ so $r = (2mV/eB^2)^{1/2}$;
(b) 10%.

67. $0.5mv^2 = (1\times10^7)(1.6\times10^{-19})$, so $v = 2.2 \times 10^7$ m/s. Then $r = mv/2eB = 0.38$ m.

68. $n = BI/qtV = 1.26 \times 10^{28}$ m^{-3}.

69. $\mu = I(\pi r^2)$ gives $I = 1.02 \times 10^9$ A.

70. $K = (erB)^2/2m$ leads to $r = (2mK)^{1/2}/eB = 3.77$ m.

71. (a) $r = mv\sin80^o/eB = 0.123$ m;
(b) $d = (v\cos80^o)(2\pi m/eB) = 0.137$ m

72. $F = evB\sin\theta$, so $v = 1.06 \times 10^7$ m/s and K = 321 eV.

73. $0.5mv^2 = eV$ gives $v = 1.68 \times 10^7$ m/s. From $-E\underline{i} = -(-v\underline{j}) \times \underline{B}$, we find $\underline{B} = (E/v)(-\underline{k}) = -5.24 \times 10^{-4}\underline{k}$ T. (B_y is arbitrary)

74. (a) $K = (qrB)^2/2m = 0.110$ MeV; (b) $T = 2\pi m/qB = 0.164$ μs.

75. (a) $p = qrB$, so $p_p/p_d = 1$; (b) $K = (qrB)^2/2m$, $K_p/K_d = 2$.

76. $F = ILB\sin\theta = 1.07$ N, due East

77. (a) $\mu = NIA = 0.748$ A.m^2;
(b) Maximum $\tau = \mu B = 5.99 \times 10^{-2}$ N.m

78. $I = \kappa\phi/NAB = 48$ μA.

Problems

1. (a) $\tau = -\mu B\sin\theta = Id^2\theta/dt^2$> for small θ, $\sin\theta \approx \theta$,
$d^2\theta/dt^2 + (\mu B/I)\theta = 0$ which indicates SHM
(b) $2\pi(I/\mu B)^{1/2}$

2. (a) $-ILB = ma$, so $v = v_o - (ILB/m)t$;
(b) $\Delta x = v_o^2/2a = mv_o^2/2ILB$

3. $\underline{F} = \int I\, d\underline{L} \times \underline{B} = I\,(\int d\underline{L}) \times \underline{B} = I\, \Delta\underline{L} \times \underline{B}$

4. $\mu = NIA$ where $A = \pi r^2 = \pi(L/2\pi N)^2$, thus $\mu \propto 1/N$.
μ is a maximum for $N = 1$.

5. (a) For an elemental ring $dI = dq/T = (\sigma 2\pi r\, dr)\omega/2\pi$
Thus, $d\mu = dI(\pi r^2) = \sigma\omega\pi r^3\, dr$; then $\mu = \int d\mu = \sigma\omega\pi R^4/4$
(b) $\tau = \mu B = \sigma\omega\pi BR^4/4$

6. $F_o = m\omega_o^2 r$; $F_o + F_B = m\omega_1^2 r = F_o + e(\omega r)B$
$(\omega_1^2 - \omega_o^2) \approx 2\omega_1\, \Delta\omega = +\, e\omega_1 B/m$. Thus, $\Delta\omega = +eB/2m$

7. $mg = 2kx_o$, so $k = 1.23$ N/m
$mg = 2k(x_o - 0.01) + ILB$, thus $B = 0.02k/IL = 1.53 \times 10^{-2}$ T

8. (a) $B = mv/er = 85.4$ G
(b) $T = 2\pi r/v = 4.19$ ns
It takes T/12 to turn through 30°, so $t = 3.49\times10^{-10}$ s
(c) $x = r\sin 30^o = 1$ cm; $y = r(1 - \cos 30^o) = 0.27$ cm

CHAPTER 30

Exercises

1. $\mu_0 I_1 I_2 c/2\pi$ [1/a - 1/(a + b)] to the right

2. (a) $B_1 = \mu_o(4\ A)/2\pi(0.08) = 10\ \mu T$; $B_2 = 40\ \mu T$
   $B_{Tx} = -B_1\cos 53^o + B_2\cos 37^o = 26\ \mu T$
   $B_{Ty} = B_1\sin 53^o + B_2\sin 37^o = 32\ \mu T$
   (b) $4/d = 12/(0.1 + d)$, thus d = 5 cm, to the left of I_1
   (c) $F/L = \mu_o I_1 I_2/2\pi r = 96\ \mu N/m$, repulsion.

3. (a) $B_x = (B_1 + B_2)\cos\theta$; $B_y = (B_1 - B_2)\sin\theta$, where $\theta = 33.7^o$.
   $\underline{B} = (6.92\underline{i} - 1.54\underline{j}) \times 10^{-5}$ T;
   (b) $\underline{F} = I\ \underline{L} \times \underline{B} = (4.6\underline{i} + 20.8\underline{j}) \times 10^{-5}$ N

4. $r = \mu_o I/2\pi B = 8$ cm, west of the wire.

5. $B = \mu_o I/2\pi r = 5 \times 10^{-4}$ T

6. $F = \mu_o I_1 I_2/2\pi d$: $F_1 = 80\ \mu N/m$, repulsion; $F_2 = 60\ \mu N/m$, attraction. $F_x = (F_1 + F_2)\cos 60^o = 70\ \mu N/m$;
   $F_y = (F_1 - F_2)\sin 60^o = 17.3\ \mu N/m$

7. $B_e = 0.5$ G due N, $B_w = \mu_o I/2\pi d = 0.06$ G due W.
   $\tan\theta = B_w/B_e$ leads to $\theta = 6.8^o$ W of N

8. $B = \mu_o I/2\pi r = 5\times10^{-5}$ T in the -z direction.
   Let $F = evB = 8\times10^{-18}$ N and note that charge is negative.
   (a) $-F\underline{j}$; (b) $+F\underline{i}$; (c) zero

9. From Eq. 30.8, with z = 0, $B = \mu_o I/2a$ at center of loop.
   $B = \mu_o I/4a + 2(\mu_o I/4\pi a) = (\mu_o I/2a)(1/2 + 1/\pi) =$
   $5.14\times10^{-7} I/a$, out

10. From Eq. 30.8, with z = 0, $B = \mu_o I/2a$ at center of loop.
    $B = \mu_o I/2\pi a + \mu_o I/2a = (\mu_o I/2a)(1/\pi + 1) = 8.28\times10^{-7} I/a$, out

11. From Eq. 30.8, with z = 0, $B = \mu_o I/2a$ at center of loop.
    $(\mu_o I/4)(1/a - 1/b)$, into page

12. From Example 30.2: $B = (\mu_o I/4\pi d)(\sin\alpha_1 + \sin\alpha_2)$

Here $\sin\alpha_1 = \sin\alpha_2 = (L/2)/[L^2/4 + d^2]1/2$

Thus $B = (\mu_0 I/2\pi d)L/(L^2 + 4d^2)^{1/2}$

13. From Example 30.2: $B = (\mu_0 I/4\pi R)(\sin\alpha_1 + \sin\alpha_2)$.
$R = \ell/2$, and $\alpha_1 = \alpha_2 = 45^o$, so $\sin\alpha_1 = \sin\alpha_2 = 1/(2)^{1/2}$.
For four wires, find $B = 2(2)^{1/2}\mu_0 I/\pi\ell$.

14. $d\ell = 10^{-3}\underline{k}$ m. Note that the unit vector is $\underline{r}/r$
(a) $\underline{r} = L(\underline{i} + \underline{k})$; $d\underline{B} = 8.84\times10^{-7}\underline{j}$ T
(b) $\underline{r} = L(\underline{i} + \underline{j} + \underline{k})$; $d\underline{B} = 4.81(-\underline{i} + \underline{j})\times10^{-7}$ T
(c) $\underline{r} = L(\underline{j} + \underline{k})$; $d\underline{B} = -8.84\times10^{-7}\underline{i}$ T
(d) $\underline{r} = L(\underline{i} + \underline{j})$; $d\underline{B} = 8.84(-\underline{i} + \underline{j})\times10^{-7}$ T
(e) $\underline{r} = L\underline{j}$; $dB = -25\times10^{-7}\underline{i}$ T

15. (a) Plot $B = (\mu_0 Ia^2)/2(a^2 + z^2)^{3/2}$.
(b) At the center $B = \mu_0 I/2a$. We find $2a^3 = (a^2 + z^2)^{3/2}$ or $z^2 = (2^{2/3} - 1)a^2$, thus $z = 0.766a$

16. $B_c = N\mu_0 I/2R$. If $\tan\theta = B_c/B_e$, then $I = 2RB_e \tan\theta/\mu_0 N$.

17. (a) Let $B_o = \mu_0 I/2\pi r$ where $I = 1$ A and $r = L/(2)^{1/2}$
Find $B_x = 0$ and $B_y = 10B_o\cos45^o$. $\underline{B} = -1.33 \times 10^{-5}\underline{j}$ T;
(b) $\underline{F} = -e\underline{v} \times \underline{B} = 8.53 \times 10^{-18}\underline{k}$ N

18. (a) $B_y = 0$, $B_x = 2(\mu_0 I/2\pi r)(a/r) = \mu_0 Ia/\pi(a^2 + x^2)$;
(b) $x = 2a$

19. (a) Assume both currents coming out: $B_x = 0$,
$B_y = 2(\mu_0 I/2\pi r)(x/r) = \mu_0 Ix/\pi(a^2 + x^2)$;
(b) Set $dB/dx = 0$ to find $x = \pm a$

20. From Eq. 30.8, with $z = 0$, $B = \mu_0 I/2a$ at center of loop.
$B = \mu_0 I/2a = 8\times10^{-5}$, leads to $I = 5.09$ A

21. From Eq. 30.8, with $z = 0$, $B = \mu_0 I/2a$ at center of loop.
For straight wires use Example 30.2: Note $R = a\cos45^o$.
$B = (\mu_0 I/4\pi R)(\sin\alpha_1 + \sin\alpha_2)$ with $\alpha_1 = \alpha_2 = 45^o$.
$B = (\mu_0 I/a)[1/4 + 1/\pi] = 7.14 \times 10^{-7} I/a$

22. From Eq. 30.8, with $z = 0$, $B = \mu_0 I/2a$ at center of loop.
For straight wires use Example 30.2:

$B = (\mu_0 I/4\pi R)(\sin\alpha_1 + \sin\alpha_2)$, where R = a or 2a.
$B = (\mu_0 I/a)[1/4 + 1/\pi(5)^{1/2} + 1/4\pi(5)^{1/2}] = 0.428\mu_0 I/a$.

23. From Eq. 30.8, with z = 0, $B = \mu_0 I/2a$ at center of loop.
(a) $B = (\mu_0 I/4)(1/a + 1/b) = 3.14 \times 10^{-5}$ T;
(b) $\mu = IA = I\pi(a^2 + b^2)/2 = 0.254$ A.m^2

24. $\underline{B} = (\mu_0 I/2\pi d)(-\underline{i} - 2\underline{k})$ where I = 20 A. $\underline{B} = -80\underline{i} - 160\underline{k}$ μT

25. $B = \mu_0(N/L)I$, so $N = BL/\mu_0 I = 265$

26. Length of wire $L = (60)(2\pi r) = 37.7$ m;
Resistance $R = \rho L/A = 0.204\ \Omega$;
$I = 1.5/0.204 = 7.35$ A; then $B = N(\mu_0 I/2a) = 2.77 \times 10^{-3}$ T

27. $\int \underline{B}.d\ell$ is not zero, even though I = 0.

28. For r < a, B = 0. For r > b, $B = \mu_0 I/2\pi r$.
For a < r < b, $I_{enc} = I(r^2 - a^2)/(b^2 - a^2)$,
thus $B = \mu_0 I(r^2 - a^2)/2\pi r(b^2 - a^2)$

29. (a) Parallel to the plate;
(b) Ampere's law: $B(2L) = \mu_0 JLt$, thus $\mu_0 Jt/2$

30. From Eq. 30.12, $B = \mu_0 NI/2\pi r$. (a) 8 mT; (b) 5.14 mT

31. Inside: See Example 30.5: $\mu_0 Ir/(2\pi R^2) = (0.25)(\mu_0 I/2\pi R)$
Thus r = R/4 = 0.5 mm
Outside: $\mu_0 I/2\pi r = (0.25)(\mu_0 I/2\pi R)$, so r = 4R = 8 mm

32. $\underline{B}$ is normal to $d\ell$ for the radial sections. For circular arcs: $\int \underline{B}.\,d\ell = (\mu_0 I/2\pi a)(\pi a/2) + (\mu_0 I/2\pi b)(3\pi b/2) = \mu_0 I$.

33. $B = \mu_0 I/2\pi r$. (a) 1.25 mT; (b) 3 μT

34. $F/\ell = \mu_0 I_1 I_2/2\pi r = 0.96$ mN/m, attractive.

35. (a) $I_1/6 = I_2/2$, so $I_2 = 2$ A along +y;
(b) $I_1/10 = I_2/2$, so $I_2 = 1.2$ A along -y.

36. $B = \mu_0(N/2\pi r)I \approx \mu_0 nI$ if r is large.

37. $B = \mu_0 nI = 0.251$ T.

38. Need $I_1 = I_2$ for zero force on 1 A. For each 2-A wire, need I_1 and I_2 coming OUT. Net force on one 2-A wire (dropping $\mu_o/2\pi$ factor): $(2)(1)/2d + (2)(2)/4d - 2I_1/d - 2I_2/3d = 0$, so $I = 0.75$ A.

39. Say I_1 is OUT. Dropping $\mu_o/2\pi$ factor, net force on I_1: $2I_1/d - I_1/d + I_1I_2/2d - 2I_1/3d = 0$, so $I_2 = -2/3$ A (IN).

40. $\mu_o I(\underline{i} - \underline{j})/2\pi d$

41. (a) $I_1/x = I_2(10 - x)$, so $x = 3.57$ cm;
(b) $I_1/d = I_2(10 + d)$, so $d = 12.5$ cm or $x = -12.5$ cm.

42. Say $F_o = \mu_o I^2/(2\pi d)$, then $F_x = F_o - F_o/2 = F_o/2$;
$F_y = -F_o - F_o/2 = -3F_o/2$, so $\underline{F} = (1.28\underline{i} - 3.84\underline{j}) \times 10^{-4}$ N/m.

43. Say $B_o = \mu_o I/2\pi d$, then $B_x = B_o + B_o/2 = 3B_o/2$; $B_y = B_o - B_o/2 = B_o/2$, thus $\underline{B} = 48\underline{i} + 16\underline{j}$ μT.

44. $B = \mu_o NI/2\pi r = 1.56$ mT.

45. $[\mu_o I/2\pi \ell](1/5^{1/2} + 3/13^{1/2}) = 2.56\times10^{-7}\ I/\ell = 34.1$ μT, Out

46. $B = \mu_o NI/2R$, so $N = 24$ turns.

47. From Eq. 30.8: $B(z) = \mu_o Ia^2/2(a^2 + z^2)^{3/2}$.
For $B(z) = 0.5B(0)$, find $(a^2 + z^2) = 2^{2/3}\ a^2$, so $z = 0.766a$.

48. $[\mu_o \epsilon_o] = (N/A^2)(C^2/N.m^2) = s^2/m^2$. Find $v = c = 3 \times 10^8$ m/s.

49. (a) $B_x = (\mu_o/2\pi)[5/0.2 + 8/0.2]\cos 60^o = 6.50$ μT
$B_y = (\mu_o/2\pi)[5/0.2 - 8/0.2]\sin 60^o = -2.60$ μT
(b) On 1 m: $\underline{F} = I\underline{L} \times \underline{B} = 5(\underline{k})\times(6.5\underline{i} - 2.6\underline{j})\times10^{-6} =$
$13.0\underline{i} + 32.5\underline{j}$ μN.

Problems

1. $I = q/T = qv/2\pi R$. From $mv^2/R = qvB$, we have $v = qRB/m$
Thus $I = q^2B/2\pi m$. From Eq. 30.8, with $z = 0$, $B = \mu_o I/2a$ at center of loop, so $B = \mu_o q^2 B/4\pi mR$. (See Eq. 30.13: $B = \mu_o qv/4\pi R^2$.)

2. (a) From Eq. 30.8:

$B = (\mu_0 NIR^2/2)\{[R^2+(R/2 + x)^2]^{-3/2}] + [R^2+(R/2 - x)^2]^{-3/2}]\}$

(b) When $x = 0$, $B = \mu_0 NI(5/4)^{-3/2}$.

3. (a) $dI = \sigma\omega r\ dr$; (b) $dB = \mu_0\ dI/2r = \mu_0\sigma\omega\ dr/2$;
(c) $B = \int dB = \mu_0\sigma\omega R/2$.

4. (a) Use Example 30.2 and note that
$B_y = 4B\cos\theta$. $B_y = (2\mu_0 I/\pi R)\sin\alpha\ \cos\theta$; where $R^2 = y^2 + L^2/4$,
$\sin\alpha = (L/2)/(R^2 + L^2/4)^{1/2} = (L/2)/(y^2 + L^2/2)^{1/2}$
$\cos\theta = (L/2)/(y^2 + L^2/4)^{1/2}$
$B_y = (\mu_0 IL^2/2\pi)(y^2 + L^2/4)^{-1}(y^2 + L^2/2)^{-1/2}$
(b) When $y >> L$, $B_y \approx \mu_0 IL^2/2\pi y^3 = 2k'\mu/y^3$
where $k' = \mu_0/4\pi$ and $\mu = IA = IL^2$.

5. $dB_x = (\mu_0\ dI/2\pi R)(D/R)$, where $R^2 = x^2 + D^2$, and $dI = Idx/w$.
$B = (\mu_0 ID/2\pi w) \int dx/(x^2 + D^2) = (\mu_0 I/\pi w)\ \tan^{-1}(w/2D)$

6. $dB = \mu_0\ dI/2\pi x$ where $dI = I\ dx/w$;
$B = (\mu_0 I/2\pi w) \int dx/x = (\mu_0 I/2\pi w)\ \ln[(L + w)/L]$

7. See Eq. (i) of Example 30.4.
Typical values: 0.034 at −5 cm, 0.498 at 0, 0.960 at +5 cm

8. (a) Consider a elementary sector with angular width $d\theta$. The field at at distance r is $dB = \mu_0 dI/2\pi r$. If I is the total current through the tube, then $dI = Ird\theta/2\pi a$, thus $dB = \mu_0 Id\theta/4\pi^2 a\ \alpha\ d\theta$ but independent of r. Thus pairs of segments defined by intersecting lines contribute equal and opposite fields at any point inside the tube.
(b) For a circular path inside tube: $(2\pi r)B = 0$, so $B = 0$

9. From Eq. (ii) of Example 30.5, $B = \mu_0 Ir/2\pi R^2$ inside a wire, where $\underline{B}$ is perpendicular to $\underline{r}$. The current that would have flowed through the cavity is $I' = a^2 I/R^2$, thus the field associated with the cavity would be $B' = \mu_0 I'r'/2\pi a^2 = \mu_0 Ir'/2\pi R^2$. The total field is the sum of the contributions of a wire of radius R and a wire of radius a with a current I' in the opposite direction. (Note that $B' = 0$ at the center of the hole, so $B = B_0 = \mu_0 Id/2\pi R^2$.) Let $\underline{s}$, $\underline{s}'$ and $\underline{D}$ be equal in magnitude but normal to $\underline{r}$, $\underline{r}'$ and $\underline{d}$ respectively The sum: $\underline{B}_T = \underline{B} + \underline{B}' = (\mu_0 I/2\pi R^2)(\underline{s} - \underline{s}') = (\mu_0 I/2\pi R^2)\underline{D}$.

10. $dB_x = (\mu_0 dI/2\pi R)\sin\theta$, where $dI = Id\theta/\pi$.
Thus $B = (\mu_0 I/2\pi^2 R) \int \sin\theta\ d\theta = \mu_0 I/\pi^2 R$.

CHAPTER 31

Exercises

1. $\Delta\phi = BA(\cos 120^o - 1) = -2.52$ mWb

2. (a) $\phi = B(\pi r^2)\cos 60^o = 1.41\text{x}10^{-3}$ Wb
   (b) $\Delta\phi = -2\phi = -2.82\text{x}10^{-3}$ Wb

3. $\phi = BA\cos\theta = \mu_o nIA\cos\theta$.
   $\xi = N\ \Delta\phi/\Delta t = \mu_o NnIA\cos\theta\ \Delta I/\Delta t = 40.2\ \mu V$

4. (a) $\xi = BLv$, thus $v = IR/BL = 32$ m/s
   (b) $F = ILB = 2.5\text{x}10^{-2}$ N

5. (a) $\phi = BA = (3.2t - 2.4t^2)\text{x}10^{-4}$ Wb;
   (b) $I = -\ N(d\phi/dt)/R = 1.33$ mA

6. $\phi = BA\cos 50^o$; and $\xi = N\ d\phi/dt = NA\cos 50^o(\Delta B/\Delta t) = 18.2$ mV

7. $\xi = -\ N\ d\phi/dt = -NA_{coil}\ dB/dt = -\mu_o NnA_{coil}\ dI/dt$.
   Find $\xi = -30.9\cos(60\pi t)$ mV

8. $\xi = -Nd\phi/dt = -N(\pi r^2)\ dB/dt = -0.682\ \cos(2\pi ft)$

9. (a) $\Delta\phi = (\pi d^2/4)\Delta B = -1.96$ mWb; (b) $\xi = AdB/dt = 98.2$ mV;
   (c) Counterclockwise

10. (a) $\xi = BLv = 0.9$ V; (b) $F = ILB = (\xi/R)LB = 3.38\text{x}10^{-2}$ N
    (c) $I^2R = 0.675$ W; (d) $Fv = 0.675$ W

11. (a) $\xi = BLv = 3.24$ V, so $I = \xi/R = 1.2$ A
    (b) $F = ILB = 0.13$ N; (c) $Fv = 3.9$ W; (d) $I^2R = 3.9$ W

12. $\xi = N(\pi r^2)dB/dt = 2.51dB/dt$. Need $\xi^2/R = 2$ W,
    so $dB/dt = +3.56$ T/s

13. $R = \rho L/A = \rho(20\text{x}2\pi\text{x}0.05)/(\pi d^2/4) = 0.136\ \Omega$
    $\xi = NA(dB/dt)$ and $P = \xi^2/R = 7.26$ mW

14. (a) $x = A\cos(\omega t)$ where $\omega = (2k/m)^{1/2}$.
    $\xi = -d\phi/dt = BL\omega A\sin(\omega t) = 0.17\sin(14.1t)$
    (b) $\xi_o = 0.17$ V occurs at $t = T/4 = 0.111$ s

15. (a) $\Sigma F_y = mg - ILB = 0$, where $I = \xi/R = (BLv)/R$.
Thus $v_T = mgR/(BL)^2$
(b) $dU_g/dt = mgv_T = (mg/BL)^2R$. Then, $P = I^2R = (mg/BL)^2R$

16. $\phi = BA$, $\Delta\phi = -2AB$.
$\xi = -N\,\Delta\phi/\Delta t = 2NAB/\Delta t$ and $\xi = IR = (\Delta Q/\Delta t)R$.
Equating these expressions for ξ we find $\Delta Q = 2NAB/R$.

17. $\xi = NA\,dB/dt = N(\pi r^2)(0.2 - t)$
$P = \xi^2/R = 3.38$ mW

18. $\phi = B(\pi r^2)$, so $d\phi/dt = 2\pi rB\,dr/dt = 24.1$ mV

19. $\xi_o = NAB\omega$, thus $\omega = 130$ rad/s

20. $\xi_o = NAB\omega$, where $\omega = \pi/2$ rad/s. Find $\xi_o = 56.5$ mV

21. (a) $\xi_o = NAB\omega = 0.201$ V
(b) $\tau = \mu B = NIAB = N(\xi/R)AB = 7.15 \times 10^{-4}$ N.m

22. (a) $\xi = NAB\omega\sin(\omega t) = 29.9$ mV
(b) $\tau = \mu B\sin(\omega t) = NIAB\sin(\omega t)$, since $I = \xi/R$,
$\tau = (NAB)^2\omega\sin^2(\omega t)/R = 2.84\text{x}10^{-5}$ N.m
(c) $P = \tau\omega = 3.57\text{x}10^{-4}$ W; (d) $P = \xi^2/R = 3.57\text{x}10^{-4}$ W

23. $(2\pi d)E = A\,dB/dt = (\pi d^2)(C)$, thus $F = eE = eCd/2$

24. $\int \underline{E}.d\ell$ is not zero even though $dB/dt = 0$.

25. Find $(2\pi r)E = \mu_o An\,dI/dt$, where $dI/dt = 12t$, and $A = \pi r^2$.
$E = 0.5\mu_o rn\,dI/dt$
(a) 6.03×10^{-5} V/m; (b) 1.21×10^{-4} V/m

26. $(2\pi r)E = \mu_o(\pi r^2)n\,dI/dt$, thus $dI/dt = 2E/\mu_o rn = 199$ A/s

27. (a) $V = \xi = BLv = 0.81$ V; (b) Zero

28. From Example 31.7, $\xi = \omega BR^2/2 = 3.18$ mV

29. From Example 31.7, $\xi = \pi fBR^2$, thus $f = 119$ rev/s $= 7150$ rpm

30. From $\xi = -NAdB/dt$ and $P = \xi^2/R$ find $dB/dt = \pm 1.62$ T/s.

31. $\xi = -NAdB/dt = NAB_0 \exp(-t/\tau)/\tau$.

32. $\xi = -NAdB/dt$, $dB/dt = -\xi/NA = -0.399$ T/s. From $B = 0.32 - 0.399t$, find $t = 0.802$ s.

33. $B = \mu_0 n 1 I$; $|\xi| = N_2 AdB/dt = N_2 A\mu_0 n_1 dI/dt$; Find $N_2 = 50$ turns.

34. (a) $\xi = NAB\omega \sin(\omega t)$, so peak $I = (NAB\omega)/(R + r) = 0.014\omega$.
$P_R = I^2R = 12$, so $\omega = 151$ rad/s.
(b) Maximum $\tau = \mu B = NIAB = 8.86 \times 10^{-2}$ N.m.

35. $\xi = 0.5\omega BR^2 = 0.141$ mV

36. $\Delta\phi = -BA = -4.8 \times 10^{-4}$ Wb.

Problems

1. $\xi = EL = BLv$, thus $E = Bv = (\mu_0 I/2\pi r)v$
$\Delta V = \int E_r\, dr = (\mu_0 Iv/2\pi) \ln[(L + d)/d]$

2. $I = BLv/R = (\mu_0 I/2\pi x)Lv/R = \mu_0 ILv/2\pi R(d + a) = 0.1$ mA
OR: Use $\phi = (\mu_0 I/2\pi)\ln[(a + x)/a]$ and $|\xi| = |d\phi/dt|$.

3. (a) From $F = ma$, we find $a = -(ILB)/m$, where $I = \xi/R = (BLv)/R$. Thus, $dv/dt = -(LB)^2 v/mR$, so
$\int dv/v = -\int (LB)^2/mR\, dt$. With $\tau = mR/(LB)^2$; find
$v = v_0 \exp(-t/\tau)$
(b) $dx/dt = v_0 \exp(-t/\tau)$, so
$\Delta x = \int dx = v_0[-\tau \exp(-t/\tau)] = v_0\tau$
(c) Energy $= \int P\, dt = \int \xi^2/R\, dt = (BLv_0)^2/R \int \exp(-2t/\tau)\, dt$
$= (BLv_0)^2\tau/2R = 1/2\, mv_0^2$.

4. Area $A = x^2 \tan\theta = (vt)^2 \tan\theta$, $\phi = BA$
$|\xi| = d\phi/dt = 2Bv^2 t \tan\theta$

5. $V = 4\pi r^3/3$, so $dV/dt = (4\pi r^2)(dr/dt) = 10^{-4}$ m^3/s
$\phi = \pi r^2 B$; $\xi = d\phi/dt = 2\pi rB\, dr/dt = 333\ \mu V$

6. $I = \xi/R = (BLv\cos\theta)/R$; and $F = ILB$
Along incline $\Sigma F = mg\sin\theta - F\cos\theta = 0$
Find $v_T = (mgR \sin\theta)/(BL\cos\theta)^2$

7. $d\phi = B\, dA = (\mu_0 I/2\pi x)(c\, dx)$, thus
$\phi = \int d\phi = (\mu_0 Ic/2\pi)\ln[(b + a)/a]$;
$\xi = d\phi/dt = \mu_0 \omega I_0 c/2\pi\ \ln[(a + b)/a]\ \cos(\omega t)$

8. $I = (\xi_0 - BLv)/R$; $F = ILB = m\, dv/dt$
Thus $dv/dt = 0$, for $v_T = \xi_0/BL$

9. (a) $(2\pi r)E = (\pi r^2)dB/dt$, so
$E = (r/2)dB/dt$. Since $\cos\theta = L/2r$,
$E\cos\theta = (L/4)dB/dt$.
(b) $(E\cos\theta)(4L) = L^2\, dB/dt$

10. (a) $I = \xi/R = -N(d\phi/dt)/R = +(NAB_0/R\tau)\exp(-t/\tau)$
(b) $Q = \int I\, dt = -(NAB_0/R\tau)[\tau \exp(-t/\tau)] = NAB_0/R$

11. $dp/dt = d(erB_{orb})/dt = e(r/2)dB_{av}/dt$, thus $B_{orb} = B_{av}/2$

CHAPTER 32

Exercises

1. (a) $L = \mu_o n^2 A\ell = 152\ \mu H$;
   (b) $dI/dt = \xi/L = 26.3$ A/s

2. (a) $L = \xi/(dI/dt) = \mu_o N^2 A/\ell$, thus $N = 47.5$
   (b) $B = \mu_o nI = 7.16 \times 10^{-4}$ T

3. (a) $\phi = LI/N = 4.8\ \mu Wb$; (b) $|\xi| = L\ dI/dt = 42$ mV

4. (a) $(LI_0/\tau)\ e^{-t/\tau}$; (b) $L(2bt - a)$; (c) $-\omega LI_0 \cos(\omega t)$

5. $L = N\phi/I = 375\ \mu H$; then $|\xi| = L\ dI/dt = 9.38$ mV

6. $L = \xi/(dI/dt) = 4.5 \times 10^{-4}$ H; then $\phi = LI/N = 33.8\ \mu Wb$

7. $L = \xi/(dI/dt) = 93.8$ mH

8. $L = (\mu_o l/2\pi)\ \ln(b/a) = 9.32\ \mu H$

9. $M = \mu_o N_1 n_2 A_2 = \mu_o(5)(360/0.2)(\pi)(1.7 \times 10^{-2})^2 = 8.56$ mH

10. $\xi = M\ dI/dt = 1$ V

11. (a) $M = \mu_o N_1 n_2 A_2 = 80.4\ \mu H$; (b) $\xi = M\ dI/dt = 402\ \mu V$

12. (a) $M = N_2\phi_{21}/I_1 = 1.5 \times 10^{-9}$ H;
    (b) $\xi = M\ dI_2/dt = 6 \times 10^{-8}$ V

13. (a) $d\phi = B\ dA = (\mu_o NI/2\pi r)(h\ dr)$, thus
    $\phi = \mu_0 h N_1 I_1/2\pi\ \ln(b/a)$;
    (b) $M = N_2\phi_{21}/I_1 = (\mu_o h N_1 N_2/2\pi)\ \ln(b/a)$

14. $M = N_1\phi_{12}/I_2$, and $\phi_{12} = B_2 A_1 \cos 60^o = (\mu_o n_2 I_2)A_1\ \cos 60^o$
    Thus $M = 0.5\mu_o n_2 N_1 A_1 = 19\ \mu H$

15. (a) $\phi_{11} = L_1 I_1/N_1 = 0.60$ mWb; (b) $\phi_{12} = MI_2/N_1 = 0.39$ mWb;
    (c) $\phi_{21} = MI_1/N_2 = 0.14$ mWb; (d) $\xi_{11} = L_1\ dI_1/dt = 80$ mV;
    (e) $|\xi_{12}| = M\ dI_2/dt = 12.6$ mV; (f) $\xi_{21} = M\ dI_1/dt = 28$ mV

16. (a) $dI/dt = (\xi/L)$, so $I = \xi t/L$
    (b) $I = (\xi/R)[1 - (1 - Rt/L)] = \xi t/L$

17. (a) $I = (\xi/R)(1 - e^{-t/\tau}) = 0.279$ A;
(b) $\xi_L = L\,dI/dt = \xi\,\exp(-t/\tau) = 10.3$ V;
(c) From $0.8 = (1 - e^{-t/\tau})$, $t = -\tau\ln(0.2) = 536$ ms

18. (a) $3T_{1/2} = 3(0.693L/R) = 0.693$ s
(b) $\xi_L = L\,dI/dt = \xi\,\exp(-t/\tau) = 1.5$ V

19. (a) $dI/dt = (\xi/L)\exp(-t/\tau)$. At $t = 0$, $dI/dt = \xi/L = 6$ A/s;
(b) $T_{1/2} = 0.693L/R = 0.231$s;
(c) $\xi/R = (\xi/L)t$, so $t = L/R = 0.333$ s

20. (a) Infer that $T_{1/2} = 0.025$ s, then
$\tau = L/R = T_{1/2}/0.693$, thus $R = 0.166\ \Omega$
(b) $0.4 = (1 - e^{-t/\tau})$, thus $-Rt/L = \ln(0.6)$, so $L = 0.392$ H

21. (a) $IR + LdI/dt = 13$ V; (b) 11 V

22. (a) $0.4 = (1 - e^{-t/\tau})$, thus $\tau = 78.3$ ms.
Find $t = -\tau\ \ln(0.2) = 126$ ms
(b) $L = R\tau = 0.94$ H

23. (a) $T_{1/2} = 0.693L/R = 5.5$ ms; (b) $I/I_o = 1 - e^{-5} = 99.3\%$

24. (a) $I_3 = 0$, $I_1 = I_2 = \xi/(R_1 + R_2)$;
(b) $I_2 = 0$, $I_1 = I_3 = \xi/R_1$
(c) $I_1 = 0$, $I_3 = -I_2 = \xi/R_1$; (d) $V_2 = I_2R_2 = \xi R_2/R_1$

25. $R = \rho L/A = \rho N(2\pi a)/(\pi d^2/4)$ where $N = 180$ turns,
$a = 2\text{x}10^{-2}$ m, $d = 10^{-3}$ m. Find $R = 0.49\ \Omega$
$L = \mu_o n^2 A\ell = 2.84\text{x}10^{-4}$ H. Finally $\tau = L/R = 0.58$ ms

26. (a) $LI^2/2 = 300$ J;
(b) $L = N_1\phi_{11}/I_1 = 3.2$ mH. Then, $U = LI^2/2 = 3.6$ mJ

27. (a) $B^2/2\mu_o = 3.98$ mJ/m^3;
(b) $L = \mu_o n^2 A\ell$; so $u = LI^2/2A\ell = 0.5\mu_o n^2 I^2$; $I = 79.6$ mA

28. $L = (\mu_o l/2\pi)\ \ln(b/a) = 2.77\text{x}10^{-7}$ H for $l = 1$ m.
$U = LI^2/2 = 5.54\text{x}10^{-7}$ J for 1 m.

29. (a) $P_R = (0.632I_o)^2R = 9.59$ W;

(b) P_L = LI dI/dt = $I_o{}^2L(1 - e{-1}_{)(Re^{-1}/L)}$
= ξ^2/R (0.632)(0.37) = 5.59 W;
(c) P_s = Iξ = 15.2 W

30. (a) τ = 0.417 ms, so ξ_L = - LdI/dt = $-\xi_o\exp(-t/\tau)$ = -3.63 V
(b) $P_R = I_o{}^2(1 - e^{-2.4})^2R$ = $(0.606\ A)^2(60\ \Omega)$ = 22.0 W
(c) P_L = LI dI/dt = (0.606)(3.63) = 2.2 W
(d) P_s = Iξ = (0.606)(40) = 24.2 W

31. I^2R = LI dI/dt, thus IR = L dI/dt:
$(1 - e^{-t/\tau}) = e^{-t/\tau}$, thus t = τ ln2 = 5.54 ms

32. From U = $LI^2/2$, L = 0.15 H

33. L = $\mu_o n^2 A\ell$, U = $LI^2/2$, thus u = $0.5\mu_o n^2I^2$ and I = 75.2 mA

34. (a) $\exp(-t/\tau) = 1 - I/I_o$ = 0.55, thus τ = L/R = t/ln(0.55) and L = 0.2 H.
(b) U = $LI^2/2$ = 16 mJ

35. (a) u = $B^2/2\mu_o$ = $(\mu_o nI)^2/2\mu_o$ = $0.5\mu_o n^2I^2$
(b) U = $LI^2/2$ = uAℓ, thus L = $\mu_o n^2A\ell$

36. (a) ξ = -L dI/dt = $-\omega LI_o\cos(\omega t)$ = -59 V
(b) P_L = LI dI/dt = 26.4 W

37. (a) $f_o = [2\pi(LC)]^{-1/2}$ = 563 Hz;
(b) $Q_o{}^2/2C = LI_o{}^2/2$, so $I_o = \omega_o Q_o$ = 0.212 A;
(c) T/8 = 1.78 ms/8 = 0.222 ms

38. (a) Take C = 25 nF. Infer that T = 4×10^{-4} s = $2\pi(LC)^{1/2}$, thus L = 0.162 H
(b) U = $LI_o{}^2/2 = Q_o{}^2/2C$ = 8 mJ

39. C = $1/L\omega_o{}^2$ = 1.98 pF to 16.7 pF

40. (a) $\omega'^2 = \omega_o{}^2 - (R/2L)^2 = 6.25\times10^6$, thus ω' = 2500 rad/s
(b) R = $2\omega_o L = 2(L/C)^{1/2}$ = 28.3 Ω

41. For $R^2 \ll 4L/C$, $\omega' \approx \omega_o$
$Q \approx Q_o\exp(-Rt/2L)\cos(\omega_o t)$

$I \approx \omega_0 Q_0 \exp(-Rt/2L) \sin(\omega_0 t)$
$U = Q^2/2C + LI^2/2 \approx Q_0{}^2/2C \exp(-Rt/L);$

42. (a) $\omega_0{}^2 = 1/LC$, so $\omega_0 = 5\text{x}10^4$ rad/s
(b) $\omega' = 0.999\omega_0$. From Eq. 32.20:
$R^2 = 4L^2\omega_0{}^2(1 - 0.999^2)$, thus $R = 179\ \Omega$

Problems

1. (a) $\xi = \xi_1 + \xi_2$, $dI/dt = dI_1/dt = dI_2/dt$
Thus $L = L_1 + L_2$
(b) $\xi = \xi_1 = \xi_2$; $dI/dt = dI_1/dt + dI_2/dt$.
Thus $1/L = 1/L_1 + 1/L_2$

2. $\xi = (L_1 + L_2)dI/dt + (M_{12} + M_{21})dI/dt$
Thus $L = L_1 + L_2 \pm 2M$. Sign depends on way coils are wound.

3. $\phi = 2(\mu_0 I\ell/2\pi)\int dx/x = (\mu_0 I\ell/\pi)\ln[(d - a)/a]$.
$L/\ell = \phi/I$

4. $B = \mu_0 Ir/2\pi a^2$, $u = B^2/2\mu_0$. $U = \int u\, dV = \int u(2\pi r\ell)dr$
Thus $U = \mu_0 I^2\ell/16\pi = LI^2/2$, so $L/\ell = \mu_0/8\pi$.
(The use of $L = \phi/I$ is tricky.)

5. $d\phi = BdA = (\mu_0 I/2\pi x)(cdx)$, thus $\phi = (\mu_0 Ic/2\pi)\ \ln[(b + a)/a]$
$L = \phi/I = (\mu_0 c/2\pi)\ \ln[(b + a)/a]$

6. $\phi = (\mu_0 NIh/2\pi)\int dr/r = LI/N$, so $L = (\mu_0 N^2 h/2\pi)\ln(b/a)$

7. (a) $U_1 = (Q_0{}^2/2C)\exp(-Rt/L)$; $U_2 = (Q_0{}^2/2C)\exp[-R(t + T)/L]$
$\Delta U = U_1[1 - \exp(-RT/L)] \approx U_1\ (RT/L) = U_1(2\pi R/\omega L)$
Thus, $\Delta U/U \approx 2\pi/Q$ where $Q = \omega L/R$.
(b) $Q = 2\pi/0.02 = 314$;
(c) $Q = (L/C)^{1/2}/R$, so $C = 0.73\ \mu F$

8. (a) $[1 - \exp(-t/\tau)] = 0.5$, so $t = -\tau\ln(0.5) = 1.54$ ms
(b) $P \propto I^2$, so $[1 - \exp(-t/\tau)]^2 = 0.5$,
so $t = -\tau\ln(0.707) = 2.73$ ms
(c) $U = LI^2/2$, so $t = 2.73$ ms as in (b)

9. $\phi = BA = \mu_0 NIA/\ell$; $L = \mu_0 N^2 A/\ell$, $N = \ell/2a$

$L = \mu_o \ell A/4a^2 = \mu_o \ell(\pi r^2)/4a^2.$

$R = \rho\ell/A = \rho(N2\pi r)/\pi a^2 = \rho\ell r/a^3$

$\tau = L/R = \mu_o \pi ar/4\rho$

10. $I_3 = I_1 - I_2;\ I_2R_2 = L\ dI_3/dt$

Loop rule: $\xi - I_1R_1 - I_2R_2 = 0$

$\xi - (I_3 + L/R_2\ dI_3/dt)R_1 - L\ dI_3/dt = 0$

$\xi - I_3R_1 = \tau R_1\ dI_3/dt$, where $\tau = L(R_1 + R_2)/R_1R_2$.

$dI_3/(\xi/R_1 - I_3) = dt/\tau$. At $t = 0$, $I_3 = 0$, thus

$I_3 = (\xi/R_1)[1 - \exp(-t/\tau)]$. Note:

$I_2 = [\xi/(R_1+R_2)]\ \exp(-R_2t/L)$

11. $P = I_o^2 e^{-2t/\tau} R;\ E = \int P\ dt = LI_o^2/2$

12. $Q = Q_o\exp(-Rt/2L)\cos(\omega' t)$

$I = -\omega' Q_o\ \exp(-Rt/2L)[\sin(\omega' t) + (R/2\omega' L)\cos(\omega' t)]$

Express $[\ldots] = [\sin(\omega' t)\cos\delta + \cos(\omega' t)\sin\delta]$

Let $\tan\delta = R/2\omega' L$, then $\cos\delta \approx 1$, and $\sin\delta \approx \tan\delta$

Thus $I \approx -A\ \sin(\omega' t + \delta)$

13. $L_1 = \mu_o n_1{}^2 A_1 \ell_1;\ L_2 = \mu_o n_2{}^2 A_2 \ell_2;$

$M = \mu_o N_1 n_2 A_1 = \mu_o N_2 n_1 A_2,\ N = n\ell$

Find $M^2 = L_1L_2$

14. $I = -dQ/dt = Q_o\ e^{-Rt/2L}[-(R/2L)\cos(\omega' t) - \omega'\sin(\omega' t)]$

In I^2R, the average value of $\sin^2(\omega' t)$ and $\cos^2(\omega' t)$ is 0.5 and of $\cos(\omega' t)\sin(\omega' t)$ is zero. Thus

$P_{av} = Q_o{}^2\ e^{-Rt/L}\ [R^2/8L^2 + \omega'^2/2]R = 0.5\omega_o{}^2 Q_o{}^2 Re^{-Rt/L}$

CHAPTER 33

Exercises

1. (a) $i_o = v_o/\omega L = 7.96$ A; (b) $f_2 = F_1/0.3 = 200$ Hz

2. (a) $i_o = v_o\omega C = 1.10$ A;
   (b) $f_2 = 1.3\text{x}50 = 65$ Hz

3. (a) $\omega L = 1/\omega C$, thus $f_o = 1/2\pi\ (LC)^{-1/2} = 5.03$ kHz;
   (b) $f = (5)^{1/2}f_o = 11.25$ kHz; (c) $f = f_o/(5)^{1/2} = 2.25$ kHz

4. $L = X_L/\omega = 0.1$ H; then $i_o = v_o/\omega L = 5.40$ A

5. (a) $Q_o = i_o/\omega = v_oC = 1.7$ mC; (b) $i_o = \omega Q_o = 0.64$ A

6. (a) $i_o = v_o/\omega L = 0.921$ A; (b) 0
   (c) $0.5 = \cos(\omega t)$, thus $\omega t = \pi/3;\ 5\pi/3$
   $i = i_o\sin(\omega t) = \pm i_o(0.866) = \pm 0.798$ A
   (d) $p = i_ov_{oL}\sin(2\omega t)/2 = 23.0$ W

7. (a) $i_o = v_o\omega C = 1.3$ A; (b) zero;
   (c) $-0.5 = \cos(\omega t)$, $\omega t = 2\pi/3,\ 4\pi/3$
   $i = i_o\ \sin(\omega t) = \pm 1.13$ A;
   (d) $p = -i_ov_{oC}\sin(2\omega t)/2 = -13.2$ W

8. (a) $\omega = 1/CX_C = 1/11C$, thus $\omega L = L/11C = 3.03\ \Omega$
   (b) $f_o = 1/2\pi(LC)^{1/2} = 4590$ Hz

9. (a) $i = i_o\sin(\omega t) = 2.39\sin(0.2\pi) = 1.40$ A,
   $p = i_ov_{oL}\sin(2\omega t)/2 = 68.2$ W;
   (b) $i_o = v_o/\omega L$, so $f = v_o/2\pi i_oL = 66.3$ Hz

10. (a) $i = i_o\sin(\omega t) = 1.06$ A; $p = -i_ov_{oC}\sin(2\omega t)/2 = -62$ W
    (b) $f_o = i_o/2\pi v_oC = 111$ Hz

11. $Z^2 = R^2 + 1/(\omega C)^2$:
    $(10.8)^2 = R^2 + 1.665\text{x}10^{-7}/C^2$;
    $(18.8)^2 = R^2 + 6.333\text{x}10^{-7}/C^2$
    Solve to find $C = 44.4\ \mu F$, $R = 5.67\ \Omega$

12. Apply $Z^2 = R^2 + \omega^2L^2$ and solve: $R = 13\ \Omega$, $L = 40$ mH

13. $i_o = 0.5$ A $= v_o/Z$, $Z^2 = R^2 + 10^4(5L - 2)^2$
Find L = 75.6 mH, 724 mH

14. $I = 60\omega C = 0.6$ A, $V_{RL}{}^2 = I^2(R^2 + \omega^2L^2) = 45^2$, thus
$R^2 + \omega^2L^2 = 75^2$. Find L = 0.14 H.

15. (a) From $V = IZ$, $V^2 - (IR)^2 = I^2(\omega L - 1/\omega C)^2$;
find I = 2.42 A;
(b) IR = 50, so R = 20.6 Ω;
(c) $V_L = I\omega L = 194$ V;   (d) $V_C = I/\omega C = 303$ V

16. (a) $X_L = 80$ Ω;   (b) $X_C = 50$ Ω;
(c) $V_{LC} = I(\omega L - 1/\omega C)$, so I = 8/3 A
(d) $120 = (8/3)[R^2 + 30^2]^{1/2}$, thus R = 33.5 Ω
(e) $P = I^2R = 238$ W

17. (a) I = 0.8 A, $Z = V/I = [10^4 + (\omega L - 1/\omega C)^2]^{1/2} = 300$ Ω;
(b) $Z^2 - R^2 = 8\text{x}10^4 = (\omega L - 1/\omega C)^2$, find L = 0.192 H;
(c) $\tan\phi = (8)^{1/2}$, so $\phi = 70.5^o$

18. $\omega L = 20$; $1/\omega C = 8$, thus L/C = 160. Also, $\omega_o{}^2 = 1/LC$
Find L = 1.01 mH, C = 6.29 μF.

19. (a) $I = V_R/R = 3$ A $= V/Z$, thus Z = 40 Ω
$Z^2 = R^2 + (\omega L - 1/\omega C)^2$, C = 49 μF;
(b) $f = 1/2\pi(LC)^{1/2} = 113$ Hz

20. (a) $Z^2 = 25^2 + (120.6 - 147.4)^2$, so Z = 36.6 Ω
(b) $I = 170/Z(2)^{1/2} = 3.28$ A
(c) $\text{Tan}\phi = (\omega L - 1/\omega C)/R$ leads to $\phi = -47^o$

21. $\tan\phi = (\omega L - 1/\omega C)/R = 0.577$; so $\omega^2L - 23.1\omega - 1/C = 0$
Find f = 264 Hz

22. Z = 24 Ω, $24^2 = R^2 + (\omega L - 1/\omega C)^2$;
$\phi = -45^o$, so $\tan\phi = -1 = (\omega L - 1/\omega C)/R$
Thus $Z^2 = 2R^2$, so R = 17 Ω, then L = 0.959 H

23. (a) $f_o = 1/2\pi(LC)^{1/2} = 919$ Hz;
(b) $I_{max} = V/R = 3.13$ A at f_o;

Given $I = V/Z = I_{max}/2$, thus $Z = 2R$ leads to
$(\omega L - 1/\omega C)^2 = 3R^2$ leads to $f = 622$ Hz, 1360 Hz

24. $Z^2 = R^2 + (\omega L - 1/\omega C)^2$; $\omega = 1200\pi$ rad/s
$(200)^2 = (120)^2 + (\omega L - 1/\omega C)^2$
$1/C = 1200\pi(96\pi + 160)$; thus $C = 0.575\ \mu F$, $1.87\ \mu F$

25. (a) $Z^2 = 20^2 + 17^2$, $Z = 26.2\ \Omega$.
$\tan\phi = -17/20$, so $\cos\phi = 0.762$;
(b) $P = I^2R = (100/Z)^2R = 291$ W

26. (a) $P = I^2R = (V/Z)^2R = 47.6$ W
$P = IV\cos\phi$, thus $\cos\phi = 0.364$

27. (a) $X_L = 20\ \Omega$, $X_C = 50\ \Omega$;
(b) $\tan\phi = (X_L - X_C)/R$, thus $\phi = -63.4^o$;
(c) $P = (V/Z)^2R = 267$ W;
(d) $\cos\phi = P/IV = 0.447$

28. $800 = 8(120)\cos\phi$, $\cos\phi = 0.833$

29. Given $R = 24\ \Omega$, $\tan\phi = (X_L - X_C)/R = 1.327$, thus
$X_L - X_C = 31.8$, thus $Z = (24^2 + 31.8^2)^{1/2} = 39.9\ \Omega$
$P = (V/Z)^2R = 151$ W

30. (a) $v_{av} = 1$ V; (b) $V = [(8 + 1)/3]^{1/2} = (3)^{1/2}$

31. $P = IV\cos\phi = V^2\cos\phi/Z$, but
$R = Z\cos\phi$, so
$P = (V\cos\phi)^2/R$

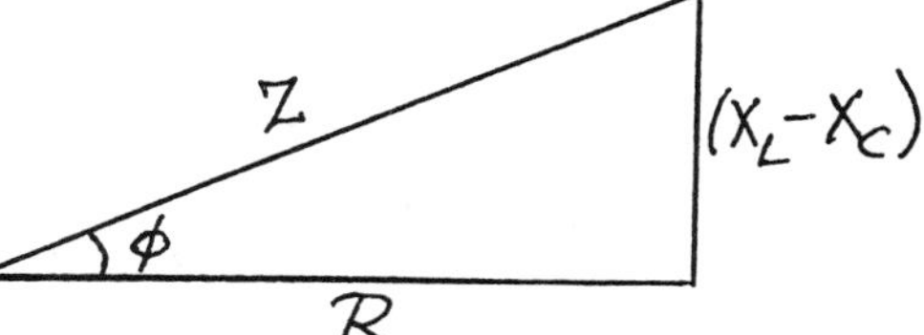

32. (a) $P_1 = V^2/R$, so $R = 14.4\ \Omega$.
$P_2 = P_1/2 = (V/Z)^2R$, thus $Z^2 = 2R^2$.
So $Z^2 = 2R^2 = R^2 + (\omega L)^2$; thus $L = R/\omega = 38.2$ mH
(b) $\tan\phi = 1$, so $\phi = 45^o$

33. $Z = 45.7\ \Omega$ and $\phi = -1.02$ rad. $v_o = i_oZ = 2.74$ V
$v = 2.74\sin(320t - 1.02)$ V

34. $I_1V_1 = I_2V_2$: so $I_1 = 33.3$ A
$P = I_1^2R = 1330$ W

35. $I_1V_1 = I_2V_2$, so $I_2 = 4500/V_2$
(a) $I_2 = 0.9$ A, $P = (0.9)^2(20) = 16.2$ W;
(b) $I_2 = 0.225$ A, $P = 1.01$ W

36. (a) $240\cos 12^o = 235$ W; (b) $(0.9)(235) = 211$ W
(c) $211 = I_2V_2\cos\phi_2$, $I_2 = 211/(24)(0.75) = 11.7$ A

37. (a) $N_1 = 5\text{x}80 = 400$;
(b) $I_2 = 12$ A, $I_1V_1 = I_2V_2$, so $I_1 = 2.4$ A

38. $V_2/V_1 = N_2/N_1$, $V_2 = (120)/8 = 15$ V
$I_2/I_1 = N_1/N_2$; $I_2 = 2.4\text{x}8 = 19.2$ A

Problems

1. $P = (V/Z)^2R = V^2R/[R^2 + (L/\omega)^2(\omega^2 - \omega_o{}^2)^2]$
$= \omega^2V^2R/[\omega^2R^2 + (\omega^2 - \omega_o{}^2)^2L^2]$

2. Set $dP/d\omega = 0$ to find $\omega = \omega_o$.

3. (a) $v_{av} = 2$ V;
(b) $(v^2)_{av} = (2/T)\int (at)^2\, dt = 16/3$, so $v_{rms} = (16/3)^{1/2}$ V

4. Phasors $\underline{i} = \underline{i}_R + \underline{i}_C + \underline{i}_L$
$i_o{}^2 = i_{oR}{}^2 + (i_{oL} - i_{oC})^2 = v_o{}^2[1/R^2 + (1/X_L - 1/X_C)^2]$
$Z = v_o/i_o = [1/R^2 + (1/X_L - 1/X_C)^2]^{-1/2}$

5. $F_{av} = (1/T)\int F\, dt$. With $v = -4 + 4t$, $v^2 = 16 - 32t + 16t^2$
$v_{av}2 = 1/2\ [16t - 16t^2 + 16t^3/3] = 16/3$.
$v_{rms} = 4/(3)^{1/2}$ V

6. $[R^2 + (\omega^2L - 1/C)^2/\omega^2] = Z^2$
$\omega^2R^2 + [\omega^4L^2 - 2\omega^2L/C + 1/C^2] = \omega^2Z^2$
$\omega^4L^2 + \omega^2(R^2 - 2L/C - Z^2) + 1/C^2 = 0$
For a quadratic, $x_1x_2 = +c/a$, so $\omega_L\omega_H = \omega_o{}^2$

7. $V_{out}/V_{in} = IR/IZ = R/(R^2 + \omega^2L^2)^{1/2}$
(a) 84.6 V; (b) 15.7 V; (c) 1.59 V. A "low-pass" filter.

8. $V_{out}/V_{in} = I\omega L/IZ = \omega L/(R^2 + \omega^2L^2)^{1/2}$
(a) 53.2 V; (b) 98.8 V; (c) 99.9 V. A "high-pass" filter.

9. $Z = v_o/i_o = 40\ \Omega$; thus $40^2 = R^2 + (X_L - X_C)^2$
Since $\phi < 0$, take $(X_L - X_C) = -38$, thus $X_L = 14\ \Omega$
Find $L = 37.1$ mH

10. (a) $Q = \int I\, dt = -(i_o/\omega)\cos(\omega t) = -(v_o/\omega Z)\cos(\omega t)$
Thus, $Q_o = v_o/\omega Z$
(b) $dQ_o/dt = 0$ leads to $\omega_{max} = (\omega_o{}^2 - R^2/2L^2)^{1/2}$

11. (a) $v_{0R} = 7.32$ V, $v_{0C} = 114$ V, $v_{0L} = 14.6$ V;
(b) $v_{oRC} = i_o(R^2 + 1/\omega^2C^2)^{1/2} = 114$ V;
(c) $v_{oLC} = i_o(\omega L - 1/\omega C) = 99.7$ V

12. $V_{out}/V_{in} = IR/IZ = R/(R^2 + 1/\omega^2C^2)^{1/2} = 1/(1 + 1/R^2\omega^2C^2)^{1/2}$

13. $V_{out}/V_{in} = (1/\omega C)/Z = 1/(\omega C)(R^2 + 1/\omega^2C^2)^{1/2}$

CHAPTER 34

Exercises

1. (a) $[\epsilon_0] = C^2/N.m^2$; $[\mu_0] = N/A^2$; so $[\epsilon_0\mu_0] = (s/m)^2$
   (b) $[E] = N/C$; $[B] = N/A.m$; $[1/\mu_0] = A^2/N$
   $[EB/\mu_0] = (N.m)/m^2.s = W/m^2$

2. (a) $[E] = N/C$; $[cB] = (m/s)(N/A.m) = N/C$
   (b) $[\epsilon_0\, d(EA)/dt] = (C^2/N.m^2)(N/C)(m^2/s) = C/s$
   (c) From 1b, $[S] = [EB/\mu_0] = W/m^2$, so $[S/c] = W.s/m^3 = N/m^2$

3. $dE/dt = (1/d)\, dV/dt$, $\phi = EA$; so
   $I_D = (\epsilon_0 A/d)\, dV/dt = 2.9 \times 10^{-7}$ A

4. (a) 3 A; (b) $dV/dt = I_D/C = I_D d/\epsilon_0 A = 3.78\text{x}10^{11}$ V/s

5. $I_D = (\epsilon_0 A/d)(dV/dt) = C\, dV/dt$

6. $I_D = (\epsilon_0 A/d)(dV/dt) = 92.7\text{x}10^{-10}$ A

7. From Example 34.1: $B = (\epsilon_0\mu_0/2d)(dV/dt)r$;
   Peak $dV/dt = \omega V_0$. Find $B = 6.28 \times 10^{-13}$ T

8. (a) Ampere's law: $(2\pi r)B = \mu_0 I_D$
   (b) $I(r) = I_D(r/R)^2$, so $B = \mu_0 I/2\pi r = \mu_0 I_D r/2\pi R^2$

9. From Example 34.1 and In-chapter exercise 1:
   (a) $B = \mu_0\epsilon_0(dE/dt)r/2 = \mu_0 Ir/2A = \mu_0 Ir/2\pi R^2 = 5\text{x}10^{-8}$ T;
   (b) $B = \mu_0\epsilon_0(dE/dt)R^2/2r = \mu_0 IR^2/2Ar = \mu_0 I/2\pi r = 8 \times 10^{-8}$ T

10. $\underline{B} = (E/c)\underline{j} = 7.0\text{x}10^{-8}\underline{j}$ T

11. (a) 1.26 cm, 23.9 GHz; (b) $E_z = 60\sin(500x + 1.5\text{x}10^{11}t)$ V/m

12. $B_x = -E_0/c\, \sin(ky + \omega t)$; $B_y = B_z = 0$

13. $u_{av} = \epsilon_0 E^2/2 = B_0^2/2\mu_0$. (a) 150 V/m; (b) 0.5 μT

14. (a) $u_{av} = \epsilon_0 E_0^2/2 = 1.11\text{x}10^{-8}$ J/m^3
    (b) $B_0 = E_0/c = 1.67\text{x}10^{-7}$ T along +z axis.
    (c) $\underline{S}_{av} = u_{av}\, c\underline{i} = 3.33\underline{i}$ W/m^2

15. (a) $u_{av} = S_{av}/c = 3.33 \times 10^{-6}$ J/m^3;

(b) $U = SAt = S(\pi R^2)(3600\ s) = 4.6 \times 1020\ J$

16. $S_{av} = P/4\pi r^2 = E_o{}^2/2\mu_o c$. Find $E_o{}^2 = 1500/r^2$;
(a) 1.55×10^{-3} V/m, 5.16×10^{-12} T
(b) 1.14×10^{-6} V/m, 3.80×10^{-15} T

17. (a) $B_o = E_o/c = 3.33 \times 10^{-8}$ T;
(b) $P = S_{av}A = (E_o{}^2/2\mu_o c)(4\pi r^2) = 60$ W

18. $F = SA/c = 0.446$ mN

19. $F = SA/c$ and $S = P/A$, so $F = P/c$.
(a) 3.33×10^{-12} N; (b) 3.33×10^{-6} N

20. Pressure $= S/c = P/4\pi r^2 c = 6.63\times10^{-15}$ N/m^2
(For perfectly refecting, twice this value.)

21. $S = B_o{}^2 c/2\mu_o = P/4\pi r^2$, thus $P = 4\pi r^2 c B_o{}^2/2\mu_o = 150$ kW

22. $F = SA/c = P/c = 3.33\times10^{-6}$ N; $a = F/m = 3.33\times10^{-8}$ m/s^2

23. (a) $S = E_o{}^2/2\mu_o c = 5.31\times10^{-15}$ W/m^2
(b) $S = P/4\pi r^2$, find $r = 3.87 \times 10^8$ m

24. (a) $S = B_o{}^2 c/2\mu_o = P/4\pi r^2$, thus $r = 25.8$ m;
(b) $E_o = B_o c = 3$ V/m

25. (a) $F/A = 2S/c = 3.33 \times 10^{-8}$ N/m^2; (b) $F = 2SA/c = 8\times10^{-9}$ N

26. (a) $S = P/4\pi r^2 = E_o{}^2/2\mu_o c$, find $E_o = 6$ V/m.
(b) $B_o = E_o/c = 2\times10^{-8}$ T

27. $P = SA = S(\pi d^2/4) = 19.6$ mW

28. 10 kW/(0.18 kW/m^2) = 55.6 m^2

29. $S = EB/\mu_o = E^2/2\mu_o c + cB^2/2\mu_o$, but $c^2 = 1/\mu_o\epsilon_o$
$S = (\epsilon_o E^2 + B^2/\mu_o)c/2$.

30. (a) $S = P/4\pi r^2 = 9.55\times10^{-2}$ W/m^2
(b) $u = S/c = 3.18\times10^{-10}$ J/m^3
(c) $F = 2SA/c = 6.37\times10^{-14}$ N

Problems

1. (a) $1/2\ \mu_0\epsilon_0 a(dE/dt)$; (b) $S = EB/\mu_0 = \epsilon_0 aE(dE/dt)/2$
   (c) $P = SA = S(2\pi ad) = \epsilon_0\pi da^2E\ dE/dt$

2. $\xi = NAdB/dt = NA\omega B_0\cos(\omega t)$; $\xi_0 = NAB\omega = 0.455\ \mu V$

3. (a) $P = V^2/R = 720$ W
   (b) $S = EB/\mu_0 = (V/\mu_0 L)(\mu_0 I/2\pi r) = 3.82 \times 10^4\ W/m^2$
   (c) $P = S(2\pi rL) = 720$ W;

4. $P = SA$, so $F = SA/c = P/c$. Then, $\Delta v = (F/m)\Delta t = SA\Delta t/mc$.

5. $F/A = (S/c)[f + 2(1 - f)] = (S/c)(2 - f)$

6. $F_G = GmM/r^2 = 4\pi\rho GMR^3/3r^2$; $F_R = SA/c = (P/4\pi r^2 c)(\pi R^2)$
   $F_G = F_R$ leads to $R = 3P/16\pi\rho cGM = 470$ nm

7. $S = P/4\pi r^2 = B_0^2 c/2\mu_0$, thus $B_0 = 1.29$ nT;
   $\xi_0 = NAB_0\omega = 2.87\text{x}10^{-4}$ V

8. Since $p = U/c$, $F_y = dp_y/dt = (dU/dt)\cos\theta/c$
   But $dU/dt = SA'$, so $F_y = SA'\cos\theta/c$. Also,
   $A' = A\cos\theta$. Thus $F_y/A = (S/c)\cos^2\theta = u\cos^2\theta$

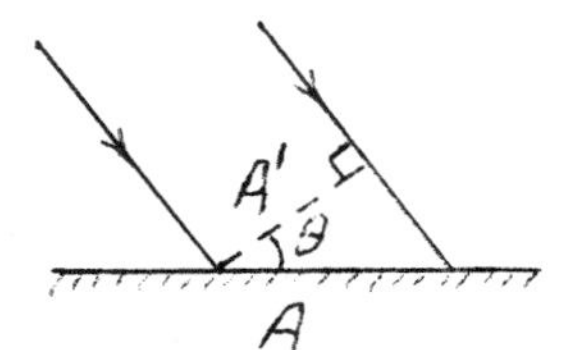

9. (a) $I_D = C\ dV/dt = 0.01$ A;
   (b) $V = IR = 4000$ V, so $\Delta t = 2$ s

10. $\tau_{net} = Fr = (SA/c)(d/2) = \kappa\theta$; $\theta = S(\pi r^2)d/2\kappa c = 7.54^o$

11. $F = (SA/c)[0.6 + 2\text{x}0.4] = 1.4SA/c = 2.9\text{x}10^{-7}$ N;
    $\Delta p = F\ \Delta t = 8.71 \times 10^{-5}$ kg.m/s

12. (a) $F = 2SA/c = 6.67$ mN; (b) $t = mv/F = 1.74$ d.

CHAPTER 35

Exercises

3. There are $(360/\theta) - 1 = 5$ images. Use the idea of a "virtual" mirror that extends behind the real mirrors.

4. Each ray is deflected by ϕ.

5. Let the mirrors lie in the xy, xz and yz planes. If the original direction is along $(\underline{i} + \underline{j} + \underline{k})$, three reflections will result in $(-\underline{i} - \underline{j} - \underline{k})$.

6. (a) $n_1\lambda_1 = n_2\lambda_2$, so $n_2 = 1.5$; (b) $v = c/n = 2\text{x}10^8$ m/s

7. $\sin\theta_i = 1.4\sin 32^o$, so $\theta_i = 47.9^o$
   Angle required: $(90 - 47.9) + (90 - 32) = 100.1^o$

8. $\sin\theta_i = n \sin(90 - \theta_i) = 1.52\cos\theta_i$.
   Thus $\tan\theta_i = 1.52$ and $\theta_i = 56.7^o$.

9. Within water $\Delta x_1 = 3\tan 30^o = 1.73$ m,
   $1.33\sin 30^o = \sin\theta_r$, $\theta_r = 41.7^o$,
   $\tan 41.7^o = \Delta x_2/(1 \text{ m})$, thus $\Delta x_2 = 0.89$ m; $\Delta x_1 + \Delta x_2 = 2.62$ m

10. Slope = n = 1.33

11. $d = t \sin(\theta - r)$. For small θ, $\sin x \approx x$,
    Snell's law takes the form: $\theta = nr$, thus
    $d \approx t(\theta - r) = t\theta(1 - 1/n) = t\theta(n - 1)/n$.

12. $n\sin 68^o = 4/3$, so $n = 1.43$ and $v = c/n = 2.1\text{x}10^8$ m/s

13. $(4/3)\sin\theta_c = 1$, so $\theta_c = 48.5$. $r = 2\tan 48.5^o = 2.27$ m

14. (a) Refracted ray emerges along interface just prior to total internal reflection.
    (b) $n_1\sin\theta_1 = n_2$. Need $n_2 < n_1$.

15. For $i = 90^o$, $\sin i = 1.5\sin r = 1$; thus maximum $\sin r = 2/3$, or minimum $\cos r = (5)^{1/2}/3$.
    For TIR, $1.5\sin\theta_c = 1 = 1.5\cos r$, so we need $\cos r > 2/3$.
    This condition is satisfied since $(5)^{1/2} > 2$.

16. After reflection, the angle of incidence at the bottom surface is 30°. From $n\sin 30^o = \sin\theta$, thus $\theta = 48.6^o$.

17. $n = \sin[(\phi + \delta)/2]/\sin 30^o = 1.54$, $v = c/n = 1.94 \times 10^8$ m/s

18. $n_{rel} = 1.66/1.33 = \sin[(\phi + \delta)/2]/\sin 30^o$;
$(\phi + \delta)/2 = 37^o$, so $\delta = 14^o$

19. The rays are inverted. Thus an image can be made erect.

20. $D = (1.633 - 1.611)/0.620 = 0.0355$

21. (a) $n\sin 45^o = 1$, so $n = 1.41$;
(b) $n\sin 45^o = 1.33$, so $n = 1.88$

22. From Example 35.5, $\theta_r = \phi/2 = 30^o$,
$\sin\theta_i = 1.6\sin\theta_r$, so $\theta_i = 53.1^o$

23. At the first face the angle fo refraction is 28.1^o. At the second face the angle of incidence is $(60^o - 28.1^o) = 31.9^o$ (form a triangle with the normals) and emerges at 52.4^o to the normal. $\delta = (45 - 28.1) + (52.4 - 31.9) = 37.4^o$

24. For small angles, Snell's law is $i = nr$; $n(\phi - r) = i'$
$\delta = (i - r) + (i' - r') = (n - 1)\phi$

25. $\cos\alpha = n \sin r$; $\phi = r + \theta_c$, thus $\cos\alpha = n \sin(\phi - \theta_c)$

26. (a) $1/q = 1/20 - 1/15$, so $q = -60$ cm, $m = 4$
(b) $1/q = 1/20 - 1/60$, so $q = +30$ cm, $m = -0.5$

27. (a) $1/q = -1/20 -1/15$, so $q = -8.57$ cm, $m = 0.57$;
(b) $1/q = -1/20 - 1/40$, so $q = -13.3$ cm; $m = 0.33$

28. (a) Concave; (b) $m = 1.8 = -q/40$, thus $q = -72$ cm
(c) $1/f = 1/40 - 1/72$, so $f = 90$ cm

29. $m = -0.4 = -q/p$, so $q = 24$ cm
$1/f = 1/60 + 1/24$, so $R = 2f = 34.3$ cm

30. (a) $m = 2.5 = -q/p$, so $q = -2.5p$.
$1/p - 1/2.5p = 1/f$, thus $p = 18$ cm.
(b) $q = +2.5p$, so $1/p + 1/2.5p = 1/f$, so $p = 42$ cm.

31. $0.4 = -q/p$, so $q = -0.4p$. So $1/p - 1/0.4p = -1/30$, $p = 45$ cm

32. $m = -3.2 = -q/p$, so $q = 3.2p = 70.4$ cm
$1/p + 1/q = 1/f$, leads to $f = 16.8$ cm

33. $m = 1/3 = -q/p$, so $p = -3q$. Since $f = -8$ cm,
$1/q - 1/3q = -1/8$, thus $q = -16/3$ cm and $p = 16$ cm

34. $0.4 = -q/3.2$, so $q = -1.28$ cm, then
$1/3.2 - 1/1.28 = 1/f$, $f = -2.13$ cm

35. $q = 1.4p = 28$ cm, then $1/20 + 1/28 = 1/f$, so $f = 11.7$ cm,
$q = -1.4p = -28$ cm, then, $1/20 - 1/28 = 1/f$, so $f = 70$ cm

36. (a) $q/p = 0.6$, so $q = 36$ cm;
then $1/60 + 1/36 = 1/f$, so $R = 2f = 45$ cm
(b) $q/p = 1.25$, so $q = 75$ cm; $R = 66.7$ cm
(c) $q/p = -1.8$, so $q = -108$ cm, $R = 270$ cm

37. (a) The image lies in the focal plane.
$\tan\alpha = y/f = r/d$, so $y = fr/d$;
(b) $y = 7.61$ cm

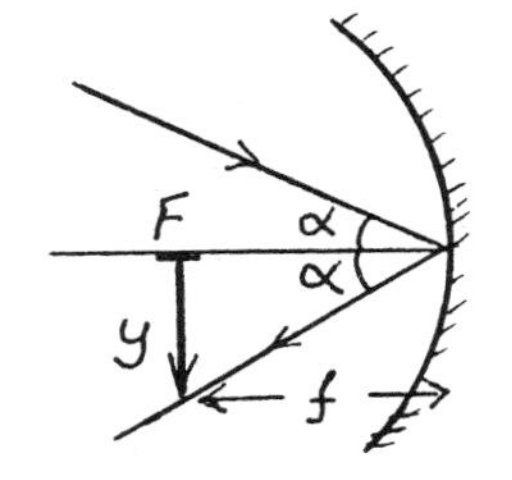

38. (a) 1.28 s; (b) 8.28 min

39. (a) 9.47×10^{15} m; (b) 1.57×10^{-5} ly

40. $c = (1.5\times10^{11}\text{ m})/(660\text{ s}) = 2.27\times10^{8}$ m/s

41. For round trip $\Delta t = 2d/c = 2(3.5\times10^{4}\text{ m})/c$
Prism must rotate 1/8 rev in Δt, i.e. $1/8\Delta t = 536$ rev/s

42. The wheel must rotate 1/360 rev in $\Delta t = 2d/c$. Thus,
rotational frequency is $1/360\Delta t = 104$ rev/s

44. $n_1\sin\theta_1 = n_2\sin\theta_2 = n_3\sin\theta_3$, so $\theta_3 = 19.5^\circ$. $\Delta\theta = 10.5^\circ$.

45. $n\sin45^\circ = 1.5\sin38^\circ$, so $n = 1.306$. Then
$v = c/n = 2.30 \times 10^{8}$ m/s.

46. See Fig. 35.26. $\sin(i) = 1.61\sin15^\circ$, so $i = 24.63^\circ$.
Then $\delta = 2(i - r) = 19.3^\circ$.

47. $\sin 45^{o} = n\sin(r)$, so $r = 27.1^{o}$. Then $1.55\sin 17.86^{o}) = \sin\theta$, so $\theta = 28.4^{o}$.

48. $1/10 - 1/14 = 1/f$, so $f = 35$ cm.
Then, $1/20 + 1/q = 1/35$, so $q = -46.7$ cm. Virtual.

49. (a) $1/p - 1/60 = 1/60$, so $p = 30$ cm. (b) $m = -q/p = +2$.

50. $\sin 75^{o} = 1.33\sin\theta_1$, so $\theta_w = 46.57^{o}$. At water-glass boundary: $1.33\sin 43.43^{o} = 1.5\sin\theta_g = 1.0\sin\theta_a$, so $\theta_a = 66.1^{o}$.

51. $m = -q/p = +3$, so $q = -3p$. From, $1/p + 1/q = 1/36$, find $p = 24$ cm.

52. $m = -q/p = 1/3$, so $q = -p/3$ and $1/p - 3/p = -1/24$, find $p = 48$ cm.

53. $m = -q/p = 4$, so $q = -4p$. From $1/p - 1/4p = 1/f$, find $f = 24$ cm.

54. From $\sin 40^{o} = n\sin\theta$ find $\theta_4 = 22.78^{o}$ and $\theta_7 = 23.53^{o}$. Along the second face $d_4 = 2.4\tan\theta_4$ and $d_7 = 2.4\tan\theta_7$, so $\Delta d = 0.037$ cm. Perpendicular distance between emerging rays $= \Delta d\tan 50^{o} = 0.044$ cm

55. $1/27 + 1/15.9 = 1/10$, then $1/15 + 1/q = 1/10$, gives $q = 30$ cm.

56. $h/f = \tan 0.52^{o}$, so $h = 7.26$ cm.

57. $n\sin 60^{o} = 1$, so $n = 1.15$.

Problems

1. $\sin 5o = 1.33\sin\theta_i$, so $\theta_i = 3.75^{o}$;
$x = 10\tan 3.75^{o} = d\tan 5^{o}$, thus $d = 7.5$ cm

2. $\tan r = 2$, so $r = 63.4^{o}$; $(1.33)\sin i = \sin r$, so $i = 42.3^{o}$.
$\tan r = 2 = (4 - x)/(H - y)$, so $x/y = \tan 42.3^{o} = 0.91$;
$y = 0.92$ m

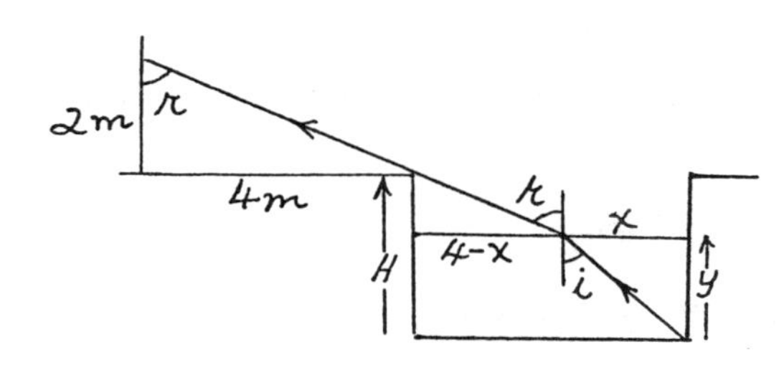

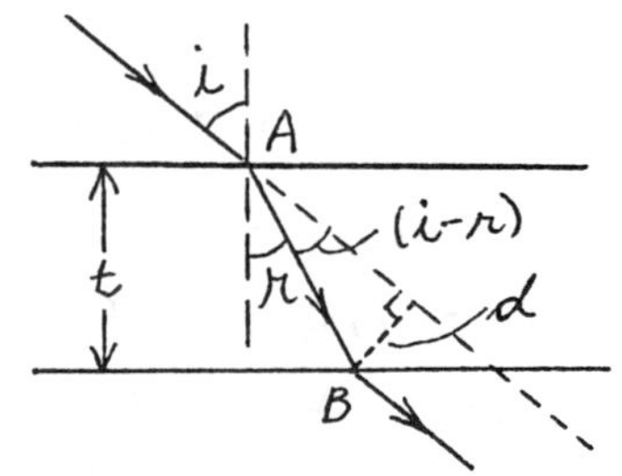

3. $AB = t/\cos r$; $d = AB \sin(i - r) = t\sin(i - r)/\cos r$

4. (a) $n\sin r = \sin 30^o$, find $\Delta r = 0.471^o$

   (b) Use result in Problem 3.

   $d_1 = 0.5066$ cm, $d_2 = 0.5256$ cm, thus $\Delta d = 0.19$ mm

6. $1/p + 1/q = 1/f$; $-1/p^2 - 1/q^2\,(dq/dp) = 0$,

   thus $dq/dp = -q^2/p^2$

7. From P35.3: $d = (t/\cos r)(\sin i.\cos r - \cos i.\sin r)$

   Use $\sin i = n \sin r$ and $\cos^2 r = (1 - \sin^2 r)$ to obtain

   $d = (t \sin i)[1 - \cos i/n(1 - \sin^2/n^2)]$

8. $n\sin\theta_c = n'$; $n_o\sin\theta = n\cos\theta_c = n[1 - (n'/n)^2]^{1/2}$

   $n_o\sin\theta = (n^2 - n'^2)^{1/2}$

9. (a) $t = d_1/c + d_2/c = (x^2 + a^2)^{1/2}/c + [(L - x)^2 + b^2]^{1/2}/c$

   (b) $c\, dt/dx = x(x^2 + a^2)^{-1/2} - (L - x)[(L - x)^2 + b^2]^{-1/2} = 0$

   This may be written $\sin\theta_1 = \sin\theta_2$, thus $\theta_1 = \theta_2$.

10. (a) $t = d_1/v_1 + d_2/v_2$

    (b) $dt/dx = x(x^2 + a^2)^{-1/2}/v_1 - (L - x)[(L - x)^2 + b^2]^{-1/2}/v_2$

    Set $dt/dx = 0$ to find $\sin\theta_1/v_1 = \sin\theta_2/v_2$.

11. (a) $\delta = 2(i - r) + (\pi - 2r) = \pi + 2i - 4r$

    (b) $d\delta/di = 2 - 4dr/di = 0$ leads to $n^2\cos^2 r = 4\cos^2 i$

    Use $\sin i = n\sin r$ and $\sin^2\theta + \cos^2\theta = 1$ to find

    $\cos i = (n^2 - 1)^{1/2}/3$.

    With $n = 4/3$, find $i = 59.4^o$, thus $r = 40.2^o$ and

    $(2i - 4r) = -42^o$. Finally, $\delta = 180^o - 42^o$.

12. (a) $\delta = 2(i - r) + 2(\pi - 2r) = 2\pi + 2i - 6r$

    (b) $d\delta/di = 2 - 6dr/di = 0$ leads to $n^2\cos^2 r = 9\cos^2 i$

    Use $\sin i = n\sin r$ and $\sin^2\theta + \cos^2\theta = 1$, to find

    $\cos i = (n^2 - 1)^{1/2}/8$

CHAPTER 36

Exercises

1. (a) $1/4 + 1/q = 1/0.05$, thus $q = 0.0506$
   $m = -q/p = -0.0127$, thus $y_I = my_O = 2.53$ cm
   (b) $m = 0.526$, so $y_I = 10.5$ cm

2. The image lies in the focal plane
   $\tan\alpha = R/D = r/f$, $r = Rf/D$
   (a) 9.06 mm; (b) 9.34 mm

3. (a) $m = -q/p = -5$, so $p = 0.4$ m;
   (b) $1/p + 1/q = 1/f$, so $f = 0.33$ m

4. (a) $m = -q/p = 4$, so $p = 4$ cm
   (b) $1/p + 1/q = 1/f$, so $f = 5.33$ cm

5. (a) $m = -q/p = -1/3$, so $p = 18$ cm;
   (b) $1/p + 1/q = 1/f$, so $f = 4.5$ cm

6. $m = -q/p = -55.6$, so $p = 0.126$ m
   $1/0.126 + 1/7 = 1/f$, so $f = 0.124$ m

7. (a) $1/q = 1/5 - 1/200$, so $q = 5.13$ cm
   (b) $1/q = 1/5 - 1/50$, so $q = 5.56$ cm

8. $q = +2p$, $1/p + 1/2p = 1/15$, so $p = 22.5$ cm
   $1/p - 1/2p = 1/15$, so $p = 7.5$ cm

9. (a) $q = 2p/3 = 8$ cm; (b) $1/8 + 1/12 = 1/f$, so $f = 4.8$ cm

10. (a) From $q = +2.5p$, find $p = 49$ cm;
    (b) From $q = -2.5p$, find $p = 21$ cm.

11. (a) $q = 0.4p$, then $1/p + 1/0.4p = 1/f$, so $p = 70$ cm;
    (b) $q = -0.4p$, then $1/p - 1/0.4p = 1/f$, so $p = -30$ cm

12. (a) $q = -p/5$, then $1/p - 5/p = -1/20$, so $p = 80$ cm
    (b) For real, erect and 150%, $q = -1.5p$.
    Then, $1/p - 2/3p = -1/20$, so $p = -6.67$ cm

13. $1/20 + 1/q_1 = 1/10$, so $q_1 = 20$ cm, thus $p_2 = -10$ cm

$-1/10 + 1/q_2 = -1/15$, so $q_2 = 30$ cm

14. $1/12 + 1/q_1 = 1/10$, so $q_1 = 60$ cm, thus $p_2 = -45$ cm
$-1/45 + 1/q_2 = 1/20$, so $q_2 = 13.8$ cm

15. $1/40 + 1/q_1 = 1/8$, so $q_1 = 10$ cm; thus $p_2 = 10$ cm
$1/10 + 1/q_2 = 1/12$, so $q_2 = -60$ cm.
$m_T = m_1 m_2 = (-1/4)(+60/10) = -1.5$

16. $-1/f_1 + 1/q_2 = 1/f_2$, $q_2 = f_{eff} = f_1 f_2/(f_1 + f_2)$

17. See Exercise 16 above: $14 = f_1 f_2/(f_1 + f_2)$
Since $f_1 = 10$ cm, find $f_2 = -35$ cm

18. (a) $M = 0.25/0.057 = 4.39$;
(b) $1/q = 1/6 - 1/5.7$, so $q = -114$ cm

19. (a) $-1/40 + 1/p = 1/4$, thus $p = 3.64$ cm, $m = -q/p$:
$y_I = m y_O = 11$ mm;
(b) $M = 0.25/p = 6.87$

20. (a) Let $N = 0.25$ m: $-1/N + 1/p = 1/f$; thus $M = N/p = 1 + N/f$
(b) $N/f = 1.4$, thus $f = 0.179$ m

21. (a) $1/p - 1/0.25 = 1/0.1$, so $p = 7.14$ cm
(b) $y_I = (25/7.14)(2) = 7.0$ mm

22. $M = -(\ell/f_o)(0.25/f_E) = 400$; thus $f_E = 2$ cm

23. $1/p_E - 1/40 = 1/3$, so $p_E = 2.79$ cm. Find $\ell = 13.7$ cm,
then $q_O = 14.7$ cm.
$1/p_O + 1/14.7 = 1/0.8$, so $p_O = 0.846$ cm,
$M = -(q_O/p_O)(25/p_E) = 157$

24. $1/0.625 + 1/q_O = 1/0.6$, so $q_O = 15.0$ cm.
$d = 17.4$ cm and $\ell = 14.4$ cm.
$M_\infty = -(\ell/f_O)(25/f_E) = 250$.

25. $M_\infty = f_O/f_E = 12$

26. $M_\infty = 180/5 = 36$

27. $M_\infty = 36/f_E = 8$, so $f_E = -4.5$ cm

28. (a) $M_\infty = 50$
(b) $1/p_E - 1/40 = 1/10$, so $p_E = 8$ cm; $M = f_O/P_E = 62.5$.

29. $f_O + f_E = 65$ cm; and $f_O/F_E = 25$.
Find $f_E = 2.5$ cm, $f_O = 62.5$ cm

30. $f_E = 5$ cm, so $M_\infty = 20/5 = 4$.

31. $M_\infty = 16.8/0.035 = 480$

32. $1/p_E - 1/0.4 = 1/f_E$, so $p_E = 0.0863$ m
$M = f_O/p_E = 1.8/0.0863 = 20.9$

33. (a) $1/12 + 1/q_O = 1/0.24$, so $q_O = 24.5$ cm and $p_E = -8.5$ cm.
$-1/8.5 + 1/q_E = -1/8$, thus $q_E = -136$ cm from objective;
(b) Need $p_E = -8$ cm, so separation should be 16.5 cm

34. $f = 35.7$ cm; then $1/0.25 - 1/NP = 1/f$, so $NP = 83.3$ cm

35. $f = -40$ cm or $P = -2.5$ D

36. Bifocal: $f_1 = -4.0$ m or $P_1 = -0.25$ D for far point
Near point: $P_2 = 1/0.25 - 1/0.4 = 1/f_2 = 1.5$ D

37. (a) $P = 1/0.25 - 1/0.34 = 1.06$ D; (b) $P = -1/0.34 = -2.94$ D

38. $1/0.28 - 1/NP = -0.5$, so $NP = 24.6$ cm

39. $1/0.4 - 1/NP = 1.5$, so $NP = 1$ m; $P = 1/0.25 - 1 = +3$ D

40. (a) $FP = 50$ cm; (b) $1/NP - 1/0.2 = -2$, so $NP = 33.3$ cm.

41. (a) $n/q_1 = (n - 1)/R$; $-n/q_1 + 1/q_2 = 0$,
$q_2 = R/(n - 1) = 24$ cm;
(b) $1/q_2 = (1 - n)/(-R)$, thus $q_2 = R/(n - 1) = 24$ cm

42. Use Eq. 36.15: (a) $1/16 = 0.5(1/12 - 1/R_2)$, so $R_2 = -24$ cm
(b) $-1/40 = 0.5(1/12 - 1/R_2)$, so $R_2 = 7.5$ cm

43. (a) $1/p + n/q = 0.5/R$; $q_1 = 3R = 12$ cm
$-1.5/4 + 1/q_2 = 0.5/4$, thus $q_2 = 2$ cm which is
6 cm from the center;

(b) $1/16 + 1.5/q_1 = 0.5/4$, so $q_1 = 24$ cm, then,
$-1.5/16 + 1/q_2 = 0.5/4$, so $q_2 = 4.57$ cm, or
8.57 cm from the center

44. (a) $1/24 + 1.5/q = 0.5/8$, so $q_1 = 72$ cm,
$-1.4/48 + 1/q_2 = 0$, $q_2 = 32$ cm
(b) $q_1 = -14.4$ cm, $q_2 = -25.6$ cm

45. $1/f = 0.5(1/12 + 1/16)$, so $f = 13.7$ cm
$1/20 + 1/q = 1/13.7$, so $q = 43.6$ cm, and $m = -2.19$

46. $1/f = 0.5(-1/12 - 1/16)$, so $f = -13.7$ cm;
$1/20 + 1/q = -1/13.7$, so $q = -8.14$ cm. Then, $m = 0.407$.

47. (a) $1/50 - 1/20 = 1/f$, so $f = -33.3$ cm.
(b) $1/50 + 1/20 = 1/f$, so $f = 14.3$ cm.

48. (a) $1/\infty - 1/0.75 = 1/f$, so $f = -0.75$ m or -1.33 diopters.
(b) $1/25 - 1/80 = 1/f$, so $f = 0.364$ m or 2.75 diopters.

49. (a) $1/p - 1/25 = 1/10$, so $p = 7.14$ cm.
(b) $m = -q/p = +3.50$.

50. (a) $1/40 + 1/q_1 = 1/15$, so $q_1 = 24$ cm and $p_2 = -14$ cm. Then,
$-1/14 + 1/q_2 = -1/10$, so $q_2 = -35$ cm.
(b) $m_1 m_2 = (-24/40)(-35/14) = 1.5$.

51. $1/10 + 1/q_1 = 1/12$, so $q_1 = -60$ cm and $p_2 = 75$ cm. Then,
$1/75 + 1/q_2 = -1/30$, so $q_2 = -21.43$ cm.
$m_T = m_1 m_2 = (60/10)(21.43/75) = 17.14$. Size = 20.6 cm.

52. $h/f = \tan\theta$, so $h = 1.82$ cm.

53. (a) $M_\infty = -f_O/f_E = -240/12 = -20$. From Eq. 36.8,
$\beta_\infty = M_\infty \alpha = 10.4^o$.
(b) $q_O = 240$ cm, then $1/p_E - 1/25 = 1/12$, so $p_E = 8.11$ cm.
$M = -f_O/p_E = -29.6$, and $\beta = M\alpha = 15.4^o$.

54. $1/x + 1/(100 - x) = 1/20$, leads to $x = 27.6$ cm, 72.4 cm.

55. (a) $m = -q/p = +1.5$. Then, $1/p - 1/1.5p = 1/12$, so $p = 4$ cm.
(b) $m = -q/p = -1.5$. Then, $1/p + 1/1.5p = 1/12$, so $p = 20$ cm

56. $1/18 + 1/q_1 = 1/10$, so $q_1 = 22.5$ cm, so $p_2 = 7.5$ cm. Then, $1/7.5 + 1/q_2 = -1/15$, so $q_2 = -5$ cm.
$m_T = m_1 m_2 = (-22.5/18)(5/7.5) = -0.833$.

57. $m = -q/p = -2.4/1800$, so $p = 75q$. Then, $1/p + 75/p = 1/f$, gives $p = 4.18$ m.

Problems

1. (a) $q_1 = 20$ cm, then $-1/8 + 1/q_2 = 1/7$, so $q_2 = 3.73$ cm,
$m = (-20/5)(3.73/8) = -1.87$;
(b) $q_1 = 6$ cm, then $1/6 + 1/q = 1/7$, so $q_2 = -42$ cm,
$m = (-1/2)(42/6) = -3.5$

2. $1/20 + 1/q_1 = 1/10$, $q_1 = 20$ cm; $1/10 + 1/q_2 = -1/5$, so $q_2 = -3.33$ cm.
$m = (-1)(3.33/10) = -0.333$.

3. $q_1 = 20$ cm, $-1/8 + 1/q_2 = -1/15$, so $q_2 = 17.1$ cm,
$m = (-1)(17.1/8) = -2.14$;

4. (a) $1/25 + 1/q_1 = -1/15$, so $q_1 = -9.38$ cm,
$1/21.38 + 1/q_2 = 1/14$, so $q_2 = 40.6$ cm
(b) $m = (9.38/25)(-40.6/21.4) = -0.71$

5. $1/p + 1/(D - p) = 1/f$; so $p^2 - Dp + fD = 0$
Solve: $p = [D + (D^2 - 4fD)^{1/2}]/2$. Thus $\Delta p = [D(D - 4f)]^{1/2}$

6. $1/(f + x) + 1/(f + x') = 1/f$ leads to $f^2 = xx'$.

7. (a) $1/p_E - 1/25 = 1/5$, so $p_E = 4.17$ cm;
$1/20 + 1/q_O = 1/0.8$, so $q_O = 83.3$ cm.
$m_T = m_O m_E = (0.833/20)(0.25/0.0417) = 0.25$
Thus $y_I = m y_O = 1.0$ cm;
(b) $M = q_O/p_E = 20$
[Note: $\alpha = 0.04/20 = 0.002$; $\beta = y_I/q_E = 0.01/0.25 = 0.04$.
$M = \beta/\alpha = 20$.]

8. $1/15 + 1/q_1 = 1/10$; so $q_1 = 30$ cm;
$-1/10 + 1/q_2 = 1/10$; so $q_2 = -5$ cm (left of the lens)

9. $n/3 + 1/q = (1 - n)/(-3)$, so $q = -3$ cm. Note that rays emerge radially with no refraction, so image is at the center.

10. (a) Use $f_1 = f$ and $f_2 = 2f$ separated by $3f$.
    (b) Use $f_1 = -f/2$ and $f_2 = f$ separated by $f/2$.

11. $n_2/p_1 - n_1/q_1 = (n_1 - n_2)/R_1$; $n_1/q_1 + n_2/q_2 = (n_2 - n_1)/R_2$
    Thus, $1/f = 1/p_1 + 1/q_2 = (n_1/n_2 - 1)(1/R_1 - 1/R_2)$

12. From Eq. 36.15, $f = 5/(n - 1)$. So $\Delta f = 8.62 - 8.06 = 0.56$ cm

13. Amount of light is proportional to area, i.e. diameter^2.
    (a) $1/(1.4)^2 = 0.51$; (b) $1/(8/5.6)^2 = 0.49$

14. $1/f = (1/1.33 - 1)(1/12 + 1/16)$, so $f = -27.4$ cm.

CHAPTER 37

Exercises

1. With $\sin\theta = y/L$, $d = m\lambda L/y = 0.17$ mm

2. With $\sin\theta = y/L$, $y = m\lambda L/d$, find $\Delta y = 1.7$ mm

3. (a) With $\sin\theta = y/L$, $d = \lambda\omega/2y = 0.141$ mm;
(b) $d = \lambda L/y = 0.281$ mm

4. $\Delta y = \lambda L/d = 3.07$ mm

5. $\lambda = yd/4L = 583$ nm

6. (a) $d = 3\lambda L/y = 0.22$ mm; (b) $\Delta y = \lambda L/d = 5.36$ mm

7. $d\sin\theta = 4.5\lambda$, so $\lambda = 543$ nm

8. $10(560) = 9.5\lambda_2$, $\lambda_2 = 589$ nm

9. $y = m\lambda L/d$, so $480m_1 = 560m_2$, or $6m_1 = 7m_2$. Find $m_1 = 7$, and $m_2 = 6$. $y = m_1(480 \text{ nm})L/d = 1.68$ cm

10. $\tan\theta \approx \sin\theta = 5.4\text{x}10^{-3}$; $d = 7\lambda/\sin\theta = 0.763$ mm.

11. $\sin\theta_2 = \sin(\theta_1 + 10^o) = 2\sin\theta_1$. Expand $\sin(\theta_1 + 10)$ to find $\tan\theta_1 = \sin 10^o/(2 - \cos 10^o)$, thus $\theta_1 = 9.71^o$
$d = \lambda/\sin\theta_1 = 17.8$ cm

12. $\Delta y = \lambda L/d = 1/700$, so $d = 0.714$ mm

13. $\delta = 3\lambda/4 = d\sin\theta = dy/L$, so $y = 3\lambda L/4d = 2.16$ mm, upward

14. $\delta = \lambda = 2(D^2 + H^2)^{1/2} - 2D$; where $D = 4$ m and $\lambda = 1.7$ m. Find $H = 2.74$ m

15. $d\sin\theta = \lambda/2 = (0.34 \text{ m})/2$, then $y = d\tan\theta = 1.38$ m.

16. (a) $\delta = (4^2 + d^2)^{1/2} - 4 = \lambda/2$, where $\lambda = 3.58$ m. Find $d = 4.18$ m; (b) $d = \lambda/2 = 1.79$ m.

17. $\delta = (4^2 + d^2)^{1/2} - 4 = \lambda/2$, where $\lambda = 0.68$ m. Find $d = 1.7$ m

18. Take speakers to be in phase. $\delta = (4^2 + d^2)^{1/2} - 4 = 0.472$ m
(a) $\delta = \lambda$, so $f = 340/\delta = 720$ Hz

(b) $\delta = \lambda/2$, so $f = 340/2\delta = 360$ Hz

19. (a) $\delta = x - (d - x) = 2x - d = (m + 1/2)\lambda$, thus
$x = 0.5d - (m + 0.5)\lambda/2$;
(b) $\delta = 2x - d = m\lambda$, so $x = 0.5d - m\lambda/2$

20. $\Delta\phi = 2\pi\delta/\lambda$, so $\delta = 5\lambda/2\pi$.
$d \sin\theta = \delta = 0.15/2\pi$, thus $\theta = 1.7^o$.

21. (a) $d(\sin\theta - \sin\alpha)$; (b) $\theta = \alpha$;
(c) $\theta = 0$, so $\delta = -d\sin\alpha = -\lambda/2$, or $\alpha = \sin^{-1}(\lambda/2d)$

22. $\sin\theta = y/f$, so $dy/f = (m + 1/2)\lambda$, or $y = (2m + 1)\lambda f/2d$.

23. $d\sin\theta = (m + 1/2)\lambda$ leads to $m = 6$, so there are 7 dark fringes.

24. Need to find the width of the reflected beam on the screen.
$d/L = y_2/24L$, $y_2 = 24d$; $d/5L = y_1/20L$, $y_1 = 4d$, thus
$\Delta y = 20d = 8$ mm.

25. Substitute $\phi = 2\pi d\sin\theta/\lambda \approx 2\pi dy/\lambda L$ into Eq. 37.9.

26. $\cos(\phi/2) = 1/(2)^{1/2}$, thus $\phi = \pi/2 = 2\pi dy/\lambda L$.
Thus, $y = \lambda L/4d = 1.12$ mm.

27. Substitute $\phi/2 = \pi dy/\lambda L$ in to Eq.37.9 to find $I = 0.098I_o$

28. $I = 4I_o\cos^2(\pi/4) = 2I_o$

29. $\cos(\phi/2) = 1/2$, so $\phi = 2\pi/3 = 2\pi\delta/\lambda$, thus $\delta = \lambda/3 = 209$ nm

30. Plot $I = 4I_o\cos^2(2\pi\sin\theta)$ in steps of 0.1 rad:
0.655; 0.1; 0.08, 0.59; 0.98; 0.84; 0.38; 0.04; 0.04;
0.30; 0.6; 0.83, 1.0, 1.0, 1.0, 1.0.

31. $2t = (m + 1/2)\lambda/n$, thus $\lambda = 2nt/(m + 1/2) =$
Find $\lambda = 417$ nm, 509 nm, 655 nm

32. $\phi = 2\pi\delta/\lambda_F = 2\pi(2t)n/\lambda = 1583/\lambda$.
(a) 3.96 rad; (b) 2.88 rad; (c) 2.26 rad

33. $\Delta t = \lambda_F/2 = \lambda/2n = 200$ nm.

$\tan\alpha = \Delta t/d = 10-4$, thus $\alpha = 0.006^o$

34. $2t = 72\lambda$; so $t = 36\lambda$. Thus, $R = t/2 = 18\lambda = 8.64\ \mu m$.

35. (a) $t = \lambda/2n = 206$ nm; (b) $t = \lambda/4n = 103$ nm

36. (a) $2t = 42\lambda$, so $t = 1.34 \times 10^{-5}$ m
(b) $r^2 = 2Rt$, thus $R = 4.5$ m

37. $2t = m\lambda_F = 8\lambda/n$; $r^2 = 2Rt$, where $r_1 = 1.8$ cm, $r_2 = 1.64$ cm.
$r_1^2 = 16\lambda R$; $r_2^2 = 16\lambda R/n$, thus
$n = r_1^2/r_2^2 = (1.8/1.64)^2 = 1.20$

38. $2\Delta t = \Delta m\ \lambda = 240\lambda$, thus $\lambda = 667$ nm

39. $\Delta\phi = 2\pi\delta(1/\lambda_F - 1/\lambda) = 2\pi\delta(n - 1)/\lambda = 5(2\pi)$.
Since $\delta = 2t$, $n = \lambda/\delta + 1 = 1.75$

40. $\Delta\phi/2\pi = \delta/\lambda$. (a) $\delta = 200$ nm; (b) 3.27 rad.

41. (a) $\lambda/n = 430$ nm; (b) $nt/\lambda = 2.79$;
(c) $\Delta\phi_{Top} = \pi$, $\Delta\phi_{Bottom}$ $2\pi\delta/\lambda = 35.1$ rad. Net $\Delta\phi = 32$ rad.

42. $\Delta y = \lambda L/d = 2.25 \times 10^{-3}$ m, so there are 4.44 per cm.

43. $y = m\lambda L/d$, so $\lambda = 520$ nm.

44. A shift of one fringe involves one wavelength. So consider difference in numbers of wavelengths in tube:
$2(n - 1)d/\lambda = 19.5$.

45. Number of wavelengths in sheet $= 2nd/\lambda$, so $\Delta N = (n_2 - n_1)d/\lambda$ and $\Delta\phi = 2\pi(\Delta N) = 3.77$ rad.

46. $\Delta y = \lambda L/d = 2.5 \times 10^{-3}$ m, so $d = 0.690$ mm.

47. $\Delta y = 2\lambda L/d = 5.30$ mm.

48. $\Delta y = \lambda L/d = 2 \times 10^{-3}$ m, so $L = 1.95$ m.

49. $d\sin\theta = m\lambda$, so $m = d\sin 1^o/\lambda = 11.3$, so number is 11.

50. (a) $\delta = d\sin\theta$ and $\phi = 2\pi\delta/\lambda = 27.1$ rad.
(b) $\tan\theta = 0.6/120$, then $\phi = 2\pi d\sin\theta/\lambda = 19.4$ rad.

51. (a) $\phi = 2\pi d\sin\theta/\lambda$, so $\sin\theta = \phi\lambda/2\pi d$ and $\theta = 0.04^o$.
(b) $\delta = d\sin\theta$, so $\sin\theta = \delta/\lambda$ and $\theta = 0.050^o$.

52. (a) $\phi = 2\pi d\sin\theta/\lambda = 17.9$ rad; (b) $I/4I_o = \cos^2(\phi/2) = 0.8$.

53. $d\sin\theta = m\lambda$, so $(m + 1)500 = (m)600$, and $m = 5$, then $\theta = 0.215^o$.

54. $\lambda = 5.5$ m and $\delta = 5.385 - 5.016 = 0.369$ m, so $\phi = 2\pi d/\lambda = 0.422$ rad.

55. $\delta = 5.590 - 5.025 = 0.565$ m $= \lambda/2$, so $\lambda = 1.13$ m.

56. $\Delta y = 2\lambda L/d$, so $\lambda = \Delta y.d/2L = 625$ nm.

57. (a) $(5 + x) - (5 - x) = \pm 0.5\lambda, \pm 1.5\lambda, \ldots$ Thus $x = \pm 1.5$ m, ± 4.5 m.
(b) $(5 + x) - (5 - x) = 0, \pm\lambda, \pm 2\lambda, \ldots$ Thus $x = 0, \pm 3$ m.

58. (a) $\delta = (2^2 + x^2)^{1/2} - x$. For $\delta = \lambda/2$, find $x = 3.75$ m; for $\delta = 3\lambda/2$, find $x = 0.583$ m. (b) Here $\delta = \lambda$ gives $x = 1.5$ m

59. (a) $2nt = m\lambda$ so $\lambda = 549$ nm;
(b) $2nt = (m + 0.5)\lambda$, so $\lambda = 439$ nm.

60. (a) $2nt = m\lambda$, so $\lambda = 420$ nm, 480 nm, 560 nm, and 672 nm.
(b) $2nt = (m + 0.5)\lambda$, so $\lambda = 448$ nm, 517 nm, 611 nm.

61. Bright $2nt = (m + 0.5)\lambda$ and $t = d\tan\theta$. (a) For given m, $d_2/d_1 = \lambda_2/\lambda_1$, thus for 680 nm, $d = 1.92$ cm.
(b) For given $\lambda = 425$ nm, $d_2/d_1 = 1.5/0.5$, so $d_2 = 3.6$ cm.

62. $\Delta t = 0.5\lambda$ for adjacent like fringes and $\Delta t/d = T/L$. Thus $d = \lambda L/2t = 1.38 \times 10^{-3}$ m and there are 7.25 fringes per cm.

63. Dark $2t = m\lambda$ and $t = d\tan\theta$, so $\Delta d = m.\Delta\lambda/2\tan\theta = 8.59$ mm.

64. Between two like fringes $\Delta t = \lambda/2$ and $\Delta t = \Delta d.\tan\theta$ where $\Delta d = 1/6$ cm. Thus $\theta = 9.39 \times 10^{-3}$ degrees.

65. Dark $2nt = m\lambda$. Here $420(m + 1) = 425m$, so $m = 84$.
From $t = d\tan\theta$, get $d = m\lambda/2n\tan\theta = 8.52$ mm.

66. (a) $3t = m\lambda/n$; (b) $3t = (m + 0.5)\lambda/n$.

67. Missing $2nt = (m + 0.5)\lambda = 3900$ nm. For m = 6, 7, 8, and 9, find λ = 411 nm, 459 nm, 520 nm, 600 nm.

68. Missing $2nt = (m + 0.5)\lambda$. Then $621(m + 0.5) = 483(m + 1.5)$ gives m = 3. Thus, t = 869 nm.

69. $2nt = (m + 0.5)620 = (m + 1)465$, gives m = 1, so t = 355 nm.

70. With n = 1 for air, $3t = (m + 0.5)\lambda$

71. $2nt = (m + 0.5)\lambda$ leads to λ = 426 nm, 596 nm.

72. (a) $2nt = (m + 0.5)\lambda$. Then $(m + 0.5)685 = (m + 1.5)411$ gives m = 1. Thus, t = 367 nm. (b) $2nt = m\lambda$ gives λ = 514 nm.

73. $t = \lambda/2n$ = 232 nm.

Problems

1. (a) $2t = m\lambda/n$, thus $(m + 1)504 = 672m$, thus m = 3.
   Thus, $t = 3\lambda/2n$ = 630 nm;
   (b) $\lambda = 2nt/(m + 1/2)$ = 448 nm, 576 nm

2. (a) $2nt = (m + 1/2)(544) = 680m$, thus m = 2, so t = 567 nm.
   (b) Constructive: $2nt = 1360 = 3\lambda_1$, so λ_1 = 453 nm

3. $\delta = \lambda\phi/2\pi = \lambda/4 = d\sin\theta = dy/L$, thus $y = \lambda L/4d$ = 2 mm, upward

4. (a) $(m + 1/2)(522) = (m + 3/2)(406)$, thus m = 3.
   $2nt = 3.5(522)$ = 761 nm
   (b) $2nt = m\lambda$; $\lambda = 1827/m$ = 609 nm, 457 nm.

5. (a) $2nt = m\lambda$, $\lambda = 2700/m$ = 450 nm, 540 nm, 675 nm (missing);
   (b) $\lambda = 2700/(m + 1/2)$ = 415 nm, 491 nm, 600 nm (enhanced)

6. $2nt = (19 + 1/2)\lambda$, thus t = 3185 nm.
   Then, $t/d = h/L$, so $d = Lt/h$ = 1.9 cm.

7. $r^2 = 2Rt$ and $2t = (m + 1/2)\lambda = r^2/R$.
   Find $(m + 1/2) = 35.5$. Thus, 36 fringes.

8. Given $\cos^2(\phi/2) = 0.5$, thus $\cos(\pi d\sin\theta/\lambda_1) = 0.707$
so, $\pi d\sin\theta/\lambda_1 = \pi/4$.
For $\cos^2(\phi/2) = 0.64$, we have $\cos(\pi d\sin\theta/\lambda) = 0.8$,
so, $\pi d\sin\theta/\lambda_2 = 36.9\pi/180$.
Find $\lambda_2/\lambda_1 = 180/(4x36.9)$, so $\lambda_1 = 488$ nm

9. (a) $d\sin\theta = \lambda = 1.13$ m, $y = 10\tan\theta = 6.9$ m;
(b) $d\sin\theta = \lambda/2$, $3\lambda/2$ lead to $y = 3.0$ m, 16 m

10. Say constructive interference occurs for both λ's at film thicknesses t_1 and t_2.
$2t_1 = m_1\lambda_1 = (m_1 + 1)\lambda_2$; so $0.6m_1 = 589$, and $m_1 = 982$
$2t_2 = m_2\lambda_1 = (m_2 + 2)\lambda_2$; so $0.6m_2 = 589x2$, i.e. $m_2 = 2m_1$.
Find $(t_2 - t_1) = (m_2 - m_1)\lambda_1/2 = 982x589.6/2 = 0.29$ mm.

11. $\Delta\phi = 2\pi(n - 1)t/\lambda = 12(2\pi)$, thus $t = 12\lambda/(n - 1) = 13\ \mu m$

12. (a) Circular fringes
(b) $d\cos\theta = m\lambda$; $y \approx L\sin\theta = L(1 - \cos^2\theta)^{1/2}$
$y = L[1 - (m\lambda/d)^2]^{1/2}$.

13. $2(n - 1)L = 40\lambda_o$, thus $n = 20\lambda_o/L + 1 = 1.0003$

14. $\delta = 2t/\cos\theta - 2(t\tan\theta)\sin i$
$\Delta\phi = 2\pi(\delta_2/\lambda_2 - \delta_1/\lambda_1) = (2\pi/\lambda)[2nt/\cos\theta - 2t\tan\theta . n\sin\theta]$
$\Delta\phi = 2\pi m = (4\pi nt/\lambda\cos\theta)[1 - \sin^2\theta] = 4\pi nt\cos\theta/\lambda$
Finally, $2nt\cos\theta = m\lambda$.

15. $2t = \lambda_F$, or $2nt = \lambda_5 = 550$ nm. There is also a π phase change on reflection. $\phi = 2\pi\delta/\lambda = 2\pi n(2t)/\lambda$
$\Delta\phi = 2\pi(550)/400 - \pi = 5.5$ rad
$\Delta\phi = 2\pi(550)/700 - \pi = 1.8$ rad
Then, $\cos^2(\phi/2) = I/4I_o$: 0.85 for 400 nm; 0.39 for 700 nm
(a) 5.5 rad; (b) 1.8 rad;
(c) 0.85 for 400 nm, 0.39 for 700 nm

CHAPTER 38

Exercises

1. (a) $\sin\theta_1 = \lambda/a$, then $y_1 \approx L\lambda/a = 2.04$ cm.
   Width $= 2y_1 = 4.08$ cm;
   (b) $\sin\theta_2 = 2\lambda/a$, so $y_2 \approx 2L\lambda/a = 4.08$ cm, then
   $y_2 - y_1 = 2.04$ cm

2. $\theta_1 = \sin^{-1}(\lambda/a)$; where $\lambda = 6.8$ cm. Width is $2\theta_1$.
   (a) 116°; (b) 26.2^o

3. $\Delta y_2/\Delta y_1 = \lambda_2/\lambda_1$, so $\Delta y_2 = (3)(436/589) = 2.22$ cm

4. $\sin\theta \approx \tan\theta = y_1/L = (\Delta y/2)/L = 2\text{x}10^{-3}$,
   $a = \lambda/\sin\theta = 0.273$ mm

5. $\Delta y = L(\sin\theta_2 - \sin\theta_1) = L\lambda/a$. So $a = L\lambda/\Delta y = 0.045$ mm

6. (a) $f = v/\lambda = 340\text{x}4/0.76 = 1790$ Hz
   (b) $\sin\theta_1 = \lambda/a = 1/4$, leads to $\theta_1 = 14.5^o$

7. $d/a = 4$, so there are 7 fringes

8. $\sin\theta_1 = 1.22\lambda/a = 6.71\text{x}10^{-4}$; Width $= 2(L\sin\theta_1) = 0.295$ mm

9. $\theta_c = 1.22\lambda/a = 9.15\text{x}10^{-4}$; then $s = d\theta_c = 1.46$ cm

10. $\theta_c = 1.22\lambda/a = 1.83\text{x}10^{-2}$; then $s = d\theta_c = 366$ m

11. (a) $\theta_c = 1.22\lambda/a = 1.34\text{x}10^{-4}$; $d = L\theta_c = 51.6$ km;
    (b) $\theta_c = 1.49\text{x}10^{-7}$; $d = L\theta_c = 57.3$ m

12. $\theta_c = 1.22\lambda/a = 1.14\text{x}10^{-6}$; $s = d\theta_c = 0.205$ m

13. $\theta_c = 1.22\lambda/a = s/25$, thus $s = 50.8$ μm

14. $\theta_c = 1.22\lambda/a = 4.47\text{x}10^{-5}$; $d = s/\theta_c = 44.7$ km

15. (a) $\theta_c = 1.22\lambda/a = 1.2\text{x}10^{-7}$; $s = d\theta_c = 1.2 \text{ x } 10^9$ m;
    (b) $\theta_c = 1.22\lambda/a = 8.4\text{x}10^{-4}$; $s = d\theta_c = 8.4 \text{ x } 10^{12}$ m

16. (a) $\theta_c = 1.22\lambda/a = 1.59\text{x}10^{-4}$; $d = s/\theta_c = 11.3$ km
    (b) $\theta_c = 1.22\lambda/a = 2.83\text{x}10^{-7}$; $d = 1.8/\theta_c = 6.26\text{x}10^6$ m

17. (a) $d = 3.33x10-6$ m; $\sin\theta_1 = \lambda/d$
$\theta_A = 7.07^o$; $\theta_B = 11.35^o$, so $\Delta\theta = 4.3^o$
(b) $\sin\theta_2 = 2\lambda/d$; $\theta_A = 14.2^o$, $\theta_B = 23.2^o$, so $\Delta\theta = 9.0^o$;
(c) Yes

18. $\delta = d(\sin\theta \pm \sin\phi) = m\lambda$.
Sign depends on direction of incoming beam.

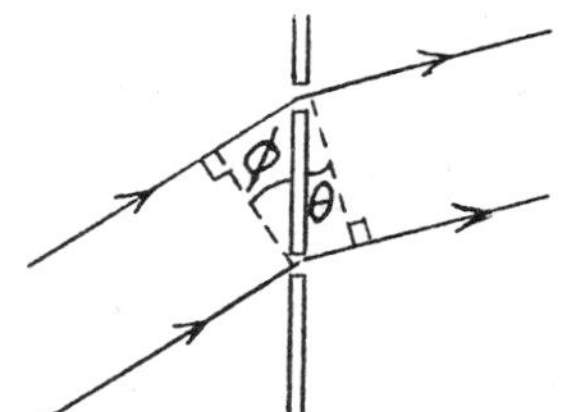

19. $\sin\theta = n\lambda/d = 0.42n$ for 700 nm. For $\theta = 90^o$, $n = 2.4$
Thus, there are two complete orders.

20. $\sin\theta = 2\lambda/d = 10^6\lambda$. Find $\theta_1 = 36.09^o$, $\theta_2 = 36.13^o$, so $\Delta\theta = 0.04^o$.

21. $d = \lambda/\sin\theta = 3.35x10^{-6}$ m
$\sin\theta_2 = 2\lambda/d = 0.293$, so $\theta_2 = 17^o$

22. $d = 2\lambda/\sin\theta = 2.61x10^{-6}$ m. $N = 2.8x10^{-2}/d = 10,700$

23. (a) $2.73E_o$; (b) $2E_o$; (c) E_o; (d) Zero

24. From the figure see that
$16\sin 50^o = E\sin 80^o$, thus
$E = 12.4$ V/m, and $\phi = 100^o$.

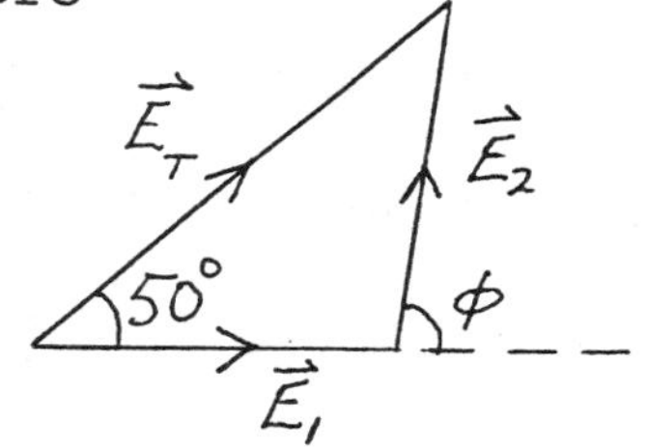

25. Closed polygon for $\phi = 2\pi/5$ rad

26. $d\sin\theta = \delta$; $\alpha = 2\pi\delta/\lambda$, thus $\theta = \sin^{-1}(\alpha\lambda/2\pi d)$.

27. (a) $\sin\theta_1 = \lambda/a$, find $\theta_1 = 0.47^o$;
(b) $\sin(\theta_1/2) = 4.1x10^{-3}$, so $\alpha/2 = \pi a\sin\theta/\lambda = 1.57$ rad
$I = I_o \sin^2(1.57)/(1.57)^2 = 0.406I_o$

28. $\sin\theta = 5.88x10^{-3}$; so $\alpha/2 = \pi a\sin\theta/\lambda = 2.12$ rad.
$I = I_o \sin^2(2.12)/(2.12)^2 = 0.162I_o$

29. E_o becomes $2E_o$, so $I \alpha E^2$ becomes $4I_o$.

30. For first order: $N = R = \lambda/\Delta\lambda = 589.3/0.6 = 982$.
$W = Nd = 982/300 = 3.27$ mm.

31. $\Delta\lambda = \lambda/Nm = 0.023$ nm

32. (a) $\Delta\lambda = 0.53$ nm, $\lambda/\Delta\lambda = 1110$; (b) $R = 2N$, so $N = 555$

33. $\sin\theta = \lambda/2d = 0.14/0.64$, so $\theta = 12.6^o$

34. (a) $\lambda = 2d\sin\theta = 0.145$ nm;
(b) $\sin\theta = \lambda/d = 0.145/0.28 = 0.518$, so $\theta = 31.2^o$

35. $\lambda = d\sin\theta = 93$ pm

36. $d = \lambda/2\sin\theta = 0.416$ nm

37. $I_1 = I_o/2$ gets through first polarizer, then
$I_2 = I_1\cos^2 60^2 = I_o/8$ passes through the second.

38. $I_1 = I_o/2$ gets through first polarizer.
$I_2 = I_1\cos^2(45^o) = I_1/2 = I_o/4$,
$I_3 = I_2\cos^2(45^o) = I_2/2 = I_o/8$

39. $\tan\theta_p = n_2/n_1$; $n_2\sin\theta_c = n_1$
Thus $n_1/n_2 = \cot\theta_p = \sin\theta_c$.

40. Use Exercise 39: $\cot\theta_p = \sin\theta_c$, so $\theta_p = 58.4^o$

41. $\tan\theta_p = 1.33$, so $\theta_p = 53.1^o$ and $\theta_i = 36.9^o$

42. $\cos^2\theta = 0.4$, so $\theta = 50.8^o$

43. $\tan\theta_p = 1.6/1.33 = 1.2$, so $\theta_p = 50.2^o$

44. When refracted and reflected rays are perpendicular.
$n\sin i = \sin r = \cos i$, thus $\tan i = 1/n$, $i = 36.9^o$.

45. $\tan\theta_p = 1.6$, so $\theta_p = 58^o = i$, then $\sin i = n\sin r$, so $r = 32^o$

46. $\sin\theta_1 = \lambda/a$ and $y = L\tan\theta = 1.86$ cm, so $2y = 3.72$ cm.

47. $\sin\theta_1 = \lambda/a$ where $\lambda = 0.550$ m. Find $\theta_1 = 43.4^o$.

48. $d/a = 4$, so $d = 0.60$ mm.

49. $d/a = 5$, thus $m = 5, 10, 15$ order of interference pattern are missing. In secondary diffraction maximum we see 4 fringes.

50. $\theta = 1.22\lambda/a = d/L$, so $d = L\theta = 376$ m.

51. $\theta = 1.22\lambda/a = d/L$, so $a = 0.488$ m.

52. (a) $\theta = 1.22\lambda/a = 2.80 \times 10^{-7}$ rad. (b) $d = L\theta = 1.32 \times 10^{14}$ m.

53. $\sin\theta = 2\lambda/d$. Find $\theta_1 = 34.35^o$ and $\theta_2 = 39.18^o$, so $\Delta\theta = 4.83^o$

54. (a) $d\sin\theta = 2\lambda$, so $d = 3.44 \times 10^{-6}$ m.
$N = 1.8$ cm/d = 5230 lines.
(b) $d\sin\theta_3 = 3\lambda$, find $\theta_3 = 28.4^o$.

55. $\sin\theta_1 = \lambda/a$ and $\sin\theta_2 = 2\lambda/a$ give $\theta_1 = 7.25 \times 10^{-3}$ rad and $\theta_2 = 1.45 \times 10^{-2}$ rad. Then $\Delta y = L(\tan\theta_2 - \tan\theta_1) = 2.32$ cm.

Problems

1. The path difference between two incoming rays is $d\cos\alpha$ and between the corresponding reflected rays it is $-d\cos\theta$. Thus, the condition for constructive interference is
$d(\cos\alpha - \cos\theta) = m\lambda$

2. See Figs. 38.14 and 38.15.
(a) $\phi = 2\pi$; $\sin\theta = \lambda/d$, $y = L\sin\theta = 20.2$ mm
(b) $\phi = \pi, 3\pi$; $\sin\theta = \lambda/2d, 3\lambda/2d$; $y = 10.1$ mm, 30.4 mm.
(c) $\phi = 2\pi/3, 4\pi/3$; $\sin\theta = \lambda/3d, 2\lambda/3d$, $y = 6.75$ mm, 13.4 mm

3. (a) $\phi = 2\pi$, $\sin\theta = \lambda/d$, $y = 20.3$ mm;
(b) $\phi = \pi/2, \pi$, $\sin\theta = \lambda/4d, \lambda/2d$; $y = 5.06$ mm, 10.1 mm

4. $d \sin\theta = n\lambda$; $d \cos\theta\, d\theta = n\, d\lambda$; thus $d\theta = nd\lambda/(d\cos\theta)$

5. Min: $d\sin(\theta + \Delta\theta) = (mN + 1)(\lambda/N)$; Max: $d\sin\theta = m\lambda$
$d[\sin\theta\cos\Delta\theta + \cos\theta\sin\Delta\theta - \sin\theta] = \lambda/N$.
With $\sin\Delta\theta \approx \Delta\theta$; find $\Delta\theta \approx \lambda/(Nd\cos\theta)$

6. $d\sin\theta = n\lambda$; $\cos\theta\, d\theta/d\lambda = n/d = \sin\theta/\lambda$, thus $d\theta/d\lambda = \tan\theta/\lambda$.

7. First polarizer transmits $I_1 = I_o/2$
Then $I_2 = I_1\cos^2 30^o$ and $I_3 = I_2\cos^2 30^o$
Find $I_3 = (0.5)(\cos^2 30^o)^2 = 0.281 I_o$

8. Add phase $\phi = 2\pi d\sin\theta/\lambda = \pi/3 \sin 11^o = 0.2$ rad or 11.5^o.

9. In Eq. 38.10, set $dI/d\alpha = 0$ to find $\tan(\alpha/2) = \alpha/2$. First solution is $\alpha = 9.0$ rad

10. From Eq. 38.10; $\sin(\alpha/2) = 0.5(\alpha/2)$. Solution is $\alpha = 2\pi a\sin\theta/\lambda = 3.79$ rad, then find $\theta = 0.04^o$.

11. From 38.16: $E_o = 2A\sin(\phi/2)$; $E_{oT} = 2A\sin(N\phi/2)$ thus $E_{oT}/E_o = \sin(N\phi/2)/\sin(\phi/2)$

CHAPTER 39

Exercises

1. $\gamma = 5/4$, so $v = 0.6c$

2. (a) $\gamma \approx 1 + v^2/2c^2$; (b) $1/\gamma = 1 - v^2/2c^2$.

3. $L_o = \gamma L = (5/4)(1.2) = 1.5$ m

4. $\gamma = 10/3$, so $1 - v^2/c^2 = 9/100$, thus $v = 0.954c$.

5. (a) $\Delta T/T_o = \gamma - 1 \approx +v^2/2c$;
   (b) $\Delta L/L_o = 1 - 1/\gamma \approx -v^2/2c$

6. (a) $\gamma = 1.005$, so $T - T_o = T_o(\gamma - 1) = 43.9$ h
   (b) $\gamma = 15.8$, so $T - T_o = T_o(\gamma - 1) = 14.8$ y

7. Use Exercise 2: $T - T_o = (\gamma - 1)T_o \approx (v^2/2c^2)T_o$
   Thus, $v/c = (2\Delta T/T_o)^{1/2} = 2.5\text{x}10^{-4}$, so $v = 75$ km/s

8. $T_o = 2.2\ \mu s$, $L_o = 400$ m. In frame S, $v = L_o/\gamma T_o$,
   so $v^2\gamma^2 = (L_o/T_o)^2$. Solve to find $v = 1.56\text{x}10^8$ m/s.

9. (a) $\gamma = 5/3$, $T = \gamma T_o = 8.33\ \mu s$;
   (b) $L_o = (0.8c)T = 2$ km; (c) $L = L_o/\gamma = 1.2$ km

10. $T - T_o = T(1 - 1/\gamma) = (500\text{ s})(-v^2/2c^2) = -4.44\text{x}10^{-10}$ s

11. $T = \gamma T_o$ and $\gamma = 2$. Find $v = 0.866c$

12. (a) $T_o = L_o/\gamma v = (4.2y \times c)/5(0.98c) = 0.857$ y
    (b) $T = \gamma T_o = 4.29$ y; (c) $L = L_o/\gamma = 0.84$ ly.

13. (a) $L_o = \gamma L = (1.02)(150) = 153$ m;
    (b) $T_o = 150\text{ m}/v = 2.5\ \mu s$; (c) $T = \gamma T_o = 2.55\ \mu s$

14. $T_o = L/v = L_o/\gamma v$, thus $\gamma v = 80c/70 = 3.43\text{x}10^8$ m/s.
    Solve to find $v = 2.26\text{x}10^8$ m/s.

15. (a) $320/0.6c = 1.78\ \mu s$;
    (b) $T = \gamma T_o = (5/4)(1.78\ \mu s) = 2.22\ \mu s$

16. (a) $L_o = \gamma L = 1.5$ km; (b) $1.5\text{ km}/0.6c = 8.33\ \mu s$
    (c) $T_o = T/\gamma = 4T/5 = 6.67\ \mu s$.

17. (a) $T_o = 120/0.98c = 0.408\ \mu s$;
(b) $T = \gamma T_o = 5.03T_o = 2.05\ \mu s$

18. (a) $\gamma = L_o/L = 4.2/3.6 = 1.166$; and $(1 - v^2/c^2) = 1/\gamma^2$
Find $v = 0.52c$
(b) $T_o = L_o/\gamma v = 24$ y; thus $\gamma v = 0.175c$. Solve to find
$v = 0.172c = 5.25x10^7$ m/s
(c) $T = 4.2c/0.172c = 24.4$ y

19. (a) $500\ km/\gamma v = 8.17$ ms; (b) $500\ km/\gamma = 490$ km

20. (a) $\gamma = 5/3$, so $T = (5/3)T_o = 43.3$ ns;
(b) $vT = 10.4$ m; (c) $vT_o = 6.24$ m

21. (a) $\gamma = 10$, so $T = 10T_o = 22\ \mu s$;
(b) $T = 10\ km/v = 33.5\ \mu s$; (c) $T_o = 3.35\ \mu s$

22. (a) $f' = (1.6/0.4)^{1/2} f_o = 2f_o = 144$ beats/min
(b) $f' = (0.4/1.6)^{1/2} f_o = 0.5f_o = 36$ beats/min

23. $f_o = 10^{10}$ Hz, and $v = 30$ m/s.
Received by car $f_1 = [(c + v)/(c - v)]^{1/2} f_o$
Received by police $f_2 = [(c + v)/(c - v)]^{1/2} f_1$
$\Delta f = f_2 - f_o = 2vf_o/(c - v) = 2$ kHz

24. $\lambda = [(c - v)/(c + v)]^{1/2} \lambda_o$, thus $(c - v)/(c + v) = 25/49$
Find $v = 0.324c$.

25. $\lambda = [(c + v)/(c - v)]^{1/2} \lambda_o = (1.2/0.8)^{1/2} \lambda_o$, so
$\lambda_o = 490$ nm

26. $f'' = [(c + v)/(c - v)]f_o = 1.22x10^9$ Hz

27. (a) $f'' = [(c + v)/(c - v)]f_o$. Note $(1 - v/c)^{-1} \approx 1 + v/c$.
Find $f'' - f_o = (2v/c)(c/\lambda) = 2v/\lambda = 4$ kHz;
(b) $f'' = [(c - v)/(c + v)]f_o$
Find $f'' - f_o = -2v/\lambda = -4$ kHz

28. $x = \gamma(x' + vt') = 697$ km; $t = \gamma(t' + vx'/c^2) = 0.018$ s

29. $\Delta x' = \gamma\ \Delta x = 600$ km; $\Delta t = \gamma(\Delta t' + v\Delta x'/c^2) = 1.2$ ms

30. $L = \gamma L_o = 5(1.2 \text{ km}) = 6 \text{ km}$

31. $\Delta x = \gamma(\Delta x' + v\Delta t') = \gamma v \Delta t' = -4 \times 10^6 \text{ m}$

32. (a) $t_A' = t_B' = 1.6 \text{ km}/c = 5.33\ \mu s$
(b) $t_A = (L/2\gamma - vt_A)/c$, so $t_A = (L/2\gamma)/(c + v) = 2.67\ \mu s$
(c) $t_B = (L/2\gamma)/(c - v) = 10.67\ \mu s$
[$\Delta t = \gamma(\Delta t' + v\Delta x'/c^2) = 1.25(5.33 + 3.2) =$
$2.67\ \mu s$; $10.67\ \mu s$.]

33. $\Delta t_1 = L/c = 333$ ms; $(L - v\Delta t_1) - v\Delta t_2 = c\Delta t_2$; so
$\Delta t_2 = (L - v\Delta t_1)/(c + v) = 37$ ms

34. (a) $\Delta t' = \gamma(\Delta t - v\Delta x/c^2) = 0$, thus $v = c^2\Delta t/\Delta x = 0.8c$;
(b) Green first

35. $\Delta t' = \gamma(\Delta t - v\Delta x/c^2) = -\gamma v\Delta x/c^2 = -7.96\ \mu s$ (G before R)
$\Delta x' = \gamma(\Delta x - v\Delta t) = 2.4$ km

36. Square Eqs. 39.17 and 39.18, then show $c^2(\Delta t)^2 - (\Delta x)^2 = c^2(\Delta t')^2 - (\Delta x')^2$. Note that $y = y'$, and $z = z'$.

37. (a) $L + v\Delta t = c\Delta t$, so $\Delta t = L/(c - v)$.
(b) $\Delta t' = \Delta t/\gamma = L/\gamma(c - v)$
OR: $\Delta x = vL/(c - v)$, then $\Delta t' = \gamma(\Delta t - v\Delta x/c^2) = L/\gamma(c - v)$

38. $v = (2)(0.96c)/[1 + (0.96)^2] = 0.999c$

39. $v_{AB} = (0.8c - 0.6c)/(1 - 0.48) = 0.385c$

40. (a) $v = 0.8c/(1 + 0.07) = 0.748c$;
(b) $v = 0.6c/(1 - 0.07) = 0.645c$

41. (a) $v_{BA} = (0.8c - 0.6c)/(1 - 0.48) = 0.385c > 0.3c$, so no;
(b) Speed of missile relative to A greater than $0.385c$

42. $0.5c = 2v/(1 + v^2/c^2)$, so $v^2 - 4cv + c^2 = 0$, then
$v = 0.268c$.

43. (a) $dm/dt = (dE/dt)/c^2 = 4.33 \times 10^9$ kg/s;
(b) $M/(dm/dt) = 1.46 \times 10^{13}$ y

44. $p = \gamma m_o v = 15.8 m_o(0.998c) = 7.9\times10^{-18}$ kg.m/s

45. $K = (\gamma - 1)m_oc^2$, where $m_oc^2 = 0.511$ MeV.
(a) $10^4 = (\gamma - 1)(0.511\text{x}10^6)$. Find $\gamma = 1.0196$, and
$v = 0.195c$;
(b) Find $\gamma = 21.57$, so $v = 0.999c$

46. (a) $K = (\gamma - 1)m_oc^2 = (15.8 - 1)(0.511\text{x}10^6 \text{ eV}) = 7.56$ MeV
(b) $p = \gamma m_ov = 15.8m(0.998c) = 4.31\text{x}10^{-21}$ kg.m/s

47. (a) $\Delta K = (\gamma_2 - \gamma_1)m_oc^2 = (5/3 - 5/4)(0.511 \text{ MeV}) = 0.213$ MeV;
(b) $\Delta K = (\gamma_2 - \gamma_1)m_oc^2 = (15.8 - 10)(0.511 \text{ MeV}) = 2.96$ MeV

48. (a) $K = (\gamma - 1)m_oc^2 = m_oc^2$, so $\gamma = 2$ and $v = 0.866c$
(b) $K = 11m_oc^2$, so $\gamma = 12$ and $v = 0.997c$.

49. $\Delta E = (0.5 \text{ m}^2)(1 \text{ kW/m}^2)(3.156\text{x}10^7 \text{ s}) = 1.58\text{x}10^{10} \text{ J} = \Delta m.c^2$
Find $\Delta W = \Delta m.g = 1.72\ \mu N$

50. $E^2 - p^2c^2 = m_o{}^2c^4$ which is an invariant

51. (a) $p^2c^2 = E^2 - m_o{}^2c^4 = 8m_o{}^2c^4$, so $p = (8)^{1/2}m_oc$;
(b) $p = \gamma m_ov$, so $\gamma v = (8)^{1/2}c$, then $v = 0.943c$

52. $\Delta E = \Delta m.c^2 = 3\text{x}10^{19}$ J, so $\Delta m = 333$ kg.

53. $\Delta K = 3m_ov^4/8c^2$; $K = m_ov^2/2$, so $\Delta K/K = 3(v/c)^2/4 = 4.0 \times 10^{-3}\%$

54. $\Delta p/p = \gamma - 1 \approx v^2/2c^2 = 0.01$, so $v = 0.14c$.

55. $\gamma = 2$, so $v = 0.866c$

56. $E = \gamma m_oc^2 = \gamma(0.511 \text{ MeV})$, so $\gamma = 97.85$.
$\beta^2 = 1 - 1/\gamma^2$, so $\beta = 0.99995$.

57. (a) $1/2\ m_o\ (0.9c)^2 = eV$, thus $V = 208$ kV;
(b) $(\gamma - 1)m_oc^2 = eV$, find $\gamma = 1.407$, then $\beta = 0.703$

58. $K_{rel} - K_{cl} = 3m_oc^2(v/c)^4/8 = 0.01(m_ov^2/2)$. Find $v = 0.115c$.

59. (a) $K = (\gamma - 1)m_oc^2 = 40$ GeV, where $m_oc^2 = 938$ MeV
Thus, $\gamma = 43.6$ leads to $v = 0.9997c$;
(b) $p = \gamma m_ov = 2.2 \times 10^{-17}$ kg.m/s

60. $E = \gamma m_o c^2 = 10$ GeV, where $m_o c^2 = 0.511$ MeV. Thus, $\gamma = 1.957 \times 10^4$. (a) $L = L_o/\gamma = 0.164$ m; (b) $T_o \approx L/c = 0.55$ ns; (c) $T = \gamma T_o = 10.8$ μs

61. $p = \gamma m_o v = m_o c$ leads to $v = 0.707c$.

62. Note $\gamma = 1.4$. (a) $L_o/\gamma = 0.857$ km; (b) $\Delta t = L_o/\gamma v = 4.08$ μs; (c) $L_o/v = 5.71$ μs,

63. $\gamma_8 = 1.667$ and $\gamma_9 = 2.294$. So $\Delta E = \Delta\gamma . m_o c^2 = 0.627 m_o c^2$.
(a) $\gamma_0 = 1$, $\gamma_1 = 1.005s$, so $\Delta\gamma = 0.005$ and $0.627/0.005 = 125$.
(b) $\gamma_5 = 1.155$, $\gamma_6 = 1.25$, so $\Delta\gamma = 0.095$ and $0.627/0.095 = 6.60$.

64. (a) $T = \gamma T_o$ so $\gamma = 7.9/2.2 = 3.59$ thus $v = 2.88 \times 10^8$ m/s.
(b) $K = (\gamma - 1)m_o c^2 = 274$ MeV.

65. (a) $\gamma = 1.005$ and $K = (\gamma - 1)m_o c^2 = 4.53 \times 10^{19}$ J.
(b) $m = E/c^2$ leads to 503 kg.

66. $K = (\gamma - 1)m_o c^2$. (a) $E = \gamma m_o c^2 = K + m_o c^2 = 1.71$ MeV;
(b) $\gamma = 1.711/0.511 = 3.348$, so $v = 2.86 \times 10^8$ m/s.
(c) $p = \gamma m_o v = 8.72 \times 10^{-22}$ kg.m/s.

67. $\Delta t = L_o/\gamma v$, so $\gamma v = 240$ m/2.76 μs $= 8.69 \times 10^7$ or $\gamma \beta = 0.29$. Find $v = 8.35 \times 10^7$ m/s.

68. (a) From Eq. 39.22, $v_{rel} = 1.3c/1.4225 = 0.914c$, so $\gamma = 2.465$ Then $L_o/\gamma = 0.406$ km. (b) $\Delta t = L_o/\gamma v = 1.48$ μs.

69. $\Delta t = L_{oP}/v + L_{oT}/\gamma v$ where L_{oP} and L_{oT} are the rest lengths of the platform and the train. With $\gamma = 1.25$, find $\Delta t = 9.11$ μs.

70. Vel. of bullet rel. to ground $= (0.6c + 0.4c)/(1 + 0.24) = c/1.24$. $v_B = 2.42 \times 10^8$ m/s and so $\gamma_B = 1.69$.
(a) $\Delta t = (10 \text{ km})/v_B = 41.3$ μs.
(b) $\Delta t = (10 \text{ km})/\gamma_B v_B = 24.4$ μs.

71. $\gamma_T = 1.09$, so $\Delta t = (10 \text{ km})/\gamma_T(0.6c) = 51$ μs.

72. (a) $E = \gamma m_o c^2 = \gamma(0.511 \text{ Mev})$, so $\gamma = 3.91 \times 10^4$.
(b) $E = pc$, so $p = E/c = 1.07 \times 10^{-17}$ kg.m/s.

Problems

1. $E^2 = \gamma^2 m_o{}^2 c^4$; $p^2c^2 = \gamma^2 m_o{}^2 v^2 c^2$
$E^2 - p^2c^2 = \gamma^2 m_o{}^2 c^4 (1 - v^2/c^2) = m_o{}^2 c^4$

2. $F = d(\gamma m_o v)/dt$, thus $F/m_o = \gamma\ dv/dt + v\ d\gamma/dt =$
$= \gamma\ dv/dt + (v/c)^2\gamma^3\ dv/dt = \gamma^3\ dv/dt$

3. Note $v^2 = c^2(1 - 1/\gamma^2)$ and $\gamma = K/m_oc^2 + 1$.
$p^2 = \gamma^2 m_o{}^2 v^2 = (\gamma^2 - 1)m_o{}^2c^2 = [(K/m_oc^2)^2 + 2K/m_oc^2]m_o{}^2c^2$
Thus, $p^2 = 2m_oK + (K/c)^2$.

4. $v_{LG} = (c/n + v)/(1 + v/nc) = (c/n)(1 + nv/c)/(1 + v/nc)$

5. (a) $\gamma_e m_e \beta_e c = \gamma_p m_p \beta_p c$; where $\gamma_e = 31.6$.
Find $\beta_p = 1.72\text{x}10^{-2}$ and $v = 5.16 \times 10^6$ m/s;
(b) $m_ec^2(\gamma_e - 1) = m_pc^2(\gamma_p - 1)$
Find $\gamma_p = 1.0167$; $v = 0.181c = 5.4 \times 10^7$ m/s

6. (a) $\Delta t' = 200 \text{ m}/c = 0.667\ \mu s$;
(b) $\Delta t = \gamma\ \Delta t' = 6.67\ \mu s$
(c) $\Delta x = v\ \Delta t = (0.995c)(6.67\ \mu s) = 1990$ m
[OR: $\Delta x = \gamma(\Delta x' + v\Delta t') = \gamma v \Delta t' = 1990$ m.]

7. (a) $\cos\theta = (dx/dt)/c$; $\cos\theta' = (dx'/dt')/c$
$\cos\theta = u_x/c = (u_x' + v)/(1 + vu_x'/c^2)c =$
$= (\cos\theta' + \beta)/(1 + \beta\cos\theta')$
(b) $\theta = 7.04^o, 15.1^o, 25.8^o$

8. $v_y = dy/dt = dy'/\gamma(dt' + vdx'/c^2) = u_y'/\gamma(1 + vu_x'/c^2)$

9. (a) $t' = L_o/\gamma v = 2.03\ \mu s$; $N = 10^3 \exp(-2.03/2.2) = 396$;
(b) $t = L_o/v = 10.2\ \mu s$; $\tau = \gamma\tau_o = 11\ \mu s$;
$N = 10^3\exp(-10.2/11) = 396$

10. $(kx - \omega t) = (k'x' - \omega' t') = k'[\gamma(x - vt)] - \omega'[\gamma(t - vx/c^2)]$
$(k - \gamma k' - \gamma v\omega'/c^2)x - (\omega - \gamma vk' - \gamma\omega')t = 0$, thus
$k = \gamma(k' + v\omega'/c^2)$; $\omega = \gamma(\omega' + vk')$

11. (a) $\tan\theta' = dy'/dx'$; $\tan\theta = dy/dx$, but $dy = dy'$.
$\tan\theta = dy'/dx = (dx'/dx)\tan\theta' = \gamma \tan\theta'$
(b) $L^2 = L_x{}^2 + L_y{}^2 = (L_o \cos\theta_o/\gamma)^2 + (L_o \sin\theta_o)^2$
$= L_o{}^2[1 - (v_o\cos\theta_o/c)^2]$

12. $\lambda' = \lambda_o(1 + v/c)^{1/2}(1 - v/c)^{-1/2} =$
$= \lambda_o(1 + v/2c + ..)(1 + v/2c + ..) = \lambda_o(1 + v/c + ..)$
Thus $(\lambda' - \lambda_o)/\lambda_o \approx v/c$.

13. (a) Each train has a contracted length in the other's frame.
$x_{AF} = 800 + 0.6ct$; $x_{BR} = -600 + 0.8ct$. Set $x_{AF} = x_{BR}$ to find
$t = 1400/0.2c = 0.233\ \mu s$
(b) $v_{BA} = (0.8 - 0.6)/(1 - 0.48) = 0.385c$, so $\gamma = 1.08$.
Length of B in frame A is $1000/\gamma = 926$ m.
$\Delta t = 1926/0.384c = 16.7\ \mu s$

14. (a) $\Delta x = \gamma(\Delta x' + v\Delta t') = 5[-10^3 + (0.98c)(4\mu s)] = 880$ m
$\Delta t = \gamma(\Delta t' + v\Delta x'/c^2) = 5[4\ \mu s - 0.98c(10^3)/c^2] = 3.67\ \mu s$
(b) $T = \Delta t - \Delta x/c = 3.67 - 2.93 = 0.74\ \mu s$

15. (a) $\gamma = 5/4$, so $L = L_o/\gamma = 800$ m for each train in the frame of the other train.
$x_{AR} = -800 + vt$; $x_{BR} = 800 - vt = 0$, so $t = 800/v = 4.44\ \mu s$.
(b) $v_{BA} = (-0.6c - 0.6c)/(1 + 0.6^2) = -0.882c$; so $\gamma = 2.13$
$L_B = 470$ m, $t' = 1470\text{ m}/0.882c = 5.6\ \mu s$

16. $u_y = u_y'/\gamma(1 + vu_x'/c^2)$; $u_x = (u_x' + v)/(1 + vu_x'/c^2)$
$\tan\theta = u_y/u_x = u_y'/\gamma(u_x' + v) = \tan\theta'/\gamma[1 + v/u'\cos\theta']$

17. $p_x = \gamma(u)m_o u$, $p_x' = \gamma(u')m_o u'$
$E = \gamma(u)m_o c^2$, $E' = \gamma(u')m_o c^2$
$\gamma(u') = (1 - u'^2/c^2)^{-1/2} = c(c^2 - u'^2)^{-1/2}$
$= [1 - c^2(u - v)^2/(c^2 - uv)^2]^{-1/2}$
$= (c^2 - uv)/[c^2(c^2 - v^2) - u^2(c^2 - v^2)]^{1/2}$
$= \gamma(1 - uv/c^2)\gamma(u)$, where $\gamma = \gamma(v)$
$E' = \gamma m_o c^2(1 - uv/c^2)\gamma(u) = \gamma(E - vp_x)$
$p_x' = \gamma m_o \gamma(u)(u - v) = \gamma(p_x - vE/c^2)$

CHAPTER 40

Exercises

1. Use Eq. 40.1. (a) 0.97 mm; (b) 0.97 μm; (c) 0.29 nm

2. Use Eq. 40.1. (a) 6170 K; (b) 8280 K

3. From Eq. 40.1, T = 4140 K to 7250 K

4. $R = \sigma(T^4 - T_o^4)$. Convert ^oC to K by adding 273.
(a) $1.5\text{x}10^6$ W/m^2; (b) 140 W/m^2; (c) 430 W/m^2

5. Power = AR = $\sigma T^4(4\pi r^2) = 3.8 \times 10^{26}$ W

6. Power = RA = $\sigma(T^4 - T_o^4)(2\pi rL) = 2280$ W

7. From Eq. 40.1, $\lambda = 9.66$ μm

8. E = hf = 0.21 eV

9. $\Delta E = P\Delta t = nhf$, thus $n = 6.03 \times 10^{29}$ each second.

10. (a) $E = hc/\lambda = (4.14\text{x}10^{-15}$ eV.s)(3x10^8 m/s)/λ =
1240x10^{-9} eV/λ where λ is in meters.
Thus, E = 1240/λ if λ is in nm.
(b) E = 1.77 to 3.10 eV

11. (a) $K_m = hf - \phi = 0.855$ eV;
(b) $hc/\lambda = 4.97\text{x}10^{-19}$ J, so N = 3x10^{-11} W/m^2/(hc/λ) =
= 6.0×10^7 m^{-2}s^{-1}

12. (a) $P = I(\pi r^2) = 9.8\text{x}10^{-18}$ W; (b) N = P/hf = 25 photons/s.

13. $K_m = hc(1/\lambda - 1/\lambda_o) = 1.33\text{x}10^{-19}$ J;
$v = (2K/m)^{1/2} = 5.4\text{x}10^5$ m/s

14. See Exercise 10, E = 1240 eV/λ, for λ in nm.
(a) 2.25 eV; (b) 4.14x10^{-7} eV; (c) 3.89x10^{-9} eV;
(d) 1.75x10^4 eV

15. (a) $f = E/h = 2.66 \times 10^{15}$ Hz;
(b) From Exercise 10: E = 1240/λ = 7.09 eV

16. From Exercise 10: λ = 1240/E = 443 nm, violet

17. $hc/\lambda = 3.61\text{x}10^{-19}$ J. $N = I/hf = 3.72 \times 10^{21}\ m^{-2}s^{-1}$

18. $E = hc/\lambda = 3.14\text{x}10^{-19}$ J; $N = P/E = 3.18\text{x}10^{15}$ photons/s.

19. (a) $K_m = hf - \phi = 0.8$ eV; (b) $eV_o = hc/\lambda - \phi$, so $\lambda = 428$ nm

20. (a) $K_m = hc/\lambda - \phi = 1.7$ eV; (b) $K_m = eV_o$, so $V_o = 1.7$ V.

21. $K_m = hf - \phi$, so $\phi = 2.34$ eV, then
$eV_o = hc/\lambda - \phi$ leads to $V_o = 3.05$ V

22. $eV_o = hf - hf_o$, thus $f_o = 1.33\text{x}10^{14}$ Hz

23. (a) $E = hc/\lambda = 3.32\text{x}10^{-19}$ J, so $N = (5\ W)/E = 1.51\text{x}10^{19}\ s^{-1}$;
(b) $NA/4\pi r^2 = 20$, where $A = \pi d^2/4$. Find $r = 652$ km

24. $K_1 = hf - \phi$; $K_2 = 1.5hf - \phi$, thus $\phi = hf_o = 3.3$ eV, so
$f_o = 7.97\text{x}10^{14}$ Hz

25. (a) $hf = 2m_oc^2 = 1.022$ MeV, so $f = 2.47 \times 10^{20}$ Hz;
(b) $p = E/c = 5.45 \times 10^{-22}$ kg.m/s

26. $E = hc/\lambda = 3.97\text{x}10^{-19}$ J
Needed power = 8E per second = $31.76\text{x}10^{-19}$ W
$I = P/A = 31.76\text{x}10^{-19}\ W/(\pi d^2/4) = 1.62\text{x}10^{-13}\ W/m^2$.
(a) $P = I(4\pi r^2) = 294$ kW; (b) $P = 3.2\text{x}10^{21}$ W

27. (a) 4.1×10^{15} V.s; (b) 5.1×10^{14} Hz

28. (a) $\lambda = 1240/E = 0.0413$ nm
$\lambda' - \lambda = \lambda_c(1 - \cos 50^o)$, so $\lambda' = 0.0422$ nm
(b) $K = hc(1/\lambda - 1/\lambda') = 1.03\text{x}10^{-16}$ J

29. $\lambda' - \lambda = 2\lambda_c$, where $\lambda = 0.0310$. so $\lambda' = 0.0359$ nm
$K = hc(1/\lambda - 1/\lambda') = 8.76\text{x}10^{-16}$ J = 5.47 keV

30. $\Delta\lambda/\lambda = 2\text{x}10^{-4}$ and $\Delta\lambda = \lambda_c(1 - \cos\theta)$, thus $\theta = 6.2^o$.

31. $E = hc/\lambda_c = 8.19\text{x}10^{-14}$ J = 0.512 MeV

32. (a) $\Delta\lambda = \lambda_c(1 - \cos\theta) = 4.89\text{x}10^{-4}$ nm
(b) $\lambda = 0.0413$ nm and $\lambda' = 0.0418$ nm.

$E = hc/\lambda' = 4.76\text{x}10\text{-}15$ J = 29.8 keV.

33. $\Delta\lambda/\lambda = 3\text{x}10^{-4}$, $\Delta\lambda = \lambda_c(1 - \cos 53^o)$, thus λ = 3.24 nm
and $E = hc/\lambda$ = 380 eV

34. (a) $\Delta\lambda = \lambda_c(1 - \cos 70^o) = 1.6\text{x}10^{-12}$ m
(b) Find λ' = 0.0816 nm, so $K = hc(1/\lambda - 1/\lambda')$ = 304 eV

35. $\Delta\lambda = \lambda_c(1 - \cos 45^o) = 7.1\text{x}10^{-13}$ m, and λ = 0.0248 nm
Find $\lambda' = 2.55\text{x}10^{-11}$ m and then $f' = 1.2 \text{ x } 10^{19}$ Hz

36. $\Delta\lambda = \lambda_c(1 - \cos\theta)$: (a) $3.26\text{x}10^{-13}$ m; (b) $2.43\text{x}10^{-12}$ m;
(c) $4.5\text{x}10^{-12}$ m

37. (a) ΔE = 12.1 eV, 10.2 eV and 1.9 eV lead to
102 nm, 122 nm, and 653 nm respectively.
(b) No lines since no ΔE is equal to energy of photon.

38. $E = -13.6/n^2$ = -0.85 eV; -0.544 eV; -0.378 eV and -0.278 eV.
Then use $\Delta E = 1240/\lambda$ (for λ in nm).
(a) 1876 nm, 1282 nm, 1097 nm; (b) 821 nm

39. From Exercise 10: $\lambda = 1240/E$ = 91.2 nm

40. $v = (ke^2/mr)^{1/2}$ where $r = 5.29\text{x}10^{-11}$ m, so $v = 2.2\text{x}10^6$ m/s
$f = v/2\pi r = 6.6\text{x}10^{15}$ Hz, so λ = 45 nm (Ultraviolet)

41. (a) $U = 2E$ = -6.8 eV; (b) $K = -E$ = 3.4 eV

42. (a) $E = -13.6Z^2/n^2$ = -122.4 eV; -30.6 eV, -13.6 eV, -7.65 eV
(a) $\lambda = 1240/\Delta E$: 10.8 nm, 11.4 nm, 13.5 nm

43. $r_n = n^2 r_1 = 5.29\text{x}10^{-11}$ m, $2.12\text{x}10^{-10}$ m, $4.76\text{x } 10^{-10}$ m

44. (a) $E = -13.6Z^2/n^2$ = -54.4 eV, -13.6 eV, -6.04 eV;
(b) 54.4 eV

45. (a) $v = (ke^2/mr)^{1/2}$ where $r = 5.29\text{x}10^{-11}$ m, so
$v = 2.19\text{x}10^6$ m/s; (b) $p = mv = 1.99 \text{ x } 10^{-24}$ kg.m/s;
(c) $a = v^2/r = 9.07 \text{ x } 10^{22}$ m/s^2

46. From Eqs. 40.19 and 40.22, $v = (kZe^2/mr)^{1/2} = nh/2\pi mr$
find $r_n = n^2 r_1/Z$.

47. From Eq. 40.19, $v = (kZe^2/n^2mr)^{1/2} = Z(2.19\text{x}10^6)/n$ m/s

48. (a) $E \alpha m$, so $E' = 207E$; (b) $r \alpha 1/m$, so $r' = r/207$

49. Eq. 40.20 becomes $E_n = -kZe^2/2r_n$ and from Exercise 46, $r_n = n^2r_1/Z$, thus $E_n = -kZ^2e^2/2n^2r_1$, where $r_1 = \hbar^2/mke^2$.

50. (a) $\lambda = c/f = 12.2$ cm; (b) $E = hf = 1.01 \times 10^{-5}$ eV.

51. $hf = eV_o + \phi = 2.53 - 0.63 = 1.90$ eV.

52. (a) hc/λ_o gives $\lambda_o = 564$ nm;
(b) From $hf = K_m + \phi$, $K_m = 0.758$ eV.

53. $hc/\lambda = K_m + hf_o$, so $f_o = 5.15 \times 10^{14}$ Hz.

54. $1/\lambda = R(1/1 - 1/2^2)$, so $\lambda = 122$ nm.

55. (a) $n = 2$ for Balmer, so $\Delta E = 13.6/2^2$ eV, which gives $\lambda = 365$ nm.
(b) $n = 1$ for Lyman, so $\Delta E = 13.6$ eV, thus $\lambda = 91.4$ nm.

56. $hf = K_m + \phi$, so $K_m = 2.02$ eV and $v = 8.42 \times 10^5$ m/s.

57. $K_m = hc(1/\lambda - 1/\lambda_o)$ leads to $v_m = 5.89 \times 10^5$ m/s.

58. $E_o = E_1 + E_2$ or $hc/\lambda_o = hc/\lambda_1 + hc/\lambda_2$, leads to $\lambda_2 = 872$ nm.

59. $1/\lambda = R(1/n^2 - 1/m^2)$. Note that $1/\lambda R = 0.889$. Then trial and error leads to $n = 1$, $m = 3$.

60. $\lambda = hc/E = 4.141 \times 10^{-11}$ m. Then $\lambda' - \lambda = (h/m_oc)(1 - \cos\theta)$ $= 1.215$ pm. Find $E' = hc/\lambda' = 29.1$ keV.

61. $E = 120$ keV and $E' = 0.95E = 114$ keV. Then $hc/E' - hc/E =$ $= (h/m_oc)(1 - \cos\theta)$ gives $\theta = 39.1^o$.

62. (a) $K = 3.08 \times 10^{-18}$ J $= hc/\lambda - hc/\lambda'$, so $\lambda' = 150.35$ pm and $\Delta\lambda = 0.35$ pm. (b) $\Delta\lambda = (h/m_oc)(1 - \cos\theta)$ gives $\theta = 31.2^o$.

Problems

1. (a) $\lambda = 1240/E = 1240/(175\text{x}10^3) = 7.09\text{x}10^{-12}$ m
(b) $\lambda' = 1240/E' = 1240/(1.3\text{x}10^5) = 9.54\text{x}10^{-12}$ m;

$\Delta\lambda = \lambda' - \lambda = 2.45\text{x}10\text{-}12 \text{ m} = \lambda_c(1 - \cos\theta)$; $\theta = 90.5^o$;

(c) From Eqs. 40.14 and 40.15: $\tan\phi = \sin\theta/(\lambda'/\lambda - \cos\theta)$

Thus, $\phi = 36.8^o$

2. $du/d\lambda = A \exp(-B/\lambda T)[-5/\lambda^6 + B/\lambda^7 T] = 0$, so
$\lambda T = B/5 = \text{constant}$.

3. $p = hf/c = \gamma m_o v$ and $E = hf = \gamma m_o c^2$. To be consistent we need $v = c$ which is not possible for an electron.

4. $E = hc/\lambda$ and $\Delta E = hc(1/\lambda' - 1/\lambda)$, thus
$\Delta E/E = \lambda(1/\lambda' - 1/\lambda) = -\Delta\lambda/\lambda' \approx -\Delta\lambda/\lambda$

5. (a) $I = 2mr^2 = 4.57 \times 10^{-48}$ kg.m^2;
(b) $I\omega_n = nh/2\pi$; so $\omega_n = 2.31\text{x}10^{13}$ n rad/s;
(c) $\Delta f = \Delta\omega/2\pi = 3.67 \times 10^{12}$ Hz, far IR

6. $p = hf/c = \gamma m_o v = 6.96\text{x}10^{-27}$ kg.m/s. Take $\gamma \approx 1$,
so $v = 4.16$ m/s.

7. $p^2\cos^2\phi = (h/\lambda - h\cos\theta/\lambda')^2$; $\quad p^2\sin^2\phi = (h\sin\theta/\lambda')^2$.
Add: $p^2 = h^2[1/\lambda^2 - 2\cos\theta/\lambda\lambda' + 1/\lambda'^2]$ (i)
For electron: $p^2c^2 = E^2 - m_o{}^2c^4 = (K + m_oc^2)^2 - m_oc^4$, so
$p^2 = K^2/c^2 + 2Km_o = h^2(1/\lambda - 1/\lambda')^2 + 2hm_oc(1/\lambda - 1/\lambda')$ (ii)
Equate (i) and (ii) to find $\Delta\lambda = (h/m_oc)(1 - \cos\theta)$

8. (a) $y = x^5(e^x - 1)^{-1}$;
$dy/dx = 5x^4(e^x - 1)^{-1} - x^5e^x(e^x - 1)^{-2} = 0$
Thus, $5(e^x - 1) = xe^x$, or $(5 - x) = 5e^{-x}$.
(b) $x = 4.965 = hc/\lambda kT$, so $\lambda_{max}T = hc/4.965k = 2.9\text{x}10^{-3}$ m.K.

9. $x = hc/\lambda kT$, so $dx = -hcd\lambda/\lambda^2 kT$ and $u = 8\pi hc/\lambda^5(e^x - 1)$
$R = \int cu/4\, d\lambda = [2\pi(kT)^4/h^3c^2] \int x^3\, dx/(e^x - 1) = \sigma T^4$.

10. $mv^2/r = ke^2/(2r)^2$; thus $K = mv^2 = ke^2/4r$, and $U = -ke^2/2r$.
$E = K + U = -ke^2/4r$.
$2mvr = n\hbar$, so equating expressions for v^2: $(n\hbar/2mr)^2 = ke^2/4mr$, find $r = n^2\hbar^2/mke^2$, then $E = -mk^2e^4/4n^2\hbar^2 = -6.8/n^2$ eV.

CHAPTER 41

Exercises

1. $\lambda = h/(2mK)^{1/2} = 4.909\text{x}10^{-19}/K^{1/2}$ where K is in joules. 1 eV = $1.602\text{x}10^{-19}$ J; find $\lambda = 1.23\ \text{nm}/K^{1/2}$ for K in eV.

2. $\lambda = h/(2meV)^{1/2} = 1.226\text{x}10^{-9}/V^{1/2}$ m = $(1.5/V)^{1/2}$ nm

3. $\lambda = (1.5/120)^{1/2} = 0.112$ nm

4. From Exercise 1, $\lambda_e = 1.23/(K)^{1/2} = 0.87$ nm
   $\lambda_p = hc/E = 622$ nm

5. $\lambda = h/mv$: (a) 0.397 nm; (b) 0.397 pm

6. $\lambda = h/(2mK)^{1/2} = 0.143$ nm

7. $\lambda = h/mv = 6\text{x}10^{-32}$ m, $a = \lambda/\sin 5^o = 7.6\text{x}10^{-30}$ m, no.

8. Photon: $E = hc/\lambda = 248$ eV;
   Electron: $p = h/\lambda$ and $K = p^2/2m = 0.061$ eV

9. $E = 1240\ \text{eV}/\lambda$ (nm): (a) 12.4 keV; (b) 1.24 GeV

10. $v = h/m\lambda = 1.2$ km/s

11. $\lambda = h/(2meV)^{1/2}$, so $V = h^2/(2me)\lambda^2 = 82$ kV

12. $D\sin\phi = \lambda = (1.5/V)^{1/2}$, thus $0.215\sin\phi = 0.152$, so $\phi = 45^o$.

13. $\lambda = 1.23/(K)^{1/2}$: (a) $\lambda = 1.23/(80)^{1/2} = 0.138$ nm;
    (b) $\lambda = 1.23/(100)^{1/2} = 0.123$ nm

14. $v = h/m\lambda = 7.27\text{x}10^6$ m/s

15. (a) $v = h/m\lambda = 1.37 \times 10^7$ m/s;
    (b) $v = (ke^2/mr_1)^{1/2} = 2.18 \times 10^6$ m/s

16. $1/2\ mv^2 = ke^2/r$, so $v = (2ke^2/mr)^{1/2}$;
    $\lambda = h/mv = (r/2mk)^{1/2}h/e = 0.32$ nm

17. $\lambda = h/(2mK)^{1/2} = 0.143$ nm. Then, $\Delta y = \lambda L/d = 2.9\ \mu\text{m}$

18. (a) $E_1 = h^2/8mL^2 = 3.29\text{x}10^{-13}$ J; $E_2 = 13.2\text{x}10^{-13}$ J
    (b) $\Delta E = 9.91\text{x}10^{-13}$ J = hf, so $f = 1.49\text{x}10^{21}$ Hz (γ ray)

19. (a) $E_1 = h^2/8mL^2 = 37.6$ eV; $E_2 = 150$ eV
(b) $\Delta E = 112$ eV $= hc/\lambda$; find $\lambda = 11.1$ nm. (UV)

20. $L = nh/(8mE)^{1/2} = 1.1$ nm

21. $\Delta E = (9 - 1)h^2/8mL^2 = 75.4$ eV, X-ray

22. $1/2\ mv^2 = h^2/8mL^2$, so $v = h/2mL = 3.64$ m/s

23. (a) $E_2 = 4E_1 = 80$ eV;
(b) $L = h/(2mE_1)^{1/2} = 0.137$ nm

24. (a) $E = h^2/8mL^2 = 6\text{x}10^{-10}$ J $= 3.8\text{x}10^9$ eV; (b) Not possible

25. $\Delta p_x = h/\Delta x = 6.6 \times 10^{-24}$ kg.m/s

26. $\Delta p_x = h/\Delta x = 3.3\text{x}10^{-24}$ kg.m/s

27. (a) $\Delta E = h/\Delta t = 6.63 \times 10^{-26}$ J; (b) $\Delta f = 10^8$ Hz

28. (a) $\Delta p = h/\Delta x = 1.66\text{x}10^{-20}$ kg.m/s;
(b) $K = (\Delta p)^2/2m = 0.516$ MeV

29. $K = 8 \times 10^{-15}$ J, and $\lambda = h/(2mK)^{1/2} = 0.128$ pm.

30. $K = 4.8 \times 10^{-19}$ J, so $\lambda = h/(2mK)^{1/2} = 0.709$ nm.

31. $k = 2\pi/\lambda = 4.72 \times 10^{10}$ rad/m, so
$p = h/\lambda = 4.98 \times 10^{-24}$ kg.m/s.

32. $E_n = n^2h^2/8mL^2\ \alpha\ n^2$. So $E_2 = 13.6$ eV.

33. $E = h^2/8mL^2 = 129$ keV.

34. $E_n = n^2h^2/8mL^2 = n^2(6.024 \times 10^{-18})$ and $f = \Delta E/h$.
Need upto $n = 3$.
$f_{12} = 2.73 \times 10^{16}$ Hz, $f_{23} = 4.55 \times 10^{16}$ Hz, $f_{13} = 7.27\text{x}10^{16}$ Hz

36. (a) $f = \Delta E/h = 5.43 \times 10^{14}$ Hz; (b) $\Delta t.\Delta E = h$, so
$\Delta f = \Delta E/h = 7.69 \times 10^6$ Hz.

37. $\lambda = h/(2mK)^{1/2} = 1.145 \times 10^{-20}/K^{1/2}$ for K in joules.
Multiply by $e^{-1/2}$: $\lambda = 2.86 \times 10^{-11}/K^{1/2}$, where K is in eV.

38. $p = mv = 1.82 \times 10^{-23}$ kg.m/s. Then, $\Delta p.\Delta x = h$, so $\Delta p = 1.33 \times 10^{-26}$ kg.m/s and $\Delta p/p = 0.073\%$.

39. (a) $\lambda = h/(2mK)^{1/2}$, so K = 151 eV. (b) $E = hc/\lambda = 12.4$ keV.

40. $p = E/c = 1.0667 \times 10^{-17}$ kg.m/s. So $\lambda = h/p = 6.21 \times 10^{-17}$ m

Problems

1. (a) At low energy, $p = (2m_oK)^{1/2}$
   (b) At high energy, $E \approx K$ and $p \approx E/c$

2. At high energy $p \approx E/c$, so $\lambda = hc/E = 6.21\text{x}10^{-15}$ m

3. (a) $p = mv = 2\text{x}10^{-24}$ kg.m/s
   (b) $\Delta x = h/\Delta p = 0.165$ nm $\approx 3.2r_B$

4. $\int A^2 \sin^2(n\pi x/L)\,dx = A^2/2 \int(1 - \cos(2\pi nx/L)\,dx = A^2L/2 = 1.$

5. $\int A^2 \sin^2(n\pi x/L)\,dx = A^2/2 \int(1 - \cos(2\pi nx/L)\,dx = 0.818$

6. At x = 0, the condition $d\psi/dx$ is continuous, $AK\exp(Kx) = kC\cos(kx)$, becomes $AK = kC$.

7. $\psi = (2/L)^{1/2} \cos(n\pi x/L)$, n = 1, 2, 3

8. $k_1 = (3)^{1/2}k_2 = 1.732k_2$. Thus, $R = (0.732/2.732)^2 = 7.2\text{x}10^{-2}$

9. The SWE is $\partial^2\psi/\partial x^2 + (2m\hbar^2)(E - m\omega^2x^2/2)\psi = 0$
   Find $\partial^2\psi/\partial x^2 = 2AB(2Bx^2 - 1)\exp(-Bx^2)$, then
   $B = m\omega/2\hbar$ and $E = B\hbar^2/m = \hbar\omega/2$

10. (a) (0, 0, 1), (0, 1, 0), (1, 0, 0)
    (b) (1, 1, 0), (1, 0, 1), (0, 1, 1)

11. $\langle x^2\rangle = (2/L) \int x^2 \sin^2(\pi nx/L)\,dx$
    $= \int x^2\,dx/L - \int x^2 \cos(2\pi nx/L)]\,dx/L$
    Integrate second term by parts to find:
    $(1/L)[x^3/3 - (L/2\pi n)^2 \cos(2\pi nx/L)(x/2)] = (1/3 - 1/2n^2\pi^2)L^2$

12. $\psi_2/\psi_1 = e^{-KL}$, thus $T = \psi_2{}^2/\psi_1{}^2 = e^{-2KL}$.
    Find $K = 3.62\text{x}10^{10}$, and $2KL = 7.24$, so
    $T = e^{-7.24} = 7.2\text{x}10^{-4}$.

CHAPTER 42

Exercises

1. (a) $(2)^{1/2}\hbar$; (b) $(12)^{1/2}\hbar$

2. $[\ell(\ell + 1)^{1/2}] = 3.46$, $\ell(\ell + 1) = 12$, so $\ell = 3$.

3. $n = 4$, $\ell = 2$, $m_\ell = 0, \pm 1, \pm 2$, $m_s = \pm 1/2$

4. $L_z = 0, +\hbar$

5. (a) $0, \pm\hbar$; (b) $90^o, \pm 45^o$

6. 0, +1, +2

7. (a) $\ell = 4$; (b) $n \geq 5$

8. (a) $\ell = 0, 1, 2$; $m_\ell = 0, +1, +2$ (9 states);
   (b) $E = -13.6Z^2/n^2 = -6.04$ eV

9. (a) $E = -13.6Z^2/n^2 = -30.6$ eV;
   (b) $\ell = 0, 1$; $m_\ell = 0, \pm 1$

10. $\cos\theta = L_z/L = m_\ell/[\ell(\ell + 1)]^{1/2} = 4/(20)^{1/2}$,
    so $\theta = 26.6^o$

11. $L/\hbar = 2.448 = [\ell(\ell + 1)]^{1/2}$, thus $\ell = 2$.
    Maximum $L_z = 2\hbar$

12. States for ℓ: 2, 6, 10, 14, 18, 22
    States for n: 2, 8, 18, 32, 50, or $2n^2$ for each n.

13. $L^2 = \ell(\ell + 1)\hbar = (L_x^2 + L_y^2) + L_z^2 =$
    $(L_x^2 + L_y^2) + m_\ell^2\hbar^2$

14. (a) $\Delta\phi$ is completely unknown; (b) L_x and L_y are unknown.

15. From Eq. 42.8, $P = 4\pi r_o^2(1/\pi r_o^3)e^{-2} = (4/r_o)e^{-2} = 0.54/r_0 =$
    $1.02\text{x}10^{10}$

16. From Eq. 42.9: $P = e^{-1}/8r_o = 8.68\text{x}10^7$

17. (a) $e^{-1}/r_o = 0.37/r_0$; (b) $4e^{-2}/r_o = 0.54/r_0$;
    (c) $16e^{-1}/r_o = 0.29/r_0$

18. (a) $9e^{-1/2}/128r_o = 8.06\text{x}10^7$; (b) $e^{-1}/8r_o = 8.68\text{x}10^7$; (c) 0.

19. $(1/\pi r_o^3)\int \exp(-2r/r_o)\, 4\pi r^2 dr =$
$= (4/r_o^3)[-r^2\exp(-2r/r_o)(r_o/2) + \int (r_o/2)\exp(-2r/r_o)(2r)dr$
$= (4/r_o^3)r_o[-r_o/2\exp(-2r/r_o)(2r) + \int (r_o/2)\exp(-2r/r_o)(2dr)$
$= [\exp(-2r/r_o)] = 1$

20. $\int P\, dr = (4/r_o^3)\int r^2\exp(-2r/r_o)dr =$
$= (4/r_o^3)[-r_o^3/2e^2 - r_o^3/2e^2 - r_o^3/4e^2] + 1 = 1 - 5/e^2 = 0.32$

21. $V = hc/e\lambda_o = 24.8$ kV

22. $\lambda_o = hc/eV = 4.98\text{x}10^{-11}$ m

23. Slope = a = 5 x 10^7 $\text{Hz}^{1/2}$

24. $\lambda_o = hc/eV = 1.24\text{x}10^{-6}/V = 1240$ nm/V

25. $f(L_\alpha) = f(K\beta) - f(K_\alpha)$ or $1/\lambda_{L\alpha} = 1/\lambda_{K\beta} - 1/\lambda_{K\alpha}$;
$\lambda_{L\alpha} = 5.59$ nm

26. $(c/\lambda)^{1/2} = a(Z - 1)$, so $\lambda_1/\lambda_2 = (Z_2 - 1)^2/(Z_1 - 1)^2$
(a) $\lambda = (41/46)^2(0.71) = 0.56$ nm;
(b) $\lambda = (41/25)^2(0.71) = 1.91$ nm

27. $\Delta E(K_\alpha) = 1240/0.71 = 1746$ eV; $\Delta E(K_\beta) = 1240/0.63 = 1968$ eV
$E_1 = -2870 - 1746 = -4616$ eV,
$E_3 = -4616 + 1968 = -2648$ eV

28. ^{23}V

29. (n, ℓ, m_ℓ, m_s): $(1, 0, 0, \pm 1/2)$, $(2, 0, 0, \pm 1/2)$,
$(2, 1, 1, \pm 1/2)$, $(2, 1, 0, \pm 1/2)$

30.

n	1	2	3
Z	1	2-5	6-14
ℓ	0	0,1	0,1,2
m_ℓ	0	0,0,+1	0,0,+1,0,+1,+2

NOBLE: Z = 1 (H); 5 (B); 14 (Si)

31. $\mu = eL/2m = (e\hbar/2m)(\ell(\ell + 1)^{1/2})\hbar$
$= (12)^{1/2}e\hbar/2m = 3.21\text{x}10^{-23}$ J/T

32. (a) $E = +\mu_B B = +3.71\text{x}10^{-24}$ J

(b) $f = 2E/h = 11.2$ GHz

33. (a) $\Delta E = hc(1/\lambda_1 - 1/\lambda_2) = 2.14$ meV;

(b) $\Delta U = 2\mu_B B = 2.14$ meV $= 3.42\text{x}10^{-22}$ J, thus B = 18.4 T

34. $F = \mu_B\ dB/dz = 1.8\text{x}10^{-25}$ N, thus $a = 6.2\text{x}10^3\ m/s^2$

$y = 1/2\ at^2 = 0.78$ mm

Problems

1. $8r_oP = x^2(2 - x)^2e^{-x} = (4x^2 - 4x^3 + x^4)e^{-x}$

$8r_o\ dP/dx = xe^{-x}(8 - 16x + 8x^2 - x^3) = 0$

With x = 5.2 find $dP/dx \approx 0$

2. $P = Ar^2\psi^2$; so $dP/dr = A\exp(-2r/r_o)(4r^3 - r^4/r_o)$,

so $r = 4r_o$.

3. Prob. $= \int P\ dr =$

$(4/r_o^3)[-r_or^2/2 - rr_o^2/2 - r_o^3/4)\exp(-2r/r_o)]$

$= 1 - 13e^{-4} = 0.76$

4. $r_{av} = (4/r_o^3) \int r^3 \exp(-2r/r_o)\ dr =$

$(r_o/2)^3(6)\int \exp(-2r/r_o)\ dr = (4/r_o^3)(3r_o^4/8) = 1.5r_o$.

5. Electron: $eV = m_oc^2(\gamma - 1)$ leads to $\gamma = 1.078$, thus

$v = 0.373c$

$\lambda_e = h/\gamma m_o v = 6.04$ pm

$\lambda_x = 1240$ nm/E = 31 pm

CHAPTER 43

Exercises

1. $R = 1.2A^{1/3}$: (a) 3.0 fm; (b) 4.6 fm; (c) 7.4 fm

2. $M = 6.0\text{x}10^{24} = 4\pi\rho R^3/3$, $R = 184$ m

3. $\rho = 3M/4\pi R^3 = 4.8 \text{ x } 10^{17}$ kg/m^3

4. $R^3 = 3M/4\pi\rho = 103.7\text{x}10^{30}$, so $R = 4.7\text{x}10^{10}$ m

5. $(A_2/A_1)^{1/3} = 2$, so $A_2/A_1 = 8$

6. $m = 235$ u $= 4\pi\rho R^3/3$, find $R = 2.6\text{x}10^{-10}$ m

7. $63.55 = 62.95x + 64.95(1 - x)$, thus $x = 0.7$, so 70% is ^{63}Cu

8. $(0.91)20 + (0.09)22 = 20.18$ u

9. $R_{Au} = 7$ fm, then $K = U = k(79e)(2e)/(R_{Au} + R_{\alpha}) = 26$ MeV

10. $\rho = 3(26e)/4\pi R^3$, where $R^3 = (1.2\text{x}10^{-15})^3(56)$
    $\rho = 1.03\text{x}10^{25}$ C/m^3

11. (a) 8.55 MeV; (b) 7.91 MeV

12. (a) BE = 31.99 MeV; BE/A = 5.33 MeV
    (b) BE = 1118 MeV, BE/A = 8.41 MeV

13. BE(^{13}C) = 97.1 MeV, BE(^{13}N) = 94.1 MeV

14. (a) 115.5 MeV; (b) 112 MeV

15. (a) BE(^{7}Li) = 39.24 MeV; BE(^{6}Li) = 31.99 MeV; so
    ΔBE = 7.25 MeV;
    (b) BE/A = 5.61 MeV

16. (a) BE(^{12}C) = 92.16 MeV; BE(^{11}B) = 76.21 MeV, so
    ΔBE = 15.95 MeV
    (b) BE/A = 7.68 MeV

17. $N_o = mN_A/M = 1.0037\text{x}10^{20}$;
    $R_o = \lambda N_o = 0.693N_o/T_{1/2} = 4.2\text{x}10^{11}$ Bq

18. $N_o = mN_A/M = 1.88\text{x}10^{19}$; $R_o = 0.693N_o/T_{1/2} = 1.05\text{x}10^{13}$ Bq

19. $N_o = R_oT_{1/2}/0.693 = 1.52x10^8$.
$N = N_o\exp(-\lambda t) = N_o\exp(-0.693/3.82) = 1.27 \times 10^8$

20. (a) $(m_{Be} - 2m_{He}) = 9.9x10^{-5}$ u; $Q = \Delta mc^2 = 92.2$ keV
(b) $(m_C - 3m_{He}) = -7.8x10^{-3}$ u. $Q < 0$, so not possible.

21. ^{11}B, yes

22. In 1 m^2, $N_o = mN_A/M = 6.69x10^{15}$.
$R_o = 0.693N_o/T_{1/2} = 5.07x10^6$ Bq, and $R = 3.7x10^4$ Bq.
Thus, $3.7x10^4 = 5.07x10^6\exp(-\lambda t)$, find t = 206 y.

23. (a) $-\lambda t = \ln(9/15)$, so $T_{1/2} = 0.693/\lambda = 3.4$ h.
(b) $N_o = R_oT_{1/2}/0.693 = 9.8 \times 10^9$

24. $N_o = mN_A/M = 2.52x10^{21}$.
$R_o = 0.693N_o/T_{1/2} = 2.3x10^9$ Bq = 0.062 Ci

25. $N_o = (1.3x10^{-12})mN_A/M = 5.22x10^{12}$.
$R_o = 0.693N_o/T_{1/2} = 20$ Bq; R = 0.75 Bq
$0.75 = 20\exp(-\lambda t)$, so $t = \ln(0.75/20)T_{1/2}/0.693 = 27{,}150$ y

26. ^{234}Th -- ^{234}Po -- ^{234}U -- ^{230}Th

27. 7 α's, 4 e^-'s

28. (a) $N_o = R_oT_{1/2}/0.693 = 7.47x10^{15}$; then
$m = MN_o/N_A = 1.63\ \mu g$.
(b) $N = N_o\exp(-\lambda t) = 3.18x10^{15}$.

29. $320 = 500\exp(-2\lambda)$, so $\lambda = 0.223\ h^{-1}$; $T_{1/2} = 0.693/\lambda = 3.11$ h

30. $N/N_o = \exp(-6.93/12.3) = 0.57$, so 57%.

31. $0.988 = \exp(-\lambda t)$; $t = T_{1/2} \ln(1/0.988)/0.693 = 8.54 \times 10^8$ y

32. In 1 g of ^{12}C number of ^{14}C atoms is $N_o = mN_A/M = 5.6x10^{10}$,
thus $R_o = \lambda N_o = 0.215$ Bq. Then,
$R = R_o\exp(-0.693x15000/5730) = 0.163R_o = 2.1$ per min (per g)

33. $N_{o1}/N_{o2} = (N_1/N_2) \exp[0.693(1/T_1 - 1/T_2)t] = 1.58\%$

34. $N_o = mN_A/M = 1.51x1016$, then $R_o = 0.693N_o/T_{1/2} = 0.265$ Bq.

35. $(M_P - Zm_e) + m_e = M_D - (Z - 1)m_e + Q/c^2$, so $Q = (M_P - M_D)c^2$.
$m_{Be}c^2 = 6536.270$ MeV, $m_{Li}c^2 = 6535.409$ MeV, so $Q = 0.861$ MeV

36. ^{206}Pb; $[m(^{210}Po) - m(^{206}Pb) - m(^{4}He)]c^2 = 5.41$ MeV

37. 6 α's, 4 e^-'s

38. $(m_n - m_p - m_e)c^2 = 939.57 - 938.28 - 0.511 = 0.78$ MeV

39. $[m(^{40}K) - m(^{40}Ca) - m_e]c^2 = 1.31$ MeV

40. α decay: $[m(^{218}Po) - m(^{214}Pb) - m(^{4}He)]c^2 = 6.11$ MeV
β^- decay: $[m(^{218}Po) - m(^{218}At)]c^2 = 0.26$ MeV

41. (a) $\Delta mc^2 = (18.005677 - 18.006956)c^2 = -1.19$ MeV;
(b) $\Delta mc^2 = (8.023830 - 8.005206)c^2 = 17.35$ MeV

42. (a) $\Delta mc^2 = (13.014786 - 13.008665)c^2 = 5.70$ MeV
(b) $\Delta mc^2 = (30.984144 - 30.987973)c^2 = -2.64$ MeV

43. $\Delta mc^2 = (10.020008 - 10.017726)c^2 = 2.13$ MeV

44. (a) ^{4}He; (b) ^{3}He; (c) n; (d) n

45. (a) ^{32}S; (b) ^{19}F; (c) ^{10}Be; (d) n

46. $\Delta mc^2 = (15.011739 - 15.011067)c^2 = 0.625$ MeV

47. $\Delta mc^2 = -0.0026302c^2 = (19.006984 - 1.008665 - m_F)c^2$
Thus, $m_F = 18.00095$ u

48. (a) $N = mN_A/M = 2.56x10^{24}$ atoms; thus $E = 4.86x10^{32}$ eV
(b) $\Delta t = 0.32E/P = 5x10^4$ s = 13.9 h

49. E = 190 MeV = $3.04x10^{-11}$ J. From NE = 4.186 J, find
$N = 1.38x10^{11}$

50. $N = mN_A/M = 2.56x10^{21}$; $R = 0.693N/T_{1/2} = 1.87x10^{-4}$ Bq
Thus in 1 day, there are 16 fissions.

51. $\Delta mc^2 = 0.186365c^2 = 173.6$ MeV

52. $E = (2x10^4)(4.2x10^9) = 8.4x1013$ J $= \Delta mc^2$, so $\Delta m = 0.93$ g

53. $0.025/10^6 = (0.5)^n$, or $4x10^7 = 2^n$; then nlog2 = 7.602 and n = 26

54. $\lambda = h/(2mK)^{1/2} = 0.143$ nm

55. $\Delta mc^2 = (0.00351\ u)c^2 = 3.27$ MeV

56. $\Delta mc^2 = (0.018884\ u)c^2 = 17.6$ MeV

57. $\Delta mc^2 = (0.004329\ u)c^2 = 4.03$ MeV

58. P = 40 MW = N(17.6 MeV)(e), so $N = 1.42x10^{18}$

59. Number of water molecules in 1 kg is $N = mN_A/M = 3.34x10^{25}$ The number of deuterium nuclei (2 per molecule) is $1.028x10^{22}$. So $E = (4.03x10^6)(1.028x10^{22})$ eV $= 6.63 \times 10^9$ J

60. ^{98}Zr

61. $^{12}C + n - {}^{13}C = 0.005310$ u = 4.95 MeV.

62. $^{12}N \rightarrow {}^{12}C + \beta^+ + \nu$; $Q = (m_X - m_Y - 2m_e)c^2 = 16.316$ MeV Find atomic mass of $^{12}N = 12.0186$ u.

63. $N_o = R_o/\lambda = 4.218 \times 10^7$. Between $t = 2T_{1/2}$ and $t = 3T_{1/2}$, the number that decay is $N_o/8 = 5.27\ x10^6$.

64. (a) mN_A/M of ^{12}C atoms gives 9.786×10^{11} atoms of ^{14}C. $R_o = \lambda N_o = 3.75$ Bq. (b) $R = R_o exp(-\lambda t) = 2.77$ Bq.

65. (a) ^{15}N; (b) $R_o = 7.4 \times 10^3$ Bq $= \lambda N_o$, so $N_o = 1.3 \times 10^6$.

66. $0.097 = exp(-\lambda t)$ leads to t = 19,300 y.

67. (a) $N_o = R_o/\lambda = mN_A/M$ where M = 131 g/mol, so $m = R_oM/\lambda N_A = 4.34 \times 10^{-13}$ g (per L);
(b) 500/2000 = 1/4 so $t = 2T_{1/2} = 16$ d.

68. 84p + 130n = 215.78375 u. Then, 7.7852 Mev x 214 = 1.78855 u so $^{214}Po = 213.9952$ u.

69. (a) 200 MeV $= 3.2 \times 10^{-11}$ J, 20 kton $= 8.36 \times 10^{13}$ J, so

number of fissions $N = 2.61 \times 10^{24}$. (b) N(235 u) = 1.02 kg.

70. $N = R/\lambda = RT_{1/2}/0.693 = 4.145 \times 10^{13}$.
Then $m = N(3.01\ u) = 2.07 \times 10^{-13}$ kg.

71. (a) ^{22}Ne; (b) $Q = (m_X - m_Y - 2m_e)c^2 = 1.82$ MeV

72. (a) $N = mN_A/M = 5.02 \times 10^{22}$ atoms of ^{12}C gives 6.53×10^{10} atoms of ^{14}C. Then, $R = \lambda N = 0.693N/T_{1/2} = 0.250$ Bq = 15 min^{-1} (per g).
(b) In 0.4 g expect $R_o = 6$ min^{-1}, but $R = 81/60 = 1.35$ min^{-1}
$R = R_o \exp(-\lambda t)$: $\ln(0.225) = -0.693t/T_{1/2}$, so $t = 1.23 \times 10^4$ y

73. (a) 16.007934 - 16.002603 = 0.005331 u = 4.97 MeV.
(b) 14.011180 - 14.003074 = 0.008106 u = 7.55 MeV

74. 8.023830 - 8.025595 = -0.001765 = -1.64 MeV.

75. $0.2 = \exp(-\lambda t)$ gives $t = 2.90 \times 10^9$ y.

76. $^{13}N \rightarrow {}^{13}O + \beta^+ + \nu$; $Q = (m_X - m_Y - 2m_e)c^2 = 1.2$ MeV.

77. (a) $R_o = \lambda N_o = 0.693N_o/T_{1/2}$, hence $N_o = 3.1 \times 10^7$;
(b) $R = R_o \exp(-\lambda t)$, so $\ln(5/65) = -\lambda t$ and $t = 1.22 \times 10^6$ s.

Problems

1. $N_o = R_oT_{1/2}/0.693 = 1.17 \times 10^{15}$
$N = N_o\exp(-0.693/29) = 0.976N_o = 1.14 \times 10^{15}$
Thus $N_o - N = 3 \times 10^{13}$

2. $dN_2/dt = \lambda N_1 = \lambda N_{o1}\exp(-\lambda t)$, thus $N_2 = N_{o1}(1 - e^{-\lambda t})$

3. $dN_2/dt = \lambda_1 N_1 - \lambda_2 N_2 = 0$, when $\lambda_1 N_1 = \lambda_2 N_2$

4. $\int t\ dN = t(-\lambda N\ dt) = -\lambda N_o \int te^{-\lambda t}\ dt$.
Let $u = t$ and $v' = -\lambda e^{-\lambda t}dt$.
Then $\tau = [te^{-\lambda t}] - \int e^{-\lambda t}\ dt = 1/\lambda$

5. Σp: $m_\alpha v_\alpha = M_D v_D$; $Q = 0.5m_\alpha v_\alpha^2 + 0.5M_D v_D^2 = K_\alpha(1 + m\alpha/M_D)$
Thus, $K_\alpha = M_D Q/(M_D + m_\alpha)$.
For ^{226}Ra, $Q = 4.87$ MeV, so $K_\alpha = 4.79$ MeV

6. Let M be the mass of a nucleus and m the mass of the neutral atom: $m_P = M_P + Zm_e$; $m_D = M_D + (Z - 1)m_e$
$Q = (M_P - M_D - m_e)c^2 = (m_P - m_D - 2m_e)c^2$

7. $R = 0.6R_o = R_o e^{-\lambda t}$, thus $\ln(0.6) = -\lambda(3.5\text{ h})$
$T_{1/2} = 0.693/\lambda = 4.75\text{ h}$

8. Since $R = 1.2A^{1/3}$;
$\Delta U = [3ke^2\text{x}10^{15}/5(1.2)][92^2/(236)^{1/3} - 56^2/(141)^{1/3} - 36^2/(92)^{1/3}] = 5.52\text{x}10^{-11}\text{ J} = 345\text{ MeV}$

9. From Problem 5, $K_\alpha = M_D Q/(M_D + m_\alpha) \approx (A - 4)Q/A$
(Approximate since we really need nuclear masses.)
For ^{236}U, $Q = 4.57$ MeV, so $K_\alpha = 4.49$ MeV

10. $Q = K_{CM}$; and $K_L = E_{Th} = 0.5m_a v_a^2$.
Σp: $m_a v_a = m_b v_b + M_Y v_Y$
$v_{CM} = m_a v_a/(m_a + M_X)$, thus
$K_{CM} = 0.5m_a(v_a - v_{CM})^2 + 0.5M_X v_{CM}^2 = 0.5m_a M_X v_a^2/(m_a + M_X)$
Thus, $E_{Th} = -(m_a + M_X)Q/M_X$ (Negative since $Q < 0$.)
For the given $^{14}N(\alpha, p)^{17}O$, find $Q = -1.19$ MeV,
so $E_{Th} = 1.53$ MeV

11. $Q = \Delta mc^2 = (-0.003224\text{ u})c^2 = -3.003$ MeV, thus $E_{Th} = 3.23$ MeV